T0211452

Learning Materials in Biosciences

Learning Materials in Biosciences textbooks compactly and concisely discuss a specific biological, bio-medical, biochemical, bioengineering or cell biologic topic. The textbooks in this series are based on lectures for upper-level undergraduates, master's and graduate students, presented and written by authoritative figures in the field at leading universities around the globe.

The titles are organized to guide the reader to a deeper understanding of the concepts covered.

Each textbook provides readers with fundamental insights into the subject and prepares them to independently pursue further thinking and research on the topic. Colored figures, step-by-step protocols and take-home messages offer an accessible approach to learning and understanding.

In addition to being designed to benefit students, Learning Materials textbooks represent a valuable tool for lecturers and teachers, helping them to prepare their own respective coursework.

More information about this series at http://www.springer.com/series/15430

Beate Brand-Saberi

Editor

Essential Current Concepts in Stem Cell Biology

 Springer

Editor
Beate Brand-Saberi
Department of Anatomy & Molecular Embryology
Ruhr-University Bochum
Bochum, Nordrhein-Westfalen, Germany

ISSN 2509-6125   ISSN 2509-6133   (electronic)
Learning Materials in Biosciences
ISBN 978-3-030-33922-7   ISBN 978-3-030-33923-4   (eBook)
https://doi.org/10.1007/978-3-030-33923-4

This Springer imprint is published by the registered company Springer Nature Switzerland AG
The registered company address is: Gewerbestrasse 11, 6330 Cham, Switzerland

Preface

Over the past two decades, stem cell research has exploded: Publications have multiplied exponentially.

The path of ground-breaking findings and innovative molecular approaches was paved with Nobel prizes. Mentioning only a few, the derivation of embryonic stem cells by Sir Martin Evans (Nobel Prize together with Professors Carpecchi and Smithies in 2007) is an outstanding example for a mutual fertilization of interdisciplinary approaches, in this case derived from developmental biology and pathology. The traditional concept of a one-way track regarding the developmental potencies during embryonic development of cells was challenged in the late 1960s by Sir John Gurdon's nuclear transfer experiments and eventually revolutionized by Yamanaka's pluripotency induction with defined factors (Nobel Prize Professor John Gurdon and Professor Shinya Yamanaka in 2012).

Clearly, stem cell biology belongs to the most innovative and competitive research fields. What is more, it has combined different disciplines in a unique way: Molecular and morphological basic sciences have met with clinical approaches, material sciences and philosophy. Reality has overtaken our imagination once more so rapidly that most of us almost forgot about implementing structured modern approaches to teach stem cell biology.

To date, there are still only very few textbooks addressing stem cell biology and none of them is as comprehensive as the present one. This book has been written by scientists who have been involved in teaching the students of the international master program – Molecular and Developmental Stem Cell Biology – which was launched at Ruhr University Bochum in 2011 and has been running with great success since then. I am very grateful to all the contributors to this textbook for their great efforts and dedication to their roles as academic teachers in spite of their competitive research projects. Our authors from Tongji University were involved in the ISAP (Internationale Studien- und Ausbildungspartnerschaften) exchange program supported by the DAAD between 2012 and 2016. It was not easy for all contributors to spare the time for writing down their valuable experience and recommendations in textbook chapters. So thank you again to all of them!

This book will be helpul to all those who strive to get into this important transdisciplinary research topic, be it master students, PhD students, MD students, postdocs or clinicians. However, as you will see, there is still some work for you to do: The book is not only intended for learning facts, but also for contemplating the white patches in the mosaic of our growing knowledge. Importantly, the chapters included here reveal that stem cell biology is rapidly expanding towards translational medical approaches, but they also address the challenges to be overcome for future progress. iPSCs, MSC, organoids, regeneration using scaffolds, extracellular vesicles, disease modelling and in depth knowledge of the multiple interacting cell types composing the tissues in our bodies are fascinating still developing aspects of stem cell biology written by experts in the field to be included into this volume. I personally hope that the book will contribute to an unbiassed interdisciplinary (and international) academic dialogue in this field. As stem cell research is often being perceived either with inappropriate hype, hope or simplistic damnation in public, a chapter on bioethics has been included as an integral part of learning about stem cells.

My thanks are due to my colleague PD Dr. Holm Zaehres for his kind advice, support and encouragement during the assembly of the chapters and final stages of this book. I would also like to express my gratitude to Dr. Amrei Strehl of the Springer Nature team and to Bibhuti Sharma of the publishing team for their constructive support and extraordinary patience. Finally and importantly, I wish to thank my family for their understanding and support.

Beate Brand-Saberi
Bochum, Germany
August 2019

Contents

Contributors

James Adjaye
Institute for Stem Cell Research and Regenerative Medicine, Heinrich Heine University Düsseldorf, Medical Faculty
Düsseldorf, Germany
James.Adjaye@med.uni-duesseldorf.de

Santoshi Biswanath
Department of Cardiac, Thoracic, Transplantation and Vascular Surgery (HTTG)
Leibniz Research Laboratories for Biotechnology and Artificial Organs (LEBAO)
Hannover Medical School (MHH)
Hannover, Germany
biswanath.santoshi@mh-hannover.de

Emiliano Bolesani
Department of Cardiac, Thoracic, Transplantation and Vascular Surgery (HTTG)
Leibniz Research Laboratories for Biotechnology and Artificial Organs (LEBAO)
Hannover Medical School (MHH)
Hannover, Germany
bolesani.emiliano@mh-hannover.de

Beate Brand-Saberi
Department of Anatomy and Molecular Embryology, Institute of Anatomy
Ruhr-University Bochum
Bochum, Germany
Beate.Brand-Saberi@rub.de

Tobias Cantz
Research Group Translational Hepatology and Stem Cell Biology, Department of Gastroenterology, Hepatology and Endocrinology, Hannover Medical School
Hannover, Germany
cantz.tobias@mh-hannover.de

Michelle Coffee
Department of Cardiac, Thoracic, Transplantation and Vascular Surgery (HTTG)
Leibniz Research Laboratories for Biotechnology and Artificial Organs (LEBAO)
Hannover Medical School (MHH)
Hannover, Germany
coffee.michelle@mh-hannover.de

Yu-Qiang Ding
Key Laboratory of Arrhythmias, Ministry of Education of China, East Hospital
Department of Anatomy and Neurobiology
Tongji University School of Medicine
Shanghai, China

Institutes of Brain Sciences, State Key Laboratory of Medical Neurobiology, and Department of Laboratory Animal Science, Fudan University
Shanghai, China
dingyuqiang@vip.163.com

Andreas Faissner
Department of Cell Morphology and Molecular Neurobiology, Faculty of Biology and Biotechnology, Ruhr-University Bochum
Bochum, Germany
andreas.faissner@ruhr-uni-bochum.de

Michael Fuchs
Department of Practical Philosophy and Ethics
Catholic Private University Linz, Linz, Austria
m.fuchs@ku-linz.at

Shaorong Gao
Tongji University, School of Life Sciences and Technology, Shanghai, China
gaoshaorong@tongji.edu.cn

Bernd Giebel
Institute for Transfusion Medicine, University
Hospital Essen, University of Duisburg-Essen
Essen, Germany
bernd.giebel@uk-essen.de

André Görgens
Institute for Transfusion Medicine, University
Hospital Essen, University of Duisburg-Essen
Essen, Germany

Clinical Research Center, Department of
Laboratory Medicine, Karolinska Institutet
Stockholm, Sweden
andre.gorgens@ki.se

Nina Graffmann
Institute for Stem Cell Research and
Regenerative Medicine, Heinrich Heine
University Düsseldorf, Medical Faculty
Düsseldorf, Germany

Hannes Klump
Institute for Transfusion Medicine
University Hospital Essen
Virchowstrasse, Essen, Germany
Hannes.Klump@uk-essen.de

Gesine Kogler
Heinrich-Heine-University, Medical Faculty
Institute for Transplantation
Diagnostics and Cell Therapeutics
Düsseldorf, Germany

University Medical Clinic
Düsseldorf, Germany
Gesine.Koegler@med.uni-duesseldorf.de

Audrey Ncube
Institute for Stem Cell Research and
Regenerative Medicine, Heinrich Heine
University Düsseldorf, Medical Faculty
Düsseldorf, Germany

Eric Bekoe Offei
Department of Anatomy and Molecular
Embryology, Institute of Anatomy
Ruhr University Bochum
Bochum, Germany

School of Veterinary Medicine
University of Ghana
Accra, Ghana

Jacqueline Reinhard
Department of Cell Morphology and Molecular
Neurobiology, Faculty of Biology and Biotech-
nology, Ruhr-University Bochum
Bochum, Germany

Lars Roll
Department of Cell Morphology and Molecular
Neurobiology, Faculty of Biology and Biotech-
nology, Ruhr-University Bochum
Bochum, Germany

Margit Schulze
Hochschule Bonn-Rhein-Sieg, Tissue
Engineering, Sankt Augustin, Germany

Lucas-Sebastian Spitzhorn
Institute for Stem Cell Research and Regenera-
tive Medicine, Heinrich Heine University
Düsseldorf, Medical Faculty
Düsseldorf, Germany

Ursula Theocharidis
Department of Cell Morphology and Molecular
Neurobiology, Faculty of Biology and Biotech-
nology, Ruhr-University Bochum
Bochum, Germany

Edda Tobiasch
Hochschule Bonn-Rhein-Sieg,
Tissue Engineering, Sankt Augustin, Germany
edda.tobiasch@h-brs.de

Christian Horst Tonk
Hochschule Bonn-Rhein-Sieg,
Tissue Engineering, Sankt Augustin, Germany

Chengzhong Wang
Mitokinin Incorporation
San Francisco, CA, USA
chengzhong_wang@hotmail.com

Yixuan Wang
Tongji University, School of Life Sciences and
Technology, Shanghai, China

Markus Witzler
Hochschule Bonn-Rhein-Sieg,
Tissue Engineering, Sankt Augustin, Germany

Wasco Wruck
Institute for Stem Cell Research and Regenera-
tive Medicine, Heinrich Heine University
Düsseldorf, Medical Faculty
Düsseldorf, Germany

Holm Zaehres
Department of Anatomy and Molecular
Embryology, Institute of Anatomy
Ruhr University Bochum
Bochum, Germany

Department of Cell and Developmental
Biology, Max Planck Institute for Molecular
Biomedicine, Münster, Germany
holm.zaehres@rub.de

Xiaoqing Zhang
Tongji University School of Medicine
Tongji University, Shanghai, China
xqzhang@tongji.edu.cn

Jianfeng Zhou
Tongji University, School of Life Sciences and
Technology, Shanghai, China

Robert Zweigerdt
Department of Cardiac, Thoracic
Transplantation and Vascular Surgery (HTTG)
Leibniz Research Laboratories for Biotechnol-
ogy and Artificial Organs (LEBAO)
Hannover Medical School (MHH)
Hannover, Germany
zweigerdt.robert@mh-hannover.de

Abbreviations

ATMP	Advanced Therapy Medicinal Products
BM MSC	Bone marrow mesenchymal/multipotent stromal cells
CB MSC	Cord blood mesenchymal/multipotent stromal cells
CB	Cord blood
CB-SC	Cord blood stromal cells
CD	Cluster of differentiation
C-MET	MET or MNNG HOS Transforming gene
CPD	Cumulative population doubling
DCBT	Double cord blood transplantation
DLK-1	Delta-like 1 homologue
DMEM	Dulbecco's Modified Eagle Medium
ECFC	Endothelial cord forming cell
EPC	Endothelial progenitor cell
ESC	Embryonic Stem Cells
FACS	Fluorescence-activated cell sorting
FCS	Fetal calf serum
FGF	Fibroblast growth factor
FLT3LG	Fms-related tyrosine kinase 3 ligand
GMP	Good manufacturing practice
GvHD	Graft-versus-host-disease
HOX	Homeobox
HPP-CFC	High Proliferative Potential-Colony Forming Cell
HSC	Hematopoietic stem cells
IL-6	Interleukin 6
iPSC	Induced pluripotent stem cells
LTC-IC	Long Term Culture-Initiating cell
MNC	Mononuclear cells
MSC	Mesenchymal/multipotent stromal cells
NC	Nucleated cells
NOD/SCID	non-obese diabetic/severe combined immunodeficiency
OC	Osteocalcin
OCT4A	Octamer binding transcription factor 4
OSX (SP7)	Osterix
Runx2	Runt domain –containing transcription factor 2

SCF	Stem cell factor
SCID	Severe combined immunodeficiency
SDF-1	Stromal cell-derived factor 1
SRC	SCID- repopulating cells
TPO	Thrombopoietin
UC MSC	Umbilical cord mesenchymal stromal cells
USSC	Unrestricted somatic stromal cells

Hematopoietic Stem Cells

Hannes Klump

What You Will Learn in This Chapter
Multipotent, hematopoietic stem cells (HSCs) as central organizers of blood cell production are the longest and probably best-known stem cell entity. In this chapter you will first
receive a brief introduction into the genesis of the stem cell theory of hematopoiesis. You
will learn which *in vitro* and *in vivo* assays can be performed to detect HSCs as well as their
gradually differentiating descendants, and what their individual informative value is. After
a short focus on when and where they develop during ontogenesis and some of the
molecular cues involved in their generation, we will get back to adult hematopoietic stem
and progenitor cells and discuss their use for organ replacement therapy in the clinical
setting. Because gene therapy of patients was first implemented in the hematopoietic system, you will learn about the current status of this rapidly evolving field. In the last paragraph, the future prospect of regenerative medicine based on *de novo* generated,
patient-tailored, autologous HSCs derived from induced, pluripotent stem cells will be
briefly discussed.

1.1   The Discovery of Hematopoietic Stem Cells

First evidence for their existence arose as a consequence from the experience with persons
in whom a complete failure of blood formation was observed after exposure to lethal doses
of ionizing radiation when the atomic bombs at the end of World War II were dropped
(Keller 1946). In animal experiments, two research groups found that intravenously trans-
fused bone marrow or spleen cells from a healthy animal were able to rescue lethally irradiated animals by restoring blood formation for a life-time (Jacobson et al. 1951; Lorenz
et al. 1951). A series of ground breaking experiments performed by James Till, Ernest
McCulloch and coworkers showed that hematopoiesis is organized by cell clones which
exist in the bone marrow, which can expand (i.e. make more of themselves) and are capable to generate myeloid, erythroid as well as lymphoid colonies in spleens after transplantation, so-called *colony forming units-spleen* (CFU-S) (◘ Fig. 1.1) (Till and McCulloch
1961; Becker et al. 1963; Wu et al. 1968) (Free access to some of the historical references
can be found at the University of Toronto Website: ▶ https://tspace.library.utoronto.ca/
handle/1807/2326). After transplantation, spleen colonies were observed between day 7
and day 12. However, the later colonies showed a different cellular composition than those
observed earlier: the early colonies mainly contained myeloerythroid cells derived from a
transplanted cell with limited potency, a lineage-restricted, myeloerythroid *progenitor*,
whereas the later colonies also contained lymphoid cells. Years later, using highly purified
cell fractions, it was demonstrated that only those day12 spleen colonies were formed by
cells on the top of a hierarchical system, multipotent hematopoietic stem cells (HSCs)
(Magli et al. 1982; Na Nakorn et al. 2002). Till & McCulloch's pioneering work set the stage
for today's research on all types of stem cells and how we functionally define them, namely
as a cell intrinsically carrying the potency to either self-renew or differentiate towards
specialized effector cells.

Based on the idea of self-renewal and multilineage differentiation, *in vitro* assays were
developed to characterize the hematopoietic tree and help understand how
HSC-differentiation towards all the different blood cell types is controlled. Donald Metcalf
and Ray Bradley were the first to demonstrate that single clonogenic bone marrow cells
(colony forming units, CFUs) were able to form myeloid colonies in semi-solid medium,

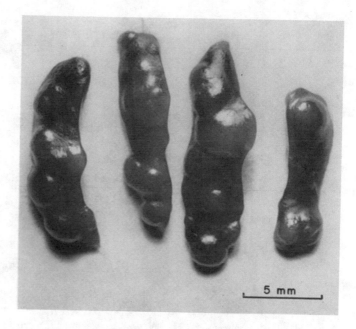

◘ Fig. 1.1 Spleens of irradiated mice 10 days after injection of 6×10^4 nucleated cells. The nodules on which the assay is based are readily seen (Original figure published by Till & McCulloch) (Till and McCulloch 1961). Each nodule contains the lymphomyeloid progeny of a multipotent hematopoietic progenitor cell clone

5 mm

in vitro, containing different mature cell types. Furthermore, they showed that soluble factors were absolutely required for this activity (Bradley and Metcalf 1966). They coined the term *colony stimulating factors* (CSF) for these soluble (glyco)peptides of which many were purified over the next years. CSFs turned out to be essential for the proliferation, differentiation and survival of hematopoietic stem and downstream progenitor cells, as well as for effector functions of the differentiated, mature cells. The first CSFs identified controlling myelopoiesis were GM-CSF (granulocyte-macrophage CSF), G-CSF (granulocyte CSF), M-CSF (macrophage CSF) and multipotential colony-stimulating factor, nowadays called interleukin-3 (IL-3). Although the colony formation cell (CFC) assay allowed to uncover key aspects of the hematopoietic hierarchy, it mainly supports the detection and quantitation of more mature, actively proliferating (and, thus, colony forming) myeloid progenitor cells. However, it is not suitable for detecting immature, slowly cycling HSCs. Thus, more complex *in vitro* assays were developed to detect and estimate the frequency of immature HSCs in a given cell population, such as the long-term culture initiating cell (LTC-IC) and cobblestone area-forming cell (CAFC) assays (◘ Fig. 1.2). Despite their value to detect more immature progenitors, these latter assays also do not support the development of all hematopoietic lineages, and therefore, conclusions about HSCs cannot be drawn. Hence, the gold standard to detect and quantify HSCs is the transplantation experiment into an appropriate, recipient animal, *in vivo*.

1.2 How to Detect Multipotent Hematopoietic Stem Cells

Ultimately, HSCs are defined by their biological activity in an organism, *in vivo*. After transplantation, HSCs have to migrate to the bone marrow (*homing*), "settle down" at the right place (the *niche*), expand and start organizing the production the whole range of blood cells (*engraftment*) for the rest of the life of the recipient (*long-term repopulation*) (Orkin and Zon, 2008). In mice, "long-term" is commonly defined as repopula-

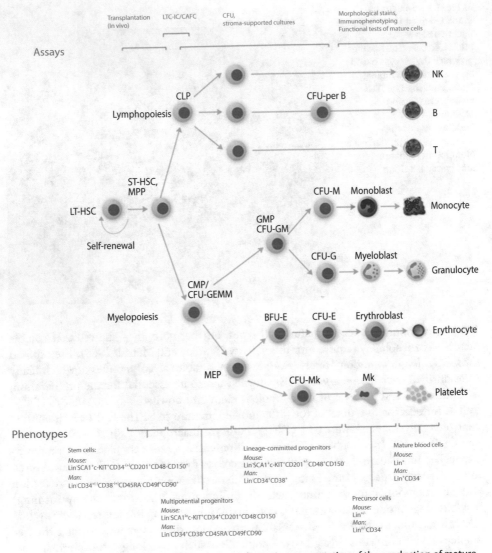

Fig. 1.2 HSC proliferation and differentiation. Schematic representation of the production of mature blood cells by the proliferation and differentiation of hematopoietic stem cells. Intermediate stages are also depicted. Transplantation assays identify repopulating stem cells. Assays for Long-Term Culture-Initiating Cells (LTC-IC) and Cobblestone Area-Forming Cells (CAFC) identify very primitive progenitor cells that overlap with stem and progenitor cells. Colony-Forming Unit (CFU) assays identify multipotential and, mainly, lineage-committed progenitor cells. LT-HSC long-term hematopoietic stem cell, ST-HSC short-term hematopoietic stem cell, MPP multipotential progenitor, CMP common myeloid progenitor, CLP common lymphoid progenitor, CFU-GEMM colony-forming unit – granulocyte/erythrocyte/ macrophage/megakaryocyte, BFU-E burst-forming unit – erythroid, CFU-E colony-forming unit – erythroid, CFU-Mk colony-forming unit – megakaryocyte, CFU-GM colony-forming unit – granulocyte/ macrophage, CFU-G colony-forming unit – granulocyte, CFU-M colony-forming unit – macrophage. The most definitive markers used to identify the various types of mouse and human hematopoietic cells are shown on the bottom. Additional markers can be used to further distinguish between subsets. Not shown are the plasmacytoid and myeloid dendritic cell (DC) lineages, which are derived from CLP and CMP, respectively. (Image taken with permission from the "Hematopoietic Stem and Progenitor Cells" mini-review; Authors: Albertus Wognum and Stephen Szilvassy from STEMCELL Technologies. URL: ▶ https:// cdn.stemcell.com/media/files/minireview/MR29068-Hematopoietic_Stem_and_Progenitor_Cells.pdf)

tion beyond 16 weeks post transplantation plus the ability to repopulate secondary mice in a serial transplantation experiment (Lemischka et al. 1986; Kent et al. 2009). Thus, HSCs *sensu stricto* are *long-term multilineage repopulating stem cells* (LT-HSCs). Based on the assumption that even one single HSC is sufficient to repopulate an entire organism, long-term and multilineage, HSCs can be quantified without any previous knowledge of their individual identity. This is done by a so-called *limiting dilution assay* (LDA), in which decreasing numbers of test cells are transplanted into predefined numbers of recipient mice. To protect the host from the immediate toxicity of the conditioning treatment (see below), more mature, committed progenitors are co-transfused which generate effector cells necessary for short-term survival (such as thrombocytes and erythrocytes) but are assumed not to compete with HSCs for their niches in the bone marrow. Based on the percentage of mice showing long-term repopulation after 4–6 months with a given number of cells transplanted, the frequency of HSCs within the test cell suspension can be calculated by Poisson statistics (Miller et al. 2008; Eaves 2015; Fazekas de St 1982). Based on this assay, the LT-HSC frequency in the bone marrow of healthy young mice is about 1:100,000 nucleated cells. It is important to keep in mind that transplanted HSCs can only engraft if the host has been prepared to accept newly incoming stem cells, because the number of niche places is limited and blocked by resident stem cells. Therefore, endogenous HSCs need to be removed to create space ("empty seats") for the incoming new HSCs, for instance by ionizing irradiation or treatment with toxic alkylating drugs such as Busulfan or Treosulfan (the process is called *conditioning*). Although transplantation is the only way to qualify and quantify HSCs present in a cell suspension, one needs to be aware of the potential limitations of the experimental approach as the transplantation setting which differs from steady state hematopoiesis in the bone marrow of a healthy organism where most of the LT-HSCs are dormant (Hofer et al. 2016).

To measure human HSPCs (human stem and progenitor cells), transplantation into humans is not possible, for obvious reasons. Thus, "humanized" immunodeficient mouse strains have been developed as surrogate xenograft models which support human HSPC engraftment and differentiation to a certain extent, for example the NOD/SCID mouse strain (Non-Obese Diabetic, Severe Combined Immunodeficiency mouse) and derivatives thereof (Shultz et al. 1995, 2005). Despite its strengths, such xenograft models have limitations when assessing human HSPC activity. For example, there is evidence that the mouse niches do not support human HSPCs very well to maintain their quiescence and stemness. Instead, human HSPCs continue to proliferate and differentiate [reviewed by (Goyama et al. 2015)]. Studies in non-human primates also raised concerns about the conclusiveness of the NOD/SCID-repopulating cell assays. In primates, a significant proportion of long-term repopulating cells did not overlap with NOD/SCID-repopulating cells (SRC) (Horn et al. 2003; Horn and Blasczyk 2007). As a matter of fact, human hematopoietic progenitors which have lost their erythro-myeloid potential, the so-called lymphoid-primed multipotent progenitor cells (LMPPs) (see ▢ Fig. 1.3), still contain SRC properties (Görgens et al. 2013; Kohn et al. 2012). Therefore, it is always imperative to test the generation of all blood lineages in a transplantation experiment to be able to draw conclusions on real HSCs. Because it is not always possible to clearly differentiate between the progeny of transplanted cells and the offspring of potentially remaining endogenous HSCs (which were not removed by the conditioning procedure), the more cautious term HSPC is often preferred to HSC when interpreting the results of transplantation experiments.

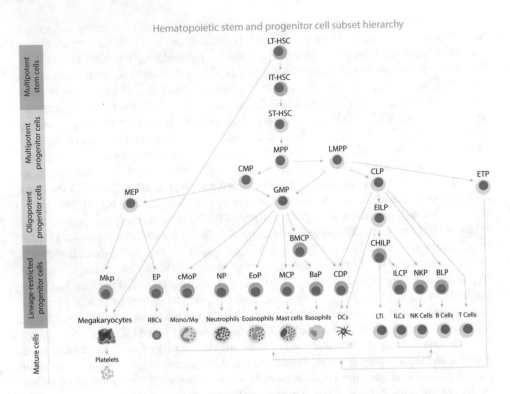

Fig. 1.3 Hematopoietic Stem and Progenitor Cell (HSPC) family tree. Classic model of mouse hematopoiesis and inferred trajectories (arrows) of HSPC differentiation leading to the production of mature blood cells. The overarching concept of this model suggests that HSC populations, which reside at the apex of a hierarchical organization of cellular relationships, give rise to several discrete intermediate progenitor populations including multipotent, oligopotent, and lineage-restricted progenitor cells. Whilst this simplified model assumes homogeneity in HSPC populations, and thus an equal ability to produce all blood cells, single cell transplantation assays as well as transcriptional profiling have revealed significant heterogeneity within subpopulations that are intrinsically biased toward the generation of certain blood lineages. For example, differentiation of HSCs to platelets may not involve transitioning to CMP or MEP intermediates and can be achieved directly via megakaryocyte-primed HSCs. It must be highlighted that models of HSPC development are continually evolving, with the current consensus indicating that step-wise commitment of HSPCs from one intermediate to the other, as traditionally depicted, may not be representative of the situation in vivo. Rather, cell intrinsic and extrinsic inputs may guide HSC development along a continuum with cells gradually differentiating and passing through a small set of intermediate stages. Definition of acronyms can be found under "Abbreviations" below the table. (Image taken with permission from the "Mouse Hematopoietic Stem and Progenitor Cell Phenotyping" wallchart; Authors: Albertus Wognum and Stephen Szilvassy from STEMCELL Technologies. URL: ▶ https://www.stemcell.com/media/files/wallchart/WA27128-Mouse_Hematopoietic_Stem_and_Progenitor_Cell_Phenotypes_and_Frequencies.pdf)

1.3　Prospective Isolation of HSCs

The enrichment of viable stem cells relies on a combination of their physical properties such as density, size and granularity and their biochemical characteristics such as membrane-bound enzymes (e.g. ATP-dependent transmembrane transport channels which actively pump drugs out of the cell), expression of distinct antigens on the cell surface ("surface markers"), or their proliferative and metabolic status (Miller et al. 2013).

Commonly employed methods to enrich hematopoietic stem and progenitor cells from a heterogeneic population are fluorescence-activated cell sorting (FACS) and magnetic activated cell separation (MACS). For FACS, the cell population is first labeled with a combination of antibodies, each coupled with a different fluorochrome, which bind to specific antigens known to be either expressed on the surface of HSCs or non-HSCs, such as differentiated effector cells ("lineage-specific markers"). Based on positive selection (expressed on HSCs) and negative depletion (not expressed on HSCs), together with a selection on small size and low granularity allows for a significant enrichment of HSCs by flow cytometry-based cell-sorting. The MACS method also relies on antibody-based selection of cells. However, the antibodies are bound to magnetic beads allowing for positive and negative separation in a magnetic field (Hu et al. 2016; Grutzkau and Radbruch 2010). In freshly isolated, primary tissues, HSCs reside in a cell population defined by a certain combination of surface antigens (termed clusters of differentiation, CD), the immunophenotype of HSCs. For example, human HSCs reside in subpopulation which does not express differentiation-specific markers (lineage negative, Lin-), but expresses CD34 (CD34+) and are negative for CD38 (CD38-). To date, they are defined as being Lin-CD34 + CD38-CD90 + CD45RA- with 1 out of 3 cells assumed to be real HSCs (Notta et al. 2011; Gorgens et al. 2013). In the mouse, HSCs reside in a fraction of bone marrow cells being Lin-Sca1 + Kit+ (LSK). In combination with the surface markers CD150 + CD244-CD48- (the so-called SLAM-code) (Kiel et al. 2005) and CD201 (the endothelial protein C receptor) (Balazs et al. 2006), a very high enrichment of bone marrow HSCs can be achieved, with approx. 1:2–1:3 being real LT-HSCs (Kent et al. 2009).

1.4 Development of HSCs in the Embryo

During embryonic development, specification of HSCs progressively occurs at different anatomic sites in the embryo before they finally colonize the bone marrow around birth. These locations depend on signals from the developing adjacent germ layers and tissues (Belaoussoff et al. 1998). In the mouse, hematopoietic progenitor activity generating only macrophages, megakaryocytes and primitive, nucleated erythrocytes can be detected as early as at 7.5 days post conception (E7.5) in extra-embryonic mesodermal yolk sac blood islands (corresponding to embryonic day, E15–18 in humans) (Palis et al. 1999; Mcgrath et al. 2015). When CFC-assays are performed with cells isolated from that region, blast colonies containing both endothelial and hematopoietic progenitors can be detected, which are formed by bipotent progenitor cells called hemangioblasts (Choi et al. 1998; Kennedy et al. 1997). From E10.5 (E30 in humans) on, HSCs capable of engrafting lethally irradiated animals and providing long-term multilineage repopulation (definitive HSCs) are detectable in the ventral part of the main vessels – the abdominal aorta (in the aorto-gonado-mesonephros region, AGM), vitelline and the umbilical arteries (Dzierzak and Speck 2008; Medvinsky and Dzierzak 1996; Muller et al. 1994; Yoshimoto et al. 2008). Slightly later, between E11.0 and 11.5, the placenta also becomes a site of HSC generation (Gekas et al. 2005; Ottersbach and Dzierzak 2005; Rhodes et al. 2008). Prior to migration to the bone marrow they colonize the fetal liver where they are assumed to expand and further mature (◻ Fig. 1.4a) However, the exact role of the fetal liver for HSC development is not fully understood yet (Rybtsov et al. 2016) (for excellent reviews on this topic, please read (Ivanovs et al. 2017, Medvinsky et al. 2011, Perlin et al. 2017).

Of all anatomical sites, the central place where first definitive HSCs are formed is the dorsal aorta (◨ Fig. 1.4b) (Muller et al. 1994; Medvinsky and Dzierzak 1996). There, they are generated by a transiently existing population of cells lining the ventral endothelial floor, the *hemogenic endothelium cells* (HECs) (Tavian et al. 2001; Bertrand et al.

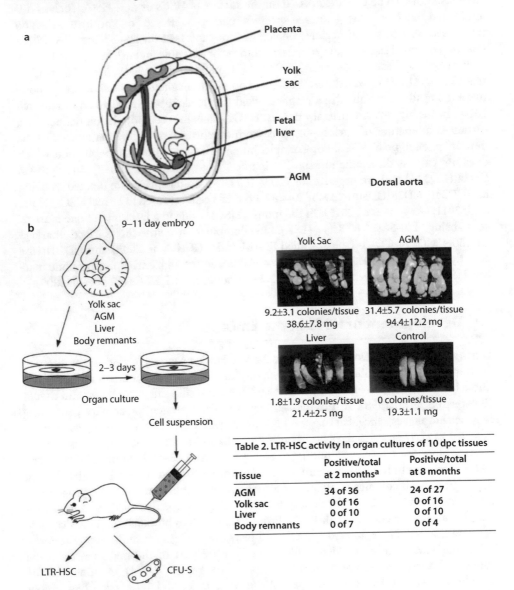

a

Placenta

Yolk sac

Fetal liver

AGM   Dorsal aorta

b   9–11 day embryo

Yolk Sac   AGM

9.2±3.1 colonies/tissue   31.4±5.7 colonies/tissue
38.6±7.8 mg   94.4±12.2 mg

Liver   Control

1.8±1.9 colonies/tissue   0 colonies/tissue
21.4±2.5 mg   19.3±1.1 mg

Yolk sac
AGM
Liver
Body remnants

2–3 days

Organ culture

Cell suspension

Table 2. LTR-HSC activity In organ cultures of 10 dpc tissues		
Tissue	Positive/total at 2 months[a]	Positive/total at 8 months
AGM	34 of 36	24 of 27
Yolk sac	0 of 16	0 of 16
Liver	0 of 10	0 of 10
Body remnants	0 of 7	0 of 4

LTR-HSC   CFU-S

◨ **Fig. 1.4**   Generation of HSCs during mouse development. **a** Scheme of a day 9–11 mouse embryo and sites of hematopoiesis. **b** Landmark experiments demonstrating that the aorto-gonado-mesonephros region is the site where first definitive HSCs are generated. In the experiments performed, cell suspensions taken from different areas of a mouse embryo were transplanted into lethally irradiated recipient mice. Colony forming units-spleen (CFU-S) are shown righthand. The presence of long-term repopulating HSCs (LT-HSCs) is shown below, which can only be detected after transplantation of AGM-derived cells. (From (Medvinsky and Dzierzak 1996)

2010; Jaffredo et al. 1998; Zovein et al. 2010; Kissa et al. 2008) which transform to hematopoietic suspension cells by process called *endothelial-to-hematopoietic transition* (EHT) (Lancrin et al. 2009; Eilken et al. 2009). EHT is triggered by signals sent from the developing tissue surrounding the aorta. The somites induce fibroblast growth factor (FGF), wingless (Wnt) and NOTCH-signalling (Clements et al. 2011; Pouget et al. 2014; Lee et al. 2014), the subendothelial gut area triggers sonic hedgehog (SHH) and bone morphogenic protein (BMP) signalling, and the sympathic nervous system acts via catecholamines (Fitch et al. 2012). Biomechanical forces originating from pulsative waves after initiation of heart beating also contribute to the induction of EHT via the nitric oxid signalling pathway (Adamo et al. 2009; North et al. 2009). All the different extrinsic cues ultimately induce an intrinsic, specific pattern of gene transcription in HECs necessary to propel the cell towards further hematopoietic specification (Souilhol et al. 2016).

The most important transcription factor controlling this critical stage of HSC specification is RUNX1. Deletion of the *RUNX1* gene is embryonic lethal as EHT of the hemogenic endothelium is blocked and, as a consequence, definitive (adult-type) HSCs are not formed (Wang et al. 1996; De Bruijn and Dzierzak 2017). However, in addition to RUNX1 a number of other transcription factors (TFs) are of key importance for the development of adult HSCs, for instance GATA2, SCL, LYL1, LMO2, FLI-1 and ERG. This heptad of transcription factors is not only critical for HSC development in the embryo but later on also regulates the function of adult HSPCs in the bone marrow. Mechanistically, they cooperatively bind to similar motifs in the genome (the individual motifs being of approx. 200 bp in size) to control the transcription of genes vital for HSPCs identity (Wilson et al. 2010, 2011). Besides the abovementioned signaling molecules, retinoic acid (RA), a morphogen necessary for an orderly embryonic development, is also required for HEC and, thus, HSC specification (Chanda et al. 2013; Gritz and Hirschi 2016). It strongly induces the expression of homeobox (*HOX*)-genes, which are well known for their importance to confer positional information to cells and tissues during embryonic development and to control stem cell self-renewal and differentiation in the adult organism (Mallo et al. 2010). Of the chromosomal regions in which the four paralogous *HOX* gene clusters are organized in mammals (clusters A, B, C, and D), only the A and B-clusters are expressed during hematopoietic cell development. Abrogation of RA-signaling leads to a reduced transcription of the *Hoxb* cluster in mice and an accompanying decrease of LT-HSC numbers in the bone marrow, emphasizing the importance of *HOX* gene expression for hematopoiesis (Qian et al. 2018). In line with these results, ectopic expression of human *HOXB4*, which is also a target of retinoic acid signaling (Cabezas-Wallscheid et al. 2017) enhances the hematopoietic potential of differentiating mouse embryonic stem cells (ESCs), *in vitro* (Helgason et al. 1996; Pilat et al. 2005; Schiedlmeier et al. 2007; Lesinski et al. 2012; Chan et al. 2008; Klump et al. 2013b; Pilat et al. 2013). It does so by turning on the transcription of genes critical for HSC specification, including the aforementioned heptad transcription factor genes and thereby strongly promotes the development of hemogenic endothelium cells (Teichweyde et al. 2018).

Many of the genes involved in HSC specification are also involved in the control of HSC self-renewal and differentiation in the adult organism. Hence, inherited or acquired mutations can lead to unbalanced hematopoietic differentiation and, by this, contribute to the development of hematologic malignancies such as leukemias and lymphomas.

1.5 Clinical Use of HSCs

When malignancies of the hematopoietic system are treated, some of the patients have to undergo a *hematopoietic stem cell transplantation* (HSCT). Although the term HSCT itself suggests that a highly defined, pure stem cell population is used, that's not really the case. Instead, cell preparations are used which contain CD34+ HSPCs, as well as many other cell types. There are three sources from which HSPCs can be obtained: mobilized peripheral blood stem cells (PBSCs), bone marrow and umbilical cord blood (UCB). Currently, PBSCs are most commonly used as they are relatively easy to obtain: after injection of mobilizing drugs such as G-CSF or Plerixafor (AMD3100, a CXCR4 antagonist) at low concentrations, a small proportion of HSPCs dislodge from their niches and enter the circulation from where they can be harvested from peripheral blood by apheresis (i.e. extracorporeal separation of blood components by centrifugation). Depending on the intended use, the collected cell preparation can additionally be subjected to a selection or depletion procedure, to enrich for CD34+ HSPCs or to deplete certain immune cells (e.g. CD19+ B-cells and CD3+ T-cells).

There are generally two forms of HSCT which are performed: *autologous HSCT*, in which a patient receives his/her own stem cells; and *allogeneic HSCT*, in which cells of a HLA-matched donor are used. Autologous HSCT is commonly performed to treat malignant hematologic diseases for which no cure yet exists, such as multiple myeloma, a malignant disease caused by uncontrolled proliferation of plasma-cells. In such cases, autologous stem cells are first harvested to allow for an intensified, bone marrow toxic (*myelotoxic*) radiochemotherapy with subsequent transplantation of the patient's own stem cells. This kind of treatment aims at improving the quality of remission (i.e. longer disease-free period), but usually does not cure the disease. In contrast, the intention of allogeneic HSCT is always curative. After intense radiochemotherapy combined with an ablation of the patient's immune-system, transplanted donor HSCs replace the patient's entire hematopoietic system. Importantly, allogeneic HSCT with unfractioned stem cell preparations is not only an organ replacement therapy but also an immuno-therapy because the co-infused immune-cells present in the graft can recognize malignant cells in the recipient as foreign and reject them, a desired effect called *graft-versus-leukemia*, GvL. However, the drawback of this therapy is an inevitable association with a rejection of the recipients healthy cells termed graft-versus-host disease (GvHD), which can become life-threatening. Therefore, an important goal of ongoing research is to find out which cell populations are responsible for GvL or GvHD and how both effects can be separated from each other, by cell enrichment/depletion and/or drug-based (Negrin 2015).

1.6 Gene Therapy of HSCs

In the case of well-characterized, inherited monogenetic diseases leading to functional defects of the hematopoietic system, such as severe combined immunodeficiencies (SCID) or hemoglobinopathies, gene therapy using autologous HSPCs is progressively becoming a practicable alternative – particularly when there is no healthy, HLA-identical sibling available as donor. Although HLA-matched unrelated donors (MUD) or haploidentical donors can be used, there is a significant risk of GvHD with all its potentially devastating consequences. Therefore, therapy with gene-modified or gene-corrected autologous HSPCs is the most reasonable treatment option in the future (Porteus 2016). At present, the most commonly used vectors for introducing transgenes into HSPCs are derived from members of the *retroviridae*, namely gammaretroviruses (derived from mouse leukemia virus, MLV) and lentiviruses (HIV-derived). Despite facing the severe adverse event of therapy-associated leukemia in the past, caused by semi-random insertions of the vector into the genome and an associated activation of neighboring proto-oncogenes (Fischer et al. 2010), many improvements of the vector architecture have contributed to a significantly increased safety profile of this technology ever since. One of the most important ones being the removal of potent enhancer sequences located in the U3 region of the long terminal repeats (LTR) resulting in self-inactivating (SIN) vectors. By the use of internal promoter elements with high tissue specificity, an additional layer of safety was introduced into the system (Kustikova et al. 2010). As a matter of fact, the numbers of promising gene therapy trials using SIN-lentiviral vector-modified autologous HSPCs is rising steadily worldwide, for instance for curing Adrenoleukodystrophy (Eichler et al. 2017; Cartier et al. 2009), Sickle Cell Anemia or Thalassemia (Sii-Felice et al. 2018; Ribeil et al. 2017; Thompson et al. 2018). The success of this kind of treatment has led to the first European market approval of gene-modified HSPCs in 2016 for the treatment of Severe Combined Immunodeficiency (ADA-SCID variant). For the production of the *advanced therapy medicinal product* (ATMP) with the trade name Strimvelis™, autologous CD34+ cells are transduced with a retroviral vector expressing the human adenosine-desaminase (ADA) coding sequence and reinfused into the immunodeficient patient (Aiuti et al. 2017) (European Medicines Agency document: ▸ https://www.ema.europa.eu/en/medicines/human/EPAR/strimvelis).

To further increase the safety of gene therapy, it would be desirable to repair HSCs by homologous recombination and expand characterized, "safe" clones to clinically relevant numbers before subsequent transplantation. However, despite the ability to purify HSCs to near homogeneity, no defined culture conditions exist allowing for their *ex vivo* expansion, yet. Even though a huge number of studies have reported the expansion of human and mouse hematopoietic stem and progenitor cells (HSPCs) with colony forming capabilities, only few studies demonstrated the expansion of a population which can repopulate immunodeficient mice post transplantation (NOD/SCID or NSG repopulating cells, SRCs), which is – despite its limitations (see ▸ Sect. 1.2) – the current gold standard to document the efficiency of HSPC-expansion protocols (Doulatov et al. 2012). Genetic engineering has also been used as an approach to expand human HSCs. Significant expansion of HSPCs has been achieved after constitutive over-expression of transcription factors, as first demonstrated in the mouse system using the human homeodomain transcription factor HOXB4 (Sauvageau et al. 1995; Antonchuk et al. 2002; Krosl et al. 2003; Klump et al. 2001, 2005). HOXB4-mediated expansion was also shown for human HSPCs (Schiedlmeier et al. 2003; Buske et al. 2002; Amsellem et al. 2003) and for non-human primate cells (Watts et al. 2010, 2012; Zhang et al. 2007), however to a less dramatic extent than in mice. Mechanistically, this transcription factor appears to regulate similar genes as in differentiating pluripotent stem cells (▸ Sect. 1.4), thereby altering the sensitivity of cells to signaling pathways known to be important for controlling stem cell self-renewal and differentiation (Schiedlmeier et al. 2007; Will et al. 2006; Klump et al. 2005). Although the use of human HSCs engineered to constitutively express a transcription factor mediating its expansion may not be safe enough for clinical application, the gained knowledge will contribute to the understanding of the fundamental mechanisms controlling self-renewal and cell fate decisions. By this, it will pave the way for the establishment of conditions allowing for selective *ex vivo* expansion of genetically unmanipulated HSCs.

1.7 Derivation of Patient-Specific, Tailored HSCs Derived from Pluripotent Stem Cells

Because of the challenges to expand HSCs without genetic manipulation, less committed, "upstream" stem cells may constitute a more feasible alternative to generate them in therapeutically useful numbers. Thus, Pluripotent Stem Cells (PSCs), such as Embryonic Stem Cells (ESCs) and induced Pluripotent Stem Cells (iPSCs) are considered a potential source for generating HSCs *in vitro*. They possess an extensive self-renewal capacity, *in vitro*, allow for efficient gene repair, clonal selection and can be differentiated towards any somatic cell type when appropriate conditions are provided (Keller 2005). Thus, reprogrammed iPSCs derived from patient cells can be considered an attractive starting point for gene correction and subsequent generation of autologous HSCs in future (◘ Fig. 1.5) (Klump et al. 2013a). Although proof-of-principle has been provided that such an approach is feasible with patient cells (Raya et al. 2009), directed differentiation towards transplantable HSCs has remained relatively inefficient. Currently, the engraftment rates after transplantation into appropriate recipient mice are very low (Ledran et al. 2008; Wang et al. 2005; Narayan et al. 2006; Tian et al. 2006). One of the most likely reasons is that the majority of the obtained cells corresponds to an immature pre-HSC incapable of efficient engraftment, indicating that some key requirements necessary for full hematopoietic specification, *in vitro*, are still ill-defined, likely owing to our incomplete knowledge of HSC development, *in vivo*. Thus, a thorough understanding of how HSCs are formed in the embryo, *in vivo*, will be critical for their development from pluripotent stem cell sources, *in vitro* (reviewed by (Rowe et al. 2016; Hotta and Yamanaka 2015).

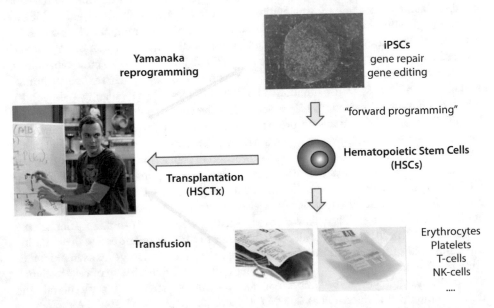

◘ **Fig. 1.5** Concept of iPSC-based generation of hematopoietic cells. After reprogramming of a patient's cells back to pluripotency, for example by the use of the Yamanaka-factors Oct4, Sox2, Klf4 and Myc (Takahashi et al. 2007), an underlying genetic defect can be repaired by gene editing, single cell clones expanded and the thoroughly characterized molecularly. Gene-corrected, presumably safe iPSCs can then be differentiated towards autologous hematopoietic stem cells (HSCs) for transplantion, or, further, to effector cells useful for transfusion purposes

┌─ **Take Home Message** ─

Hematopoietic stem cells are the longest and best studied stem cell entity for which most clinical experience exists. They will likely continue to be a pacemaker of our fundamental understanding of stem cell biology and presumably remain in a pioneering position regarding the development of new cell-based therapies. This is especially true for the area of genome editing-based gene-therapy where novel techniques such as CRISPR-Cas9 are currently under way into first clinical trials (phase I/II) for treating transfusion-dependent β-thalassemia, in Germany (Regensburg, Tübingen) and in Great Britain (London) (▶ https://clinicaltrials.gov/ct2/show/NCT03655678).

A brief introduction into theoretical and practical aspects of the basic biology of HSCs, their development during embryogenesis, assays how to detect them, their clinical application and some future prospects were given in this chapter. However, 'panta rhei' – everything flows: because of the rapid progress made in this area of research, some parts of this chapter may be outdated soon (as is true for any printed textbook). Thus, internet resources have become an invaluable complementary source of up-to-date information. Some trustworthy, recommendable information sites on HSCs are located at the U.S. National Institute of Health (▶ https://stemcells.nih.gov/info/2001report/chapter5.htm), the International Society of Stem Cell Research (ISSCR) (▶ www.isscr.org) and the European "Eurostemcell" network (▶ https://www.eurostemcell.org/resource-type/fact-sheet).

❓ Questions

1. Hematopoietic stem cells (HSCs) are defined as:
   (a) committed progenitor cells
   (b) pluripotent cells
   (c) multipotent cells
   (d) totipotent cells

2. Hematopoietic colony forming cells (CFCs) are:
   (a) multipotent stem cells
   (b) pluripotent stem cells
   (c) committed progenitor cells
   (d) the same as colony forming units spleen (CFU-S)

3. In healthy adult humans, hematopoiesis takes place at the following sites:
   (a) spleen
   (b) thymus
   (c) liver
   (d) bone marrow

4. During embryonic development, first definitive HSCs are formed in the
   (a) fetal liver
   (b) dorsal aorta
   (c) yolk sac
   (d) somites

5. The most important transcription factor for endothelial-to-hematopoietic transition in the dorsal aorta is
   (a) LMO2
   (b) ID2
   (c) RUNX1
   (d) HOXD3

✅ **Answers**

1. c
2. c
3. d
4. b
5. c

References

Adamo, L., Naveiras, O., Wenzel, P. L., Mckinney-Freeman, S., Mack, P. J., Gracia-Sancho, J., Suchy-Dicey, A., Yoshimoto, M., Lensch, M. W., Yoder, M. C., Garcia-Cardena, G., & Daley, G. Q. (2009). Biomechanical forces promote embryonic haematopoiesis. *Nature, 459*, 1131–1135.

Aiuti, A., Roncarolo, M. G., & Naldini, L. (2017). Gene therapy for ADA-SCID, the first marketing approval of an ex vivo gene therapy in Europe: Paving the road for the next generation of advanced therapy medicinal products. *EMBO Molecular Medicine, 9*, 737–740.

Amsellem, S., Pflumio, F., Bardinet, D., Izac, B., Charneau, P., Romeo, P. H., Dubart-Kupperschmitt, A., & Fichelson, S. (2003). Ex vivo expansion of human hematopoietic stem cells by direct delivery of the HOXB4 homeoprotein. *Nature Medicine, 9*, 1423–1427.

Antonchuk, J., Sauvageau, G., & Humphries, R. K. (2002). HOXB4-induced expansion of adult hematopoietic stem cells ex vivo. *Cell, 109*, 39–45.

Balazs, A. B., Fabian, A. J., Esmon, C. T., & Mulligan, R. C. (2006). Endothelial protein C receptor (CD201) explicitly identifies hematopoietic stem cells in murine bone marrow. *Blood, 107*, 2317–2321.

Becker, A. J., McCulloch, E. A., & Till, J. E. (1963). Cytological demonstration of the clonal nature of spleen colonies derived from transplanted mouse marrow cells. *Nature, 197*, 452–454.

Belaoussoff, M., Farrington, S. M., & Baron, M. H. (1998). Hematopoietic induction and respecification of A-P identity by visceral endoderm signaling in the mouse embryo. *Development, 125*, 5009–5018.

Bertrand, J. Y., Chi, N. C., Santoso, B., Teng, S., Stainier, D. Y., & Traver, D. (2010). Haematopoietic stem cells derive directly from aortic endothelium during development. *Nature, 464*, 108–111.

Bradley, T. R., & Metcalf, D. (1966). The growth of mouse bone marrow cells in vitro. *The Australian Journal of Experimental Biology and Medical Science, 44*, 287–299.

Buske, C., Feuring-Buske, M., Abramovich, C., Spiekermann, K., Eaves, C. J., Coulombel, L., Sauvageau, G., Hogge, D. E., & Humphries, R. K. (2002). Deregulated expression of HOXB4 enhances the primitive growth activity of human hematopoietic cells. *Blood, 100*, 862–868.

Cabezas-Wallscheid, N., Buettner, F., Sommerkamp, P., Klimmeck, D., Ladel, L., Thalheimer, F. B., Pastor-Flores, D., Roma, L. P., Renders, S., Zeisberger, P., Przybylla, A., Schonberger, K., Scognamiglio, R., Altamura, S., Florian, C. M., Fawaz, M., Vonficht, D., Tesio, M., Collier, P., Pavlinic, D., GEIGER, H., Schroeder, T., Benes, V., Dick, T. P., Rieger, M. A., Stegle, O., & Trumpp, A. (2017). Vitamin A-retinoic acid Signaling regulates hematopoietic stem cell dormancy. *Cell, 169*(807–823), e19.

Cartier, N., Hacein-Bey-Abina, S., Bartholomae, C. C., Veres, G., Schmidt, M., Kutschera, I., Vidaud, M., Abel, U., Dal-Cortivo, L., Caccavelli, L., Mahlaoui, N., Kiermer, V., Mittelstaedt, D., Bellesme, C., Lahlou, N., Lefrere, F., Blanche, S., Audit, M., Payen, E., Leboulch, P., L'Homme, B., Bougneres, P., Von Kalle, C., Fischer, A., Cavazzana-Calvo, M., & Aubourg, P. (2009). Hematopoietic stem cell gene therapy with a lentiviral vector in X-linked adrenoleukodystrophy. *Science, 326*, 818–823.

Chan, K. M., Bonde, S., Klump, H., & Zavazava, N. (2008). Hematopoiesis and immunity of HOXB4-transduced embryonic stem cell-derived hematopoietic progenitor cells. *Blood, 111*, 2953–2961.

Chanda, B., Ditadi, A., Iscove, N. N., & Keller, G. (2013). Retinoic acid signaling is essential for embryonic hematopoietic stem cell development. *Cell, 155,* 215–227.

Choi, K., Kennedy, M., Kazarov, A., Papadimitriou, J. C., & Keller, G. (1998). A common precursor for hematopoietic and endothelial cells. *Development, 125,* 725–732.

Clements, W. K., Kim, A. D., Ong, K. G., Moore, J. C., Lawson, N. D., & Traver, D. (2011). A somitic Wnt16/Notch pathway specifies haematopoietic stem cells. *Nature, 474,* 220–224.

De Bruijn, M., & Dzierzak, E. (2017). Runx transcription factors in the development and function of the definitive hematopoietic system. *Blood, 129,* 2061–2069.

Doulatov, S., Notta, F., Laurenti, E., & Dick, J. E. (2012). Hematopoiesis: A human perspective. *Cell Stem Cell, 10,* 120–136.

Dzierzak, E., & Speck, N. A. (2008). Of lineage and legacy: the development of mammalian hematopoietic stem cells. *Nat Immunol, 9,* 129–136.

Eaves, C. J. (2015). Hematopoietic stem cells: Concepts, definitions, and the new reality. *Blood, 125,* 2605–2613.

Eichler, F., Duncan, C., Musolino, P. L., Orchard, P. J., De Oliveira, S., Thrasher, A. J., Armant, M., Dansereau, C., Lund, T. C., Miller, W. P., Raymond, G. V., Sankar, R., Shah, A. J., Sevin, C., Gaspar, H. B., Gissen, P., Amartino, H., Bratkovic, D., Smith, N. J. C., Paker, A. M., Shamir, E., O'Meara, T., Davidson, D., Aubourg, P., & Williams, D. A. (2017). Hematopoietic stem-cell gene therapy for cerebral Adrenoleukodystrophy. *The New England Journal of Medicine, 377,* 1630–1638.

Eilken, H. M., Nishikawa, S., & Schroeder, T. (2009). Continuous single-cell imaging of blood generation from haemogenic endothelium. *Nature, 457,* 896–900.

Fazekas de St, G. (1982). The evaluation of limiting dilution assays. *Journal of Immunological Methods, 49,* R11–R23.

Fischer, A., Hacein-Bey-Abina, S., & Cavazzana-Calvo, M. (2010). 20 years of gene therapy for SCID. *Nature Immunology, 11,* 457–460.

Fitch, S. R., Kimber, G. M., Wilson, N. K., Parker, A., Mirshekar-Syahkal, B., Gottgens, B., Medvinsky, A., Dzierzak, E., & Ottersbach, K. (2012). Signaling from the sympathetic nervous system regulates hematopoietic stem cell emergence during embryogenesis. *Cell Stem Cell, 11,* 554–566.

Gekas, C., Dieterlen-Lievre, F., Orkin, S. H., & Mikkola, H. K. A. (2005). The Placenta is a Niche for Hematopoietic Stem Cells. *Dev Cell, 8,* 365–375.

Gorgens, A., Radtke, S., Mollmann, M., Cross, M., Durig, J., Horn, P. A., & Giebel, B. (2013). Revision of the human hematopoietic tree: Granulocyte subtypes derive from distinct hematopoietic lineages. *Cell Reports, 3,* 1539–1552.

Görgens, A., Radtke, S., Möllmann, M., Cross, M., Dürig, J., Horn, P. A., & Giebel, B. (2013). Revision of the human hematopoietic tree: Granulocyte subtypes derive from distinct hematopoietic lineages. *Cell Reports, 3,* 1539–1552.

Goyama, S., Wunderlich, M., & Mulloy, J. C. (2015). Xenograft models for normal and malignant stem cells. *Blood, 125,* 2630–2640.

Gritz, E., & Hirschi, K. K. (2016). Specification and function of hemogenic endothelium during embryogenesis. *Cellular and Molecular Life Sciences, 73,* 1547–1567.

Grutzkau, A., & Radbruch, A. (2010). Small but mighty: How the MACS-technology based on nanosized superparamagnetic particles has helped to analyze the immune system within the last 20 years. *Cytometry. Part A, 77,* 643–647.

Helgason, C. D., Sauvageau, G., Lawrence, H. J., Largman, C., & Humphries, R. K. (1996). Overexpression of HOXB4 enhances the hematopoietic potential of embryonic stem cells differentiated in vitro. *Blood, 87,* 2740–2749.

Hofer, T., Busch, K., Klapproth, K., & Rodewald, H. R. (2016). Fate mapping and quantitation of Hematopoiesis in vivo. *Annual Review of Immunology, 34,* 449–478.

Horn, P. A., & Blasczyk, R. (2007). Severe combined Immunodefiency-repopulating cell assay may overestimate long-term repopulation ability. *Stem Cells, 25,* 3271–3272.

Horn, P. A., Thomasson, B. M., Wood, B. L., Andrews, R. G., Morris, J. C., & Kiem, H. P. (2003). Distinct hematopoietic stem/progenitor cell populations are responsible for repopulating NOD/SCID mice compared with nonhuman primates. *Blood, 102,* 4329–4335.

Hotta, A., & Yamanaka, S. (2015). From genomics to gene therapy: Induced pluripotent stem cells meet genome editing. *Annual Review of Genetics, 49,* 47–70.

Hu, P., Zhang, W., Xin, H., & Deng, G. (2016). Single cell isolation and analysis. *Frontiers in Cell and Development Biology, 4,* 116.

Ivanovs, A., Rybtsov, S., Ng, E. S., Stanley, E. G., Elefanty, A. G., & Medvinsky, A. (2017). Human haematopoi-etic stem cell development: From the embryo to the dish. *Development, 144*, 2323–2337.

Jacobson, L. O., Simmons, E. L., Marks, E. K., & Eldredge, J. H. (1951). Recovery from radiation injury. *Science, 113*, 510–511.

Jaffredo, T., Gautier, R., Eichmann, A., & Dieterlen-Lievre, F. (1998). Intraaortic hemopoietic cells are derived from endothelial cells during ontogeny. *Development, 125*, 4575–4583.

Keller, P. D. (1946). A clinical syndrome following exposure to atomic bomb explosions. *Journal of the American Medical Association, 131*, 504–506.

Keller, G. (2005). Embryonic stem cell differentiation: Emergence of a new era in biology and medicine. *Genes & Development, 19*, 1129–1155.

Kennedy, M., Firpo, M., Choi, K., Wall, C., Robertson, S., Kabrun, N., & Keller, G. (1997). A common precursor for primitive erythropoiesis and definitive haematopoiesis. *Nature, 386*, 488–493.

Kent, D. G., Copley, M. R., Benz, C., Wohrer, S., Dykstra, B. J., Ma, E., Cheyne, J., Zhao, Y., Bowie, M. B., Zhao, Y., Gasparetto, M., Delaney, A., Smith, C., Marra, M., & Eaves, C. J. (2009). Prospective isolation and molecular characterization of hematopoietic stem cells with durable self-renewal potential. *Blood, 113*, 6342–6350.

Kiel, M. J., Yilmaz, O. H., Iwashita, T., Yilmaz, O. H., Terhorst, C., & Morrison, S. J. (2005). SLAM family receptors distinguish hematopoietic stem and progenitor cells and reveal endothelial niches for stem cells. *Cell, 121*, 1109–1121.

Kissa, K., Murayama, E., Zapata, A., Cortes, A., Perret, E., Machu, C., & Herbomel, P. (2008). Live imaging of emerging hematopoietic stem cells and early thymus colonization. *Blood, 111*, 1147–1156.

Klump, H., Schiedlmeier, B., Vogt, B., Ryan, M., Ostertag, W., & Baum, C. (2001). Retroviral vector-mediated expression of HoxB4 in hematopoietic cells using a novel coexpression strategy. *Gene Therapy, 8*, 811–817.

Klump, H., Schiedlmeier, B., & Baum, C. (2005). Control of self-renewal and differentiation of hematopoietic stem cells: HOXB4 on the threshold. *Annals of the New York Academy of Sciences, 1044*, 6–15.

Klump, H., Teichweyde, N., Meyer, C., & Horn, P. A. (2013a). Development of patient-specific hematopoietic stem and progenitor cell grafts from pluripotent stem cells, in vitro. *Current Molecular Medicine, 13*, 815–820.

Klump, H., Teichweyde, N., Meyer, C., & Horn, P. A. (2013b). Development of patient-specific hematopoietic stem and progenitor cell grafts from pluripotent stem cells, in vitro. *Current Molecular Medicine, 13*, 815–820.

Kohn, L. A., Hao, Q. L., Sasidharan, R., Parekh, C., Ge, S., Zhu, Y., Mikkola, H. K., & Crooks, G. M. (2012). Lymphoid priming in human bone marrow begins before expression of CD10 with upregulation of L-selectin. *Nature Immunology, 13*, 963–971.

Krosl, J., Austin, P., Beslu, N., Kroon, E., Humphries, R. K., & Sauvageau, G. (2003). In vitro expansion of hematopoietic stem cells by recombinant TAT-HOXB4 protein. *Nature Medicine, 9*, 1428–1432.

Kustikova, O., Brugman, M., & Baum, C. (2010). The genomic risk of somatic gene therapy. *Seminars in Cancer Biology, 20*, 269–278.

Lancrin, C., Sroczynska, P., Stephenson, C., Allen, T., Kouskoff, V., & Lacaud, G. (2009). The haemangioblast generates haematopoietic cells through a haemogenic endothelium stage. *Nature, 457*, 892–895.

Ledran, M. H., Krassowska, A., Armstrong, L., Dimmick, I., Renstrom, J., Lang, R., Yung, S., Santibanez-Coref, M., Dzierzak, E., Stojkovic, M., Oostendorp, R. A., Forrester, L., & Lako, M. (2008). Efficient hematopoietic differentiation of human embryonic stem cells on stromal cells derived from hematopoietic niches. *Cell Stem Cell, 3*, 85–98.

Lee, Y., Manegold, J. E., Kim, A. D., Pouget, C., Stachura, D. L., Clements, W. K., & Traver, D. (2014). FGF signalling specifies haematopoietic stem cells through its regulation of somitic Notch signalling. *Nature Communications, 5*, 5583.

Lemischka, I. R., Raulet, D. H., & Mulligan, R. C. (1986). Developmental potential and dynamic behavior of hematopoietic stem cells. *Cell, 45*, 917–927.

Lesinski, D. A., Heinz, N., Pilat-Carotta, S., Rudolph, C., Jacobs, R., Schlegelberger, B., Klump, H., & Schiedlmeier, B. (2012). Serum- and stromal cell-free hypoxic generation of embryonic stem cell-derived hematopoietic cells in vitro, capable of multilineage repopulation of immunocompetent mice. *Stem Cells Translational Medicine, 1*, 581–591.

Lorenz, E., Uphoff, D., Reid, T. R., & Shelton, E. (1951). Modification of irradiation injury in mice and Guinea pigs by bone marrow injections. *Journal of the National Cancer Institute, 12*, 197–201.

Magli, M. C., Iscove, N. N., & Odartchenko, N. (1982). Transient nature of early haematopoietic spleen colonies. *Nature, 295,* 527–529.

Mallo, M., Wellik, D. M., & Deschamps, J. (2010). Hox genes and regional patterning of the vertebrate body plan. *Developmental Biology, 344,* 7–15.

Mcgrath, K. E., Frame, J. M., Fegan, K. H., Bowen, J. R., Conway, S. J., Catherman, S. C., Kingsley, P. D., Koniski, A. D., & Palis, J. (2015). Distinct sources of hematopoietic progenitors emerge before HSCs and provide functional blood cells in the mammalian embryo. *Cell Reports, 11,* 1892–1904.

Medvinsky, A., & Dzierzak, E. (1996). Definitive hematopoiesis is autonomously initiated by the AGM region. *Cell, 86,* 897–906.

Medvinsky, A., Rybtsov, S., & Taoudi, S. (2011). Embryonic origin of the adult hematopoietic system: Advances and questions. *Development, 138,* 1017–1031.

Miller, C. L., Dykstra, B. & Eaves, C. J. (2008). Characterization of mouse hematopoietic stem and progenitor cells. *Current Protocols in Immunology,* Chapter 22, Unit 22B 2.

Miller, P. H., Knapp, D. J., & Eaves, C. J. (2013). Heterogeneity in hematopoietic stem cell populations: Implications for transplantation. *Current Opinion in Hematology, 20,* 257–264.

Muller, A. M., Medvinsky, A., Strouboulis, J., Grosveld, F., & Dzierzak, E. (1994). Development of hematopoietic stem cell activity in the mouse embryo. *Immunity, 1,* 291–301.

Na Nakorn, T., Traver, D., Weissman, I. L., & Akashi, K. (2002). Myeloerythroid-restricted progenitors are sufficient to confer radioprotection and provide the majority of day 8 CFU-S. *The Journal of Clinical Investigation, 109,* 1579–1585.

Narayan, A. D., Chase, J. L., Lewis, R. L., Tian, X., Kaufman, D. S., Thomson, J. A., & Zanjani, E. D. (2006). Human embryonic stem cell-derived hematopoietic cells are capable of engrafting primary as well as secondary fetal sheep recipients. *Blood, 107,* 2180–2183.

Negrin, R. S. (2015). Graft-versus-host disease versus graft-versus-leukemia. *Hematology. American Society of Hematology. Education Program, 2015,* 225–230.

North, T. E., Goessling, W., Peeters, M., Li, P., Ceol, C., Lord, A. M., Weber, G. J., Harris, J., Cutting, C. C., Huang, P., Dzierzak, E., & Zon, L. I. (2009). Hematopoietic stem cell development is dependent on blood flow. *Cell, 137,* 736–748.

Notta, F., Doulatov, S., Laurenti, E., Poeppl, A., Jurisica, I., & Dick, J. E. (2011). Isolation of single human hematopoietic stem cells capable of long-term multilineage engraftment. *Science, 333,* 218–221.

Orkin, S. H., & Zon, L. I. (2008). Hematopoiesis: An evolving paradigm for stem cell biology. *Cell, 132,* 631–644.

Ottersbach, K., & Dzierzak, E. (2005). The Murine Placenta Contains Hematopoietic Stem Cells within the Vascular Labyrinth Region. *Dev Cell, 8,* 377–387.

Palis, J., Robertson, S., Kennedy, M., Wall, C., & Keller, G. (1999). Development of erythroid and myeloid progenitors in the yolk sac and embryo proper of the mouse. *Development, 126,* 5073–5084.

Perlin, J. R., Robertson, A. L., & Zon, L. I. (2017). Efforts to enhance blood stem cell engraftment: Recent insights from zebrafish hematopoiesis. *The Journal of Experimental Medicine, 214,* 2817–2827.

Pilat, S., Carotta, S., Schiedlmeier, B., Kamino, K., Mairhofer, A., Will, E., Modlich, U., Steinlein, P., Ostertag, W., Baum, C., Beug, H., & Klump, H. (2005). HOXB4 enforces equivalent fates of ES-cell-derived and adult hematopoietic cells. *Proceedings of the National Academy of Sciences of the United States of America, 102,* 12101–12106.

Pilat, S., Carotta, S., & Klump, H. (2013). Development of hematopoietic stem and progenitor cells from mouse embryonic stem cells, in vitro, supported by ectopic human HOXB4 expression. *Methods in Molecular Biology, 1029,* 129–147.

Porteus, M. (2016). Genome editing: A new approach to human therapeutics. *Annual Review of Pharmacology and Toxicology, 56,* 163–190.

Pouget, C., Peterkin, T., Simoes, F. C., Lee, Y., Traver, D., & Patient, R. (2014). FGF signalling restricts haematopoietic stem cell specification via modulation of the BMP pathway. *Nature Communications, 5,* 5588.

Qian, P., De Kumar, B., He, X. C., Nolte, C., Gogol, M., Ahn, Y., Chen, S., Li, Z., Xu, H., Perry, J. M., Hu, D., Tao, F., Zhao, M., Han, Y., Hall, K., Peak, A., Paulson, A., Zhao, C., Venkatraman, A., Box, A., Perera, A., Haug, J. S., Parmely, T., Li, H., Krumlauf, R., & Li, L. (2018). Retinoid-sensitive epigenetic regulation of the Hoxb cluster maintains Normal Hematopoiesis and inhibits Leukemogenesis. *Cell Stem Cell, 22*(740–754), e7.

Raya, A., Rodriguez-Piza, I., Guenechea, G., Vassena, R., Navarro, S., Barrero, M. J., Consiglio, A., Castella, M., Rio, P., Sleep, E., Gonzalez, F., Tiscornia, G., Garreta, E., Aasen, T., Veiga, A., Verma, I. M., Surralles, J., Bueren, J., & Izpisua Belmonte, J. C. (2009). Disease-corrected haematopoietic progenitors from Fanconi anaemia induced pluripotent stem cells. *Nature, 460,* 53–59.

Rhodes, K. E., Gekas, C., Wang, Y., Lux, C., Francis, C. S., Chan, D. N., Conway, S., Orkin, S. H., Yoder, M., & Mikkola, H. K. A. (2008). The Emergence of Hematopoietic Stem Cells is Initiated in the Placental Vasculature in the Absence of Circulation. *Cell Stem Cell, 2,* 252–263.

Ribeil, J. A., Hacein-Bey-Abina, S., Payen, E., Magnani, A., Semeraro, M., Magrin, E., Caccavelli, L., Neven, B., Bourget, P., El Nemer, W., Bartolucci, P., Weber, L., Puy, H., Meritet, J. F., Grevent, D., Beuzard, Y., Chretien, S., Lefebvre, T., Ross, R. W., Negre, O., Veres, G., Sandler, L., Soni, S., De Montalembert, M., Blanche, S., Leboulch, P., & Cavazzana, M. (2017). Gene therapy in a patient with sickle cell disease. *The New England Journal of Medicine, 376,* 848–855.

Rowe, R. G., Mandelbaum, J., Zon, L. I., & Daley, G. Q. (2016). Engineering hematopoietic stem cells: Lessons from development. *Cell Stem Cell, 18,* 707–720.

Rybtsov, S., Ivanovs, A., Zhao, S., & Medvinsky, A. (2016). Concealed expansion of immature precursors underpins acute burst of adult HSC activity in foetal liver. *Development, 143,* 1284–1289.

Sauvageau, G., Thorsteinsdottir, U., Eaves, C. J., Lawrence, H. J., Largman, C., Lansdorp, P. M., & Humphries, R. K. (1995). Overexpression of HOXB4 in hematopoietic cells causes the selective expansion of more primitive populations in vitro and in vivo. *Genes & Development, 9,* 1753–1765.

Schiedlmeier, B., Klump, H., Will, E., Arman-Kalcek, G., Li, Z., Wang, Z., Rimek, A., Friel, J., Baum, C., & Ostertag, W. (2003). High-level ectopic HOXB4 expression confers a profound in vivo competitive growth advantage on human cord blood CD34+ cells, but impairs lymphomyeloid differentiation. *Blood, 101,* 1759–1768.

Schiedlmeier, B., Santos, A. C., Ribeiro, A., Moncaut, N., Lesinski, D., Auer, H., Kornacker, K., Ostertag, W., Baum, C., Mallo, M., & Klump, H. (2007). HOXB4's road map to stem cell expansion. *Proceedings of the National Academy of Sciences of the United States of America, 104,* 16952–16957.

Shultz, L. D., Schweitzer, P. A., Christianson, S. W., Gott, B., Schweitzer, I. B., Tennent, B., Mckenna, S., Mobraaten, L., Rajan, T. V., Greiner, D. L., et al. (1995). Multiple defects in innate and adaptive immunologic function in NOD/LtSz-scid mice. *Journal of Immunology, 154,* 180–191.

Shultz, L. D., Lyons, B. L., Burzenski, L. M., Gott, B., Chen, X., Chaleff, S., Kotb, M., Gillies, S. D., King, M., Mangada, J., Greiner, D. L., & Handgretinger, R. (2005). Human lymphoid and myeloid cell development in NOD/LtSz-scid IL2R gamma null mice engrafted with mobilized human hemopoietic stem cells. *Journal of Immunology, 174,* 6477–6489.

Sii-Felice, K., Giorgi, M., Leboulch, P., & Payen, E. (2018). Hemoglobin disorders: Lentiviral gene therapy in the starting blocks to enter clinical practice. *Experimental Hematology, 64,* 12–32.

Souilhol, C., Gonneau, C., Lendinez, J. G., Batsivari, A., Rybtsov, S., Wilson, H., Morgado-Palacin, L., Hills, D., Taoudi, S., Antonchuk, J., Zhao, S., & Medvinsky, A. (2016). Inductive interactions mediated by interplay of asymmetric signalling underlie development of adult haematopoietic stem cells. *Nature Communications, 7,* 10784.

Takahashi, K., Tanabe, K., Ohnuki, M., Narita, M., Ichisaka, T., Tomoda, K., & Yamanaka, S. (2007). Induction of pluripotent stem cells from adult human fibroblasts by defined factors. *Cell, 131,* 861–872.

Tavian, M., Robin, C., Coulombel, L., & Peault, B. (2001). The human embryo, but not its yolk sac, generates lympho-myeloid stem cells: Mapping multipotent hematopoietic cell fate in intraembryonic mesoderm. *Immunity, 15,* 487–495.

Teichweyde, N., Kasperidus, L., Carotta, S., Kouskoff, V., Lacaud, G., Horn, P. A., Heinrichs, S., & Klump, H. (2018). HOXB4 promotes Hemogenic endothelium formation without perturbing endothelial cell development. *Stem Cell Reports, 10,* 875–889.

Thompson, A. A., Walters, M. C., Kwiatkowski, J., Rasko, J. E. J., Ribeil, J. A., Hongeng, S., Magrin, E., Schiller, G. J., Payen, E., Semeraro, M., Moshous, D., Lefrere, F., Puy, H., Bourget, P., Magnani, A., Caccavelli, L., Diana, J. S., Suarez, F., Monpoux, F., Brousse, V., Poirot, C., Brouzes, C., Meritet, J. F., Pondarre, C., Beuzard, Y., Chretien, S., Lefebvre, T., Teachey, D. T., Anurathapan, U., Ho, P. J., Von Kalle, C., Kletzel, M., Vichinsky, E., Soni, S., Veres, G., Negre, O., Ross, R. W., Davidson, D., Petrusich, A., Sandler, L., Asmal, M., Hermine, O., De Montalembert, M., Hacein-Bey-Abina, S., Blanche, S., Leboulch, P., & Cavazzana, M. (2018). Gene therapy in patients with transfusion-dependent beta-thalassemia. *The New England Journal of Medicine, 378,* 1479–1493.

Tian, X., Woll, P. S., Morris, J. K., Linehan, J. L., & Kaufman, D. S. (2006). Hematopoietic engraftment of human embryonic stem cell-derived cells is regulated by recipient innate immunity. *Stem Cells, 24,* 1370–1380.

Till, J. E., & McCulloch, E. A. (1961). A direct measurement of the radiation sensitivity of normal mouse bone marrow cells. *Radiation Research, 14,* 213–222.

Wang, Q., Stacy, T., Binder, M., Marin-Padilla, M., Sharpe, A. H., & Speck, N. A. (1996). Disruption of the Cbfa2 gene causes necrosis and hemorrhaging in the central nervous system and blocks definitive hematopoiesis. *Proceedings of the National Academy of Sciences of the United States of America, 93*, 3444–3449.

Wang, L., Menendez, P., Shojaei, F., Li, L., Mazurier, F., Dick, J. E., Cerdan, C., Levac, K., & Bhatia, M. (2005). Generation of hematopoietic repopulating cells from human embryonic stem cells independent of ectopic HOXB4 expression. *The Journal of Experimental Medicine, 201*, 1603–1614.

Watts, K. L., Delaney, C., Humphries, R. K., Bernstein, I. D., & Kiem, H. P. (2010). Combination of HOXB4 and Delta-1 ligand improves expansion of cord blood cells. *Blood, 116*, 5859–5866.

Watts, K. L., Nelson, V., Wood, B. L., Trobridge, G. D., Beard, B. C., Humphries, R. K., & Kiem, H. P. (2012). Hematopoietic stem cell expansion facilitates multilineage engraftment in a nonhuman primate cord blood transplantation model. *Experimental Hematology, 40*, 187–196.

Will, E., Speidel, D., Wang, Z., Ghiaur, G., Rimek, A., Schiedlmeier, B., Williams, D. A., Baum, C., Ostertag, W., & Klump, H. (2006). HOXB4 inhibits cell growth in a dose-dependent manner and sensitizes cells towards extrinsic cues. *Cell Cycle, 5*, 14–22.

Wilson, N. K., Foster, S. D., Wang, X., Knezevic, K., Schutte, J., Kaimakis, P., Chilarska, P. M., Kinston, S., Ouwehand, W. H., Dzierzak, E., Pimanda, J. E., De Bruijn, M. F., & Gottgens, B. (2010). Combinatorial transcriptional control in blood stem/progenitor cells: Genome-wide analysis of ten major transcriptional regulators. *Cell Stem Cell, 7*, 532–544.

Wilson, N. K., Calero-Nieto, F. J., Ferreira, R., & Gottgens, B. (2011). Transcriptional regulation of haematopoietic transcription factors. *Stem Cell Research & Therapy, 2*, 6.

Wu, A. M., Till, J. E., Siminovitch, L., & McCulloch, E. A. (1968). Cytological evidence for a relationship between normal hemotopoietic colony-forming cells and cells of the lymphoid system. *The Journal of Experimental Medicine, 127*, 455–464.

Yoshimoto, M., Porayette, P., & Yoder, M. C. (2008). Overcoming obstacles in the search for the site of hematopoietic stem cell emergence. *Cell Stem Cell, 3*, 583–586.

Zhang, X. B., Schwartz, J. L., Humphries, R. K., & Kiem, H. P. (2007). Effects of HOXB4 overexpression on ex vivo expansion and immortalization of hematopoietic cells from different species. *Stem Cells, 25*, 2074–2081.

Zovein, A. C., Turlo, K. A., Ponec, R. M., Lynch, M. R., Chen, K. C., Hofmann, J. J., Cox, T. C., Gasson, J. C., & Iruela-Arispe, M. L. (2010). Vascular remodeling of the vitelline artery initiates extravascular emergence of hematopoietic clusters. *Blood, 116*, 3435–3444.

Mesenchymal Stem Cells

Christian Horst Tonk, Markus Witzler, Margit Schulze, and Edda Tobiasch

© Springer Nature Switzerland AG 2020
B. Brand-Saberi (ed.), *Essential Current Concepts in Stem Cell Biology*,
Learning Materials in Biosciences, https://doi.org/10.1007/978-3-030-33923-4_2

What You Will Learn from This Chapter

This chapter focuses on mesenchymal stem cells (MSCs), a multipotent stem cell type that has been found in a variety of tissues and organs of the human body since their discovery in 1970. Their main function is to maintain and repair the respective tissue in vivo. Mesenchymal stem cells can be easily isolated from different tissues and can undergo extensive self-proliferation prior to differentiation into various mesodermal cell types such as osteoblasts, chondrocytes, adipocytes, tenocytes, myocytes, and fibroblasts. Because of this vast differentiation potential, mesenchymal stem cells are a promising tool for regenerative medicine approaches. They could play an important role in cellular therapy, tissue replacement and regeneration in the future. Mesenchymal stem cells will be compared for their application and differentiation potential to embryonic stem cells and induced pluripotent stem cells and the limitations and challenges using scaffolds for tissue repair will be presented. In addition, legal and ethical aspects of the use of mesenchymal stem cells will be discussed.

Moreover, isolation protocols for mesenchymal stem cells from the most common sources namely bone marrow, adipose tissue and umbilical cord are included.

2.1 Localization, Characterization and Storage of Mesenchymal Stem Cells

A stem cell is an unspecialized cell that has the ability for self-renewal for a long time period with or without senescence and for differentiation into cells of various lineages (Till and McCulloch 1961; Gordon 1972), generally. Stem cells can have diverse plasticity. Totipotent stem cells from the fertilized egg can give rise to a complete organism and in addition to the extra embryonal tissues, while pluripotent stem cells like embryonic stem cells (ESCs) and induced pluripotent stem cells (iPSCs) can differentiate into any cell type from all three germ layers but can not form an organism without transferring them into a blastocyst (Hanna et al. 2010). By contrast, the most commonly known adult stem cells, the mesenchymal stem cells (MSCs) and hematopoietic stem cells (HSCs), have a more limited differentiation potential and are thus called multipotent. As the name suggests, MSCs can give rise to several cell types of the mesenchyme which is derived from the mesoderm during development. They form then cell types from bone, muscle, and cartilage but also to cells from tissue of other origin (◘ Fig. 2.1) (Pittenger et al. 1999; Pansky et al. 2007).

Mesenchymal stem cells were first discovered by Friedenstein in 1970 (1970). He and his research team were interested in isolating hematopoietic stem cells (HSCs) from bone marrow as a treatment for leukemia and thereby found an unknown cell type. This unknown cell type seemed to be responsible for the microenvironment typical for hematopoietic tissue (Friedenstein et al. 1974), and today it's known that it is a major part of the HSC niche. Due to the capacity of these cells to form stromal tissue, they named them "stromal precursors" (Friedenstein et al. 1970). Later the term "mesenchymal stromal cells" was established (Prockop 1997; Baksh et al. 2004).

In the last years, it was discussed whether mesenchymal stem cells are actually stem cells or just stromal precursors. That MSCs were first found exclusively in the connective tissue and that they influence the surrounding cells are among the strongest arguments towards conserving the name "mesenchymal stromal cells". The most prominent argument to name them mesenchymal stem cells is their capacity to differentiate into several

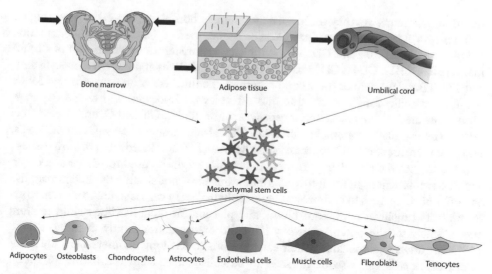

□ **Fig. 2.1** Major sources and differentiation linages of MSCs. The main isolation sources are bone marrow of the iliac crest, subcutaneous adipose tissue and Wharton's jelly from umbilical cord (shown with arrows). Mesenchymal stem cells from diverse tissues can have a different plasticity and might be pre-committed already towards different lineages (depicted with different colors of the MSCs). The acknowledged differentiation potential of MSCs is shown in the lower part of the figure

lineages. The reason that they can be found in so many tissues are that they are needed for maintenance and repair. The idea to use the same mechanism artificially to replace tissues or even organs came up. It had been expected that this will work similar to the treatment of certain cancers such as leukemia and other diseases using hematopoietic stem cell transplantation.

As mentioned, MSCs were first found in the bone marrow a semi-solid tissue (Friedenstein et al. 1970), which has high cell turnover and regeneration rate. Later, it was discovered that other tissues with high cell turnover and regeneration rate like the epidermis (Rheinwald and Green 1975; Huang et al. 2013; Salerno et al. 2017) and the intestinal tissue (Lanzoni et al. 2009) also possess MSCs. So, scientist thought that those are the prerequisites for tissues to have the cells. Much later, scientists found that MSCs could also be isolated from fat tissue where cells have a low cell turnover and a high regeneration rate (Zuk et al. 2002). Finally, MSCs were also found in tissues of organs with a low cell turnover and a low regeneration rate like the kidney (Bruno et al. 2009). Besides the mentioned tissues, there are further sources where MSCs can be isolated from, such as dental pulp (Alkhalil et al. 2015) or follicle (Haddouti et al. 2009), peripheral blood (Cao et al. 2005), hair follicle (Liu et al. 2010) and several neonatal tissues (da Silva Meirelles et al. 2006) such as umbilical cord (Lee et al. 2004), Wharton's jelly (Wang et al. 2004), amniotic fluid (Steigmann and Fauza 2007; Moraghebi et al. 2017), and placenta (In't Anker et al. 2004). The vast differentiation potential makes them a promising stem cell type for use in regenerative medicine.

In response to the recent unraveled series of MSC sources and isolation procedures from various tissue types, combined with an interest in MSCs for research and application in regenerative medicine, the Mesenchymal and Tissue Stem Cell Committee of the International Society for Cellular Therapy proposed a set of minimal criteria to define MSCs. The first criterion is that MSCs must adhere to plastic, which also simplifies the isolation procedure. As they unfortunately resemble fibroblasts, the expression (higher than 95%) of at least three specific markers on the cell surface, i.e. cluster of differentiation (CD)73, CD90, CD105 and CD133 and the lack expression (lower than 5%) of surface markers like CD14, CD34, CD45 or CD11b, CD19 or CD79α and human leucocyte antigen-DR (HLA-DR) defines the second criterion. The third criterion demands that MSCs possess the ability for differentiation into osteoblasts, adipocytes, and chondrocytes *in vitro* confirmed by specific staining usually Alizarin Red S, Oil Red O, and Alcian Blue (Horwitz et al. 2005; Dominici et al. 2006). These features should be valid for all MSCs, however there are some differences existing between MSCs isolated from diverse tissues.

In line with the promising prospect that MSCs might be used for tissue and organ repair or replacement in the future also a new business model came up. Parents can use the offer of companies to cryopreserve mesenchymal stem cells isolated from Wharton's jelly of the umbilical cord, cord blood, and placenta of their newborn child to store them. Solely for the potential use by the child or the donor family members. These MSCs are advantageous due to their early age due to e.g. long telomeres. On the other hand, those cells can also be donated to be collected in stem cell banks for future research and use.

To ensure the safety and efficacy of the cells during banking and cryopreservation, optimal conditions for cryopreservation must be selected, for further detail see Ullah and co-workers (2015).

For efficient storage there must be an optimal cryopreservation medium as well as a defined (1 °C/min) constant and strict drop in temperature as for other cells as well (Thirumala et al. 2005) using a controlled rate freezer or other appropriate storage devices (Ullah et al. 2015). Due to the possible impact with new approaches for various treatments in regenerative medicine, the Food and Drug Administration (FDA) in the U.S.A. and the European Medicines Agency (EMA) in Europe take responsibility by monitoring the large-scale banking next to supervising MSC based cell therapy products (Hourd et al. 2008).

2.2 Mesenchymal Stem Cells in Regenerative Medicine

2.2.1 The Application of Mesenchymal Stem Cells

MSCs are promising tools in the field of regenerative medicine (Tobiasch 2008), so approaches for the clinical use of mesenchymal stem cells are in progress. The number of clinical trials using mesenchymal stem cells has risen over the last years. Currently there are 711 registered clinical trials in different clinical phases (phase I, II, III, IV) worldwide, exhibiting the huge potential of MSC-based cell therapy (▶ www.clinicaltrials.gov).

Most of these clinical trials are still in the early stages phase I or II. Only few clinical trials are in phase III (36 studies) or phase IV (2 studies), the stages preceding the potential use in regenerative medicine such as tissue engineering to the replacement or reconstruction of damaged human tissues. Clinical trials using mesenchymal stem cells encompass many different diseases. The most common diseases registered for cell therapy with mesenchymal stem cells are shown in ◘ Fig. 2.2.

One example, for a clinical phase III study in Germany is lead by research team of Professor Steinhoff in Rostock. In cooperation with several leading university heart centers, they applied autologous CD133+ bone marrow stem cells intramyocardially during

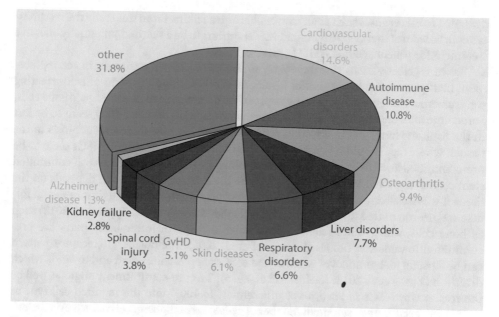

☐ Fig. 2.2 Percentage distribution of clinical trials with mesenchymal stem cells. There are 711 clinical trials for different diseases in phase I – IV ongoing in the world (► www.clinicaltrials.gov). 104 of these clinical trials address cardiovascular disorders, 77 autoimmune diseases, 67 osteoarthritis, 55 liver disorders, 47 respiratory disorders, 43 skin diseases, 36 graft versus host diseases (GvHD), 27 spinal cord injuries, 20 kidney failures and 9 Alzheimer disease. The other diseases include i.e. muscular dystrophies, aplastic anemia, osteogenesis imperfecta, Parkinson's disease and ulcerative colitis

bypass surgery to patients that had a myocardial infarction (heart attack). Following the application of the stem cells, the patients had an improved cardiac function. That led to a positive completion of the clinical trial and the establishment of a reference standard for future cell therapies of this illness (White et al. 2016; Steinhoff et al. 2017).

Another example is a clinical trial with ten patients using MSCs as therapy against spinal cord injuries. In this approach, autologous MSCs were harvested, culturally expanded and injected directly into the spinal cord. After a second injection a few weeks later, the motor power of the upper extremities was improved. Moreover, after a six-month follow-up, none of the patients experienced any permanent complication associated with MSCs transplantation (Park et al. 2012).

Normally new treatments have to enter the marketing-authorization process starting with clinical trials to create a safety profile and dosage guidelines. However, there is also individual medical treatment, where patients after getting sufficient clarification have the chance to receive a treatment which is not authorized yet. In accordance with Declaration of Helsinki from 1964, this is only possible for patients with life-threatening, long-lasting or seriously debilitating illnesses that can not be medicated with any of the currently authorized treatments. In 2011 Prasad and coworkers described the first individual curative trial for the treatment of pediatric patients with acute graft versus host disease (aGvHD) of refractory grades III-IV. Twelve patients that were resistant against steroids and other immunosuppressive therapies received a therapy with allogenic human MSCs. Seven of the patients showed a positive response to the treatment while five did not. The seven responders stayed alive almost 700 days after the treatment and eventually passed away due to other causes such as an irreversible organ failure. From this study the researchers concluded that therapy with human MSCs appears to be a safe and potentially effective treatment for patients with aGvHD (Prasad et al. 2011).

Taken together various researchers try to use mesenchymal stem cells in cell replacement therapies hoping to get similar results as with hematopoietic stem cells for treating e.g. leukemia. However, most therapy approaches do not replace the damaged tissue cells. Some time after the injection of MSCs, there is no trace left of them. They seem to be lost in the liver and lung, and in addition it is postulated that they undergo apoptosis in the tissues, where they should replace the damaged cells. Next to this some MSCs seem to be pre-committed towards specific cell types, resulting in a mixture of various pre-committed stem cells at a specific localization site. Where the mixture composition depends on the tissue it was isolated from. For example, mesenchymal stem cells isolated from dental follicle are pre-committed towards hard tissues (Haddouti et al. 2009; Zippel et al. 2012) such as bone cells, while MSCs isolated from adipose tissue of lower body parts are pre-committed towards soft tissues (Sakaguchi et al. 2005). The purinergic receptor pattern can be a useful tool to uncover whether the MSCs are pre-committed and towards which direction (Zhang et al. 2014; Kaebisch et al. 2015). This pre-commitment might as well be the reason why MSCs *in vitro* never differentiate to 100% into the required cell type as could be expected. At least a small number of cells stays in the stem cell state. Nevertheless, they seem to be beneficial in clinical trials. So, the question arises – why?

The reason might be that MSCs have a paracrine effect on the surrounding tissue (Cotter et al. 2018; Linero and Chaparro 2014). It was shown in different, previously described studies, that the administration of MSCs prevented injuries and led, in some cases, to tissue recovery (Bartholomew et al. 2002; Herrera et al. 2007; Yao et al. 2016). Next to this MSCs are hardly causing an immunogenic response. MSCs are therefore more often (53.91%) used in autologous (from the patient itself) but also (46.09%) allogenic (from a donor) approaches are promising (▸ www.clinicaltrials.gov; Haubner et al. 2015). Taken the beneficial effect of MSCs into consideration various studies nowadays take a new approach by using more than cell type for replacements. They either use MSCs differentiated towards the desired tissue type, such as cardiomyocytes and combine these cells with e.g. fibroblasts or undifferentiated MSCs.

However not only MSCs are used in regenerative medicine. Other stem cell types like embryonic stem cells (ESCs) and induced pluripotent stem cells (iPSCs) can be used as well.

2.2.2 Pros and Cons of Mesenchymal Stem Cells Compared to Embryonic Stem Cells and Induced Pluripotent Stem Cell

Next to multipotent mesenchymal stem cells, there are two pluripotent stem cell types often used in regenerative medicine: embryonic stem cells (ESCs) and induced pluripotent stem cells (iPSCs). ESCs are generated by isolating the cells of the inner cell mass of a blastocyst from an early embryo (around day five to seven). However, the generation of ESCs arises ethical issues (Evans and Kaufman 1981; Thomson et al. 1998). Since 2006 there is an artificial method creating very similar cells by reprogramming adult somatic cell using the four Yamanaka transcription factors Oct3/4, Sox2, Klf4 and c-Myc or the Thomson factors Oct3/4, Sox2, Nanog, Lin28 (Takahashi and Yamanaka 2006; Takahashi et al. 2007; Yu et al. 2007; Nakagawa et al. 2008).

Embryonic stem cells are the original and therefore the gold standard to understand early development, and also diseases where they play a role. Due to their higher plasticity compared to MSCs, some scientists claim that they are needed for the replacement of cells and tissues which can not be differentiated from less plastic cells such as MSCs. A commonly mature cell type where this is claimed are nervous. However, as ESCs also do not all differentiate *in vitro*, they also pertain to the increased risk for cancer development, particularly teratomas.

The induced pluripotent stem cell type has the advantage that it can be obtained from the patient himself and thus cause only little unwanted immunogenic reactions and hardly any ethical issues. They do have, similar to ESCs, the high potential being able to differentiate into all tissues of an organism, but related to this, also the high risk for teratoma formation. In addition, if retroviruses are used to express the mentioned transcription factors their use increase the risk for cancerogenesis due to random integration into the genome. Another risk is that this cell type is barely understood today. It was only recently unraveled that their epigenetic pattern is different to ESCs, specifically a mixture of the initial cell type and ESC.

Nevertheless, today 121 clinical studies are on the way with ESCs (in the USA) and 21 studies have been tested using iPSCs. Two of them used in 2018 iPSCs for spinal cord injury and for macular degeneration (▶ www.clinicaltrials.gov; Tsuji et al. 2018; Takagi et al. 2018).

2.3 Scaffolds for Tissue Repair in Combination with MSCs

2.3.1 Scaffold Requirement for MSC-Based Tissue Repair

Mesenchymal stem cells are affected by extracellular matrix (ECM) that defines the microenvironment that surrounds them. ECM defines geometry, topography and morphology that the cells are facing. But other molecules with similar properties such as extracellular nucleotides, growth factors and cytokines, for instance transforming growth factor-β (TGF-β), tension induced proteins (TIPs), integrins and transient receptor potential (TRP), can also regulate cytoskeleton tension successively activate a series of mechanical transduction events and thus influence stem cell fate. Mechanical forces such as shear stress and blood pressure influence stem cell proliferation and differentiation as well as chemical and physical factors such as pH or oxygen levels. ◘ Figure 2.3 gives a detailed overview on stem cells and their micro-environment (Schulze and Tobiasch 2012).

A scaffold for tissue repair should resemble or mimic original tissue as closely as possible. This means that not only the chemical structure, but also other aspects such as morphology, mechanical properties, technical functionality, stability against physiological conditions and biocompatibility have to be tuned to match the implant site. "Gold standard" for implants are usually autografts: implants taken from another site of the patient itself since they have the lowest adverse effects. However, these grafts are rarely available. Allografts, implants coming from a different donor, are also widely being used, with a limited risk of rejection. But as they also rely on limited donor material, research focuses on synthetic substitute material that should also lower risks of infections or implant rejections (Henkel et al. 2013).

Depending on the tissue that is going to be repaired, a scaffold has to fulfil quite different requirements. A bone scaffold, for instance, has to provide much more mechanical stability than a scaffold for vascular tissue and an implant for tendon repair has to provide

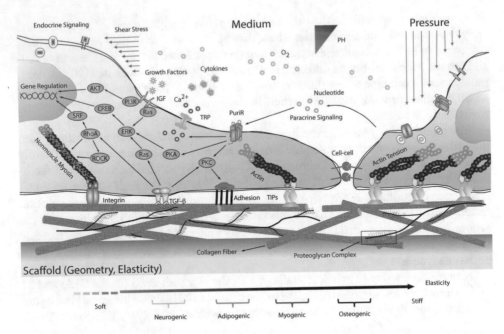

■ **Fig. 2.3** Stem cells and their natural micro-environment. There are roughly three groups into which cell-influencing factors can be categorized: physical factors such as shear forces, elasticity and topography, cellular interactions such as immune and nerve cells, nearby blood vessels and neighboring stem cells and biochemical factors such as oxygen, glucose, hormones and growth factors (Schulze and Tobiasch 2012). (Copyright 2012 Springer Verlag)

both tensile strength and flexibility. Cell attachment and proliferation is crucial for vascularization of the scaffold and a successful implantation.

This may be ensured by choosing cell-friendly, non-cytotoxic materials such as (bio-) ceramics and or (bio-)polymers that have proven to be cytocompatible. Depending on the application, scaffolds can be made of either ceramic or polymer, a composite of both or even certain metals. There have been studies where synthetic polymers (e.g. polycaprolactone, poly lactic acid), polymers from renewable sources (such as polysaccharides or collagen) or copolymers of different types of polymers were used for bone, tendon, vascular tissue and cartilage repair. Many polymers tend to be biocompatible and their flexibility makes them useful candidates with regard to a tunable morphology. However, using only a polymer as scaffold material can also have certain disadvantages. Some polymer materials have rather low cell attachment properties or are not suitable for load bearing applications due to a low stiffness and mechanical resistance. Similarly to polymers, there are different ceramic substances that are being used especially for bone scaffolds, however, most studies use some variation of calcium phosphates. This makes sense, since 70% of original bone is comprised of calcium phosphate. It has great cell attachment properties and high mechanical stiffness although being brittle. Generating scaffolds with a defined hierarchical porosity on the other hand is often challenging. Plus, pure ceramic scaffolds lack the flexibility of polymer scaffolds. This leads to a composite approach, where the advantages of both polymers and ceramics are combined, overcoming most of the individual disadvantages. Here, a polymer is used to create a flexible, porous network, while the ceramic (often calcium phosphates) are added to enhance stability and cell attachment (Schulze and Tobiasch 2012; Henkel et al. 2013; Hielscher et al. 2018).

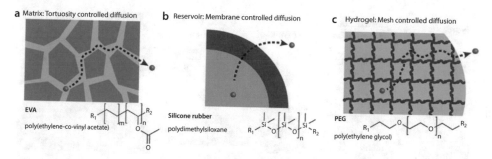

Fig. 2.4 Examples of controlled release platforms and their materials: **a** Matrix tortuosity-controlled diffusion out of porous materials; **b** Membrane-controlled diffusion from capsules and reservoirs; **c** Hydrogels (Fenton et al. 2018). (Copyright 2018 John Wiley and Sons)

Additionally, certain drugs or growth factors can be incorporated into the scaffold to accelerate cell differentiation or prevent infections and inflammation. However, drug delivery, especially a sustained release over several days remains a great challenge. Current research focuses on different methods of encapsulation and release of drugs. Basic release mechanisms include: matrix tortuosity-controlled diffusion, membrane-controlled diffusion for small molecules, hydrogel network swelling (■ Fig. 2.4) and scaffold or capsule degradation (Wong and Choi 2015; Fenton et al. 2018; Witzler et al. 2018).

2.3.2 Current Approaches in Scaffold Development

As already mentioned, current approaches in scaffold development focus mainly on composite scaffolds that serve different functions. Most of these are produced using so-called "additive fabrication techniques". This can range from simple cross-linking and sometimes mineralizing of hydrogels to foams, to electrospinning of fibers and mats and ultimately to freeform fabrication of complete scaffolds (■ Fig. 2.5). Depending on the fabrication technique, additives (such as drugs or growth factors) can be added directly in the forming process or have to be added in a second step (Grotheer et al. 2014; Velasco et al. 2015; Simon 2018).

Hydrogel formation is a simple way of creating a porous polymer network. There are natural polysaccharides such as gelatin and agarose that form hydrogels by simple heating and cooling and there are natural polymers like chitosan and collagen or synthetic polymers such as poly(acrylamide), poly(ethylene glycol) or poly(lactic acid) that have to be somehow crosslinked. By varying concentration and degree of crosslinking, the stiffness of the hydrogel can be tuned. Mineralization of these scaffolds for bone applications then takes place in a second step, either by coating the polymer with ceramic or simply by mixing-in the ceramic particles into the solution prior to gelling. Incorporation of other active ingredients can be achieved by simple adding to the solution (Paris et al. 2015; Hu et al. 2016).

Foaming or leaching processes are also used for generating porous structures. Here, porogens (such as ice crystals, liquid CO_2 or salts) are added to the polymer or ceramic paste and removed after hardening of the scaffold by varying temperature, pressure or simple washing steps (Chen et al. 2018).

Electrospinning is often used for fibers and mats or fleeces that can be used for tendon, cartilage and bone applications and are often mineralized in a second step. Polymers, both

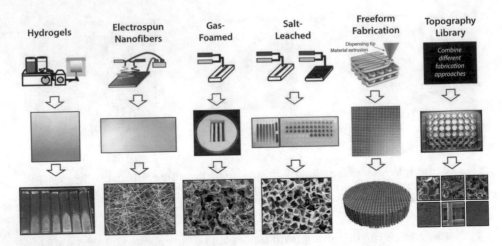

□ Fig. 2.5 Overview on different techniques for additive scaffold manufacturing. (Printed with permission of C.D. Simon Jr. (2018)). Schematic description of the technique (top line), macroscopic result (middle line) and microscale structure of the samples (bottom line)

natural and synthetic, are spun into fibers from a solution by acceleration in an electric field. Depending on the collector of the fiber, single strands or up to large mats of varying thickness can be created. It is also possible to incorporate growth factors or other active ingredients into the scaffold. The resulting porous fleeces can be seeded with cells and have been reported to show improved cell adhesion and proliferation (Rajzer et al. 2017).

Freeform fabrication or "3D printing" is on the rise for creating structures with a well-defined structure and porosity for cell in-growth. There are several possibilities to alter shape, functionality or mechanical properties by choosing different materials for separate sections of the scaffold. Ceramic parts can be added both directly to the printing mass or in a separate mineralization step. Active ingredients may be added directly to the printing feed or have to be added in a later step if the substance is heatsensitive (Grémare et al. 2018). Novel manufacturing techniques have comprehensively been reviewed focusing scaffold development for stem cell-based therapies (Schipper et al. 2017; El Khaldi-Hansen et al. 2017; Ottensmeyer et al. 2018).

2.3.3 Limitations and Challenges of Scaffolds

Current tissue repair research focuses on mimicking original tissue and its properties. While many aspects are already being considered, there are still a lot of limitations and challenges: the "perfect" biocompatible scaffold with ideal porosity, morphology, technical and mechanical properties has still not been created yet. There are also the questions of which cell types and which growth factors will ensure proper vascularization and host integration. Aside from these fundamental challenges there are other things still unknown: what are possible side effects of scaffolds, donor cells and growth factors? What is the most appropriate animal model? Which quality and functionality will the regenerated tissue have and what are its long-terms properties? And then there are of course clinical and regulatory questions: how can a patient-specific scaffold be generated? How to get FDA (and other) approval for these systems? How will the costs be covered?

2.4 Ethical and Legal Aspects of Mesenchymal Stem Cells

Interestingly, there are no specific regulations concerning research with human embryonic stem cells on the level of the United Nations (UNESCO/UN) or Europe (Council of Europe/European Union). On both levels, however, there are suggestions and regulatory efforts concerning the use of cloning methods in humans.

Due to this it is not astonishing that nearly each EU country has a different regulation. For example, the German signed the Embryo Protection Act from first January 1991 into law which prohibits the import, production and use of human ESCs for any purpose. In addition, each totipotent stem cell having the capacity to develop into an organism given circumstances is considered to be an embryo in Germany. After a long and controversial discussion, the German Bundestag agreed to a due date and some years later the due day was moved to first May 2007. Furthermore, an alteration of the specific stem cell law allows the use of imported ESCs which were generated before the due date and were produced from surplus embryos through in-vitro-fertilization. The Robert Koch Institute is the corresponding ordinance giving permission for projects with ESCs and the authority for all ethical questions related to human embryonic stem cells. In contrast in Great Britain embryos generated by using somatic cell nuclear transfer (keyword therapeutical cloning), and thereby even producing a chimera (e.g. cow egg with human nucleus) can be used for research.

Many Asian countries and Israel have the least restrictive laws. Singapore, for example, is widely considered as Asia's stem cell center because it is allowed to use embryos up to 2 weeks of age for therapeutical purposes (Dhar and Hsi-En 2009; Poulos 2018). In Israel it is legal to use an embryo up to 40 days after fertilization. The differences in the restriction and use of ESCs are enormous between different countries. Depending on the country, an ethical committee has to agree upon the project and the approval of the donor must be obtained. In China, a scientist stated in November 2018 that two twin girls were born where the genome was altered using CRISPR-Cas9 to generate a partical resistance against HIV infection by altering the CCR5 receptor. This led to a heated debate in most societies because it alters the human genome also for following generations (keyword human cloning).

MSCs on the other hand can be isolated and used worldwide with very few ethical controversies. Two different regulations have to be considered with the Regularity of the EU if stem cells are combined with a scaffold. Whereas the rule of the so called "medical devices" are easy to meet, the rule for a "medicinal product" were tight regulated leading to only few stem cell products on the market. In addition, the EU legislation created a regulatory gap between medical devices and medicinal products which hinders the approval of advanced-therapy products. However, there are only a few mesenchymal stem cell products approved so far. The EMA (Europe) authorized the first stem cell product called Holoclar which is used in the eye to replace damaged cells of the epithelium of the cornea (Pellegrini et al. 2018). Also, mesenchymal stem cells have secured conditional approval since 2012 for treating children with graft versus host diseases (GvHD). Health Canada and also MedSafe in New Zealand and the regulatory agency in Japan (Najima and Ohashi 2017) authorized the application. Osiris Therapeutics (U.S.A) completed the first major clinical phase III trial, sponsored by industry, of allogeneic, marrow-derived MSCs for treatment of steroid-refractory GvHD (NCT00366145) (Galipeau and Sensébé 2018). The Therapeutic Goods Administration in Australia admitted one mesenchymal stem cell product called Prochymal which is also used in acute GvHD in pediatric patients (Patel and Genovese 2011). The Korean Food and Drug Administration approved in total three medical treatments based on MSCs. NeuroNata-R which is only available in South Korea and Hearticellgram®-AMI (Yang 2011) are bone marrow mesenchymal stem cell therapies. NeuroNata-R has a neuroprotective effect and is indicated for amyotrophic lateral sclerosis while Hearticellgram®-AMI is used to treat acute myocardial infarction through intracoronary injection (Yang 2011). Further Cartistem, is the first allogenic stem cell drug. The mesenchymal stem cells drug is derived from allogenic umbilical cord blood and is used for treatment of knee cartilage defects in patients with osteoarthritis (Park et al. 2017).

In line with this, surgeon prefer to use stem cells with and without scaffolds using a technique which allow a "one step" procedure. During surgery, the operating physician may isolate mesenchymal stem cells from patient's bone marrow or rarer fat tissue and apply the cells back to the patient during surgery. The physician is allowed to treat the cells using an apheresis device or aspirator. The cells can also be separated by different on-site separation methods (magnetic antibodies, by filtration, or by gradient centrifugation) (Ruiz-Navarro and Kobinia 2017; Heinrichsohn 2017). Besides that, cells can be processed by activation through ultrasound or lasers. Also, mixing the mesenchymal stem cells with scaffolds or washing them before administration and using clinical syringes or other delivering tools e.g. for collection, separation, processing, delivery is allowed. This procedure is called Point-of-Care, and it allows the physician to apply a new cellular therapy to a patient. However, if the mesenchymal stem cell preparation is taken out of the operating room and the cells are expanded or treated in other ways, the application back to the patient has to follow another rule, which hampers the procedure as it is an elaborate process (■ Fig 2.6).

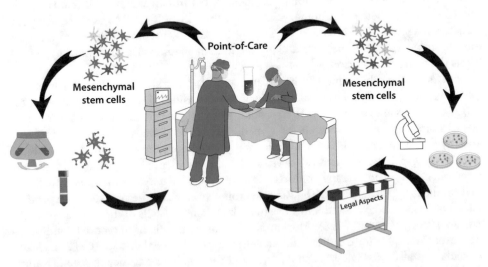

■ **Fig. 2.6** Point-of-Care cell therapy with mesenchymal stem cells. During surgery, physicians are allowed to isolate the patients' cells and apply them back. Only a few modifications and processing steps are allowed in this process (shown on the left side) namely collecting cells from the patient, separate them via density gradient centrifugation or magnetic beads, and apply the cells back to the patient using syringes or other delivering tools. If the cells have to be further processed such as being cultured or modified otherwise, they may not be given back to the patient without an official permission, where multiple obstacles have to be overcome (shown on the right side)

2.5 Perspectives for Future Use of Mesenchymal Stem Cells

MSCs are a promising tool for future use in regenerative medicine approaches. However, there are several obstacles which still have to be overcome.

Against expectations which have arisen from the success of hematopoietic stem cell treatments for cancers such as leukemia in the last 50 years, combined with the more recent findings that most tissues have adult stem cells, that MSCs will replace the loss of tissue cells in the respective organs or tissues, this is not the case (Eggenhofer et al. 2014). The cells are rapidly lost over time after transplantation into the recipient. Nevertheless, there seem to be a benefit for the patient which presumably is due to paracrine effect of the MSCs (Linero and Chaparro 2014; Baraniak and McDevitt 2010).

Another hurdle is that MSCs as other stem cells never differentiate completely *in vitro* which results in a potential risk for the recipient since they seem to be pre-committed to several lineages already in the tissue they are derived from. This might lead to unwanted differentiations within the tissue they are transplanted in.

Further, in opposite to pluripotent stem cells, they cannot differentiate unlimited in culture, so they must be replaced with newly isolates cells after some time. However, legal aspects in the EU are a major hindrance if the cells have to be cultured, expanded, or treated otherwise more than what is necessary for the isolation procedure leading to a quite difficult procedure necessary to be allowed to transplant them afterwards into a patient.

Last not least thinking about tissue or organ replacement the necessary procedure gets even more complex and thus complicated. Cells are not proliferating *in vitro* in three dimensions (and if this is unwanted as it is a sign for tumorigenicity). The extracellular matrix of the tissue or organ is needed for the cells to do so. If it cannot be obtained, it must be constructed artificially by either using natural or artificial substrates or a combination thereof to produce a biomaterial/scaffold for the cells. The scaffold must have interconnected pores to direct vascularization for nutrient and oxygen supply and disposal of metabolic waste and a lot of other features to eventually mimic the highly complex architecture of a "simple" tissue – not to forget that biomolecules and other cells than the specific tissue cell type are needed as well, such nerve cells, fibroblasts and the cells of the blood vessels namely endothelial cells and smooth muscle cells.

Nevertheless, ongoing clinical trials are encouraging to proceed using MSCs for future treatment of a variety of different diseases. Their great potential to differentiate in several lineages, their low immunogenicity, the easy accessibility, and their higher safety compared to pluripotent stem cells gives great hope that the obstacles can be overcome. More time is needed for basic research and thus a better understanding of the underlying signaling pathways within the cells and with their environment composed of other cell types, the matrix and biomolecules to make the huge step from bench to bedside safe.

2.6 Isolation Protocol of Mesenchymal Stem Cells

To get you started, step by step isolation protocols of MSCs from the most commonly used tissues are listed below. Be aware that you first need an ethical committee to give you permission if you want to use human tissue and that you need a permission for animal experiments as well if the cell source is not human.

ℹ **Protocol**

Isolation of MSCs from bone marrow (BM) (Susa et al. 2004; Secunda et al. 2015)

1. Collect 20 mL of BM and fill into a Falcon tube containing anti-coagulant citrate dextrose (ACD) in a ration 1:5.
2. Isolate the mononuclear cells by using the Ficoll-Paque gradient method: Pipet 15 mL of the Ficoll solution into a Falcon tube and overlay with 30 mL of the collected BM mixed with ACD.
   Don't mix the layers. Centrifuge at 400 x g for 30 minutes at room temperature without breaks. Five layers will be resulting after the centrifugation, namely from top to bottom: plasma phase, peripheral blood mononuclear cells (PBMCs), Ficoll, granulocytes, erythrocytes.
3. Collect the PBMC in the interphase, put them in a new Falcon tube and wash them twice with phosphate buffered saline.
4. The cell pellet can be resuspended in DMEM, supplemented with 10% fetal bovine serum and 1% antibiotics (100 U/mL streptomycin and 100 µg/mL penicillin) and the cells can be seeded on plates with a density of $15 \times 10^4/cm^2$ at 37 °C in a humidified atmosphere with 5% CO_2.

ℹ **Protocol**

Isolation of MSCs from umbilical cord (UC) (Secunda et al. 2015):

1. The cord can be gained after pregnancy must be devoid of blood. Cut it into pieces (thickness: 0.5 mm) and placed into PBS, supplemented with 1% antibiotics (100U/mL streptomycin and P100 µg/mL penicillin). proceed within 2 hours.
2. Transfer the pieces of the UC to 50 mL Falcon tubes with serum-free DMEM (Sigma) and centrifuge at 400 × g for 10 minutes at RT.
3. Discard the supernatant and add 0.1% collagenase (0.15 U/mL) solved in serum-free DMEM and incubate overnight.
4. Add the double volume of PBS and centrifuge at 400 × g for 10 minutes at RT.
5. Discard the supernatant and treat the pellet with 2.5% trypsin in PBS at 37 °C for 30 minutes.
6. Add at least threefold FBS to neutralize the trypsin and wash with serum-free DMEM culture medium.
7. Resuspend the cell pellet in 10 mL DMEM, supplemented with 10% FBS and 1% antibiotics (100 U/mL streptomycin and 100 µg/mL penicillin) and seeded on plates with a density of $15 \times 10^4/cm^2$ at 37 °C in a humidified atmosphere with 5% CO_2.

This method uses an enzymatic digest. An explant method can be used as well, were the UC pieces were culture in DMEM-low glucose with 20% FBS at 37 °C in a humidified atmosphere with 5% CO_2 and left undisturbed for 7 days to allow the migration of the cells from the explant.

ℹ **Protocol**

Isolation of MSCs from adipose tissue from liposuction material (Pittenger et al. 1999; Secunda et al. 2015):

1. Determine the amount of received fat solution.
2. Add PBS to the fat tissue (ratio 1:1), shake well and incubate for 30 minutes at RT to separate the phases.

3. Discard the bottom layer and add PBS (ratio 1:1) and add 0.1% collagenase (0.15 U/mL) in PBS to the upper phase and incubate for 60 minutes at 37 °C while shaking in the water bath.
4. Centrifuge the incubated suspension at $200 \times g$ at RT for 10 minutes.
5. Discard the supernatant and resuspend the pellet in 10 mL erolysis buffer (0.5 M EDTA, pH 8.0, 0.154 M ammonium chloride, 1×10^{-2} M potassium hydrogen carbonate) and incubate for 10 minutes at RT.
6. Centrifuge again at $200 \times g$, RT for 10 minutes and discard the supernatant afterwards.
7. Resuspend the cell pellet in 10 mL DMEM, supplemented with 10% FBS and 1% antibiotics (300 U/mL streptomycin and 300 µg/mL penicillin) and seed them at in a density of $15 \times 10^{4}/cm^2$ on plates at 37 °C in a humidified atmosphere with 5% CO_2.
8. Change medium after 24 hours to remove non-adherent cells and debris.

If necessary, the bottom layer (from 3.) can be used as well following the same procedure to increase the MSC yield.

▪ **Additional Comment**

The mesenchymal stem cells isolated from different tissues were all cultured in DMEM, supplemented with 10% FBS and 1% antibiotics (300 U/mL streptomycin and 300 µg/ mL penicillin) at 37 °C in a humidified atmosphere with 5% CO_2. Reaching > 75% confluence, the cells were trypsinised with 0.05% trypsin in 0.2% EDTA and subsequently frozen or passaged by splitting them in two or more new flasks (Freshney et al. 2007). Adapted and slightly altered methods can also be used for the isolation of MSCs from other tissues.

Take Home Message

- Mesenchymal stem cells can be found in various tissues and have a vast differentiation potential which is interesting for approaches in regenerative medicine.
- Mesenchymal stem cells are defined by a set of three minimal criteria proposed by the Mesenchymal and Tissue Stem Cell Committee of the International Society for Cellular Therapy
- Mesenchymal stem cells can be used in autologous therapy approaches, have a low immunogenicity, are easily accessible, quickly expandable and they cause no major ethical problems.
- Mesenchymal stem cells can be used in cell therapy approaches during a surgery at the Point-of-Care. If they must be expanded or treated otherwise, the application in humans causes additional obstacles which have to be overcome.
- Many clinical trials for the treatment of various diseases are ongoing. At the moment, almost all of them are in clinical trials phase I or II.
- The beneficial effect of mesenchymal stem cells in regenerative medicine seem to be due to paracrine effects since over time the cells get lost, thus not replacing the tissue cells.

- New approaches in regenerative medicine for tissue repair often use two cell types e.g. tissue cell and mesenchymal stem cells.
- Scaffold composition and properties should closely resemble original tissue.
- Scaffolds must be biocompatible and have suitable morphology, mechanical properties, technical functionality, stability against physiological conditions.
- Today, scaffolds are often produced via additive manufacturing including 3D printing and electrospinning.
- Major difficulties remain in proper vascularization, bio resorption and in transferring the scaffold into clinical trials.

Acknowledgements   This article was supported by the Bundesministerium für Bildung und Forschung (BMBF) FHprofUnt to E.T., FKZ: 13FH012PB2; EFRE co-financed NRW Ziel 2: "Regionale Wettbewerbsfähigkeit und Beschäftigung", 2007-2013, Ministerium für Innovation, Wissenschaft und Forschung (MIWF) NRW FH-Extra to E.T., FKZ: z1112fh012; DAAD PPP Vigoni to E.T., FKZ: 314-vigoni-dr and FKZ: 54669218; Bundesministerium für Bildung und Forschung (BMBF)-AIF to E.T., FKZ: 1720X06; Bundesministerium für Bildung und Forschung (BMBF) IngenieurNachwuchs to E.T., FKZ: 13FH019IX5; Ministerium für Innovation, Wissenschaft und Forschung (MIWF) NRW FH Zeit für Forschung to E.T., FKZ 005-1703-0017.

Bundesministerium für Bildung und Forschung (BMBF) IngenieurNachwuchs to S.W.; FKZ: 13FH569IX6.

References

Alkhalil, M., Smajilagić, A., & Redžić, A. (2015). Human dental pulp mesenchymal stem cells isolation and osteoblast differentiation. *Medicinski Glasnik, 12*(1), 27–32.

Baksh, D., Song, L., & Tuan, R. S. (2004). Adult mesenchymal stem cells. Characterization, differentiation, and application in cell and gene therapy. *Journal of Cellular and Molecular Medicine, 8*(3), 301–316.

Baraniak, P. R., & McDevitt, T. C. (2010). Stem cell paracrine actions and tissue regeneration. *Regenerative Medicine, 5*(1), 121–143.

Bartholomew, A., Sturgeon, C., Siatskas, M., et al. (2002). Mesenchymal stem cells suppress lymphocyte proliferation in vitro and prolong skin graft survival in vivo. *Experimental Hematology, 30*(1), 42–48.

Bruno, S., Bussolati, B., Grange, C., et al. (2009). Isolation and characterization of resident mesenchymal stem cells in human glomeruli. *Stem Cells and Development, 18*(6), 867–880.

Cao, C., Dong, Y., & Dong, Y. (2005). Study on culture and in vitro osteogenesis of blood-derived human mesenchymal stem cells. *Zhongguo Xiu Fu Chong Jian Wai Ke Za Zhi, 19*(8), 642–647.

Chen, Y., Kawazoe, N., & Chen, G. (2018). Preparation of dexamethasone-loaded biphasic calcium phosphate nanoparticles/collagen porous composite scaffolds for bone tissue engineering. *Acta Biomaterialia, 67*, 341–353.

Cotter, E. J., Wang, K. C., Yanke, A. B., et al. (2018). Bone marrow aspirates concentrate for cartilage defect of the knee: From bench to bedside evidence. *Cartilage., 9*(2), 161–170.

da Silva Meirelles, L., Chagastelles, P. C., & Nardi, N. B. (2006). Mesenchymal stem cells reside in virtually all postnatal organs and tissues. *Journal of Cell Science, 119*(11), 2204–2213.

Dhar, D., & Hsi-En, H. O. (2009). Stem cell research policies around the world. *The Yale Journal of Biology and Medicine, 82*(3), 113–115.

Dominici, M., Le Blanc, K., Mueller, I., et al. (2006). Minimal criteria for defining multipotent mesenchymal stromal cells. The International Society for Cellular Therapy position statement. *Cytotherapy, 8*(4), 315–317.

Eggenhofer, E., Luk, F., Dahlke, M. H., et al. (2014). The life and fate of mesenchymal stem cells. *Frontiers in Immunology, 5*, 148.

El Khaldi-Hansen, B., El-Sayed, F., Tobiasch, E., et al. (2017). Functionalized 3D scaffolds for template-mediated biomineralization in bone regeneration. In Atta-ur-Rahman & S. Anjum (Eds.), *Frontiers in stem cell and regenerative medicine research* (pp. 130–178). Sharjah: Bentham Science Publishers.

Evans, M. J., & Kaufman, M. H. (1981). Establishment in culture of pluripotential cells from mouse embryos. *Nature, 292*(5819), 154–156.

Fenton, O. S., Olafson, K. N., Pillai, P. S., et al. (2018). Advances in biomaterials for drug delivery. *Advanced Materials, 30*, 1705328.

Freshney, R. I., Stacey, G. N., & Auerbach, J. M. (2007). *Culture of human stem cells*. Hoboken: Wiley.

Friedenstein, A. J., Chailakhjan, R. K., & Lalykina, K. S. (1970). The development of fibroblast colonies in monolayer cultures of guinea-pig bone marrow and spleen cells. *Cell and Tissue Kinetics, 3*(4), 393–403.

Friedenstein, A. J., Chailakhyan, R. K., Latsinik, N. V., et al. (1974). Stromal cells responsible for transferring the microenvironment of the hemopoietic tissues. Cloning in vitro and retransplantation in vivo. *Transplantation, 17*(4), 331–340.

Galipeau, J., & Sensébé, L. (2018). Mesenchymal stromal cells. Clinical challenges and therapeutic opportunities. *Cell Stem Cell, 22*(6), 824–833.

Gordon, A. S. (1972). Haemopoietic cells. In D. Metcalf, M. A. S Moore (Eds.) *Frontiers of biology* (vol. 178(4064), pp. 974–975). Elsevier.

Grémare, A., Guduric, V., Bareille, R., et al. (2018). Characterization of printed PLA scaffolds for bone tissue engineering. *Journal of Biomedical Materials Research. Part A, 106*(4), 887–894.

Grotheer, V., Schulze, M., & Tobiasch, E. (2014). Trends in bone tissue engineering: Proteins for osteogenic differentiation and the respective scaffolding. In: *Protein purification – Principles and trends*. iConcept Press Ltd, Hong Kong.

Haddouti, E. M., Skroch, M., Zippel, N., et al. (2009). Human dental follicle precursor cells of wisdom teeth. Isolation and differentiation towards osteoblasts for implants with and without scaffolds. *Materialwissenschaft und Werkstofftechnik, 40*(10), 732–737.

Hanna, J. H., Saha, K., & Jaenisch, R. (2010). Pluripotency and cellular reprogramming. Facts, hypotheses, unresolved issues. *Cell, 143*(4), 508–525.

Haubner, F., Muschter, D., Pohl, F., et al. (2015). A co-culture model of fibroblasts and adipose tissue-derived stem cells reveals new insights into impaired wound healing after radiotherapy. *International Journal of Molecular Sciences, 16*(11), 25947–25958.

Heinrichsohn, F. (2017). Cellular therapy, an autologous cellular point of care approach to satisfy patient needs. *Journal of Translational Science, 3*(1).

Henkel, J., Woodruff, M. A., Epari, D. R., et al. (2013). Bone regeneration based on tissue engineering conceptions – A 21st century perspective. *Bone Research, 1*(3), 216–248.

Herrera, M. B., Bussolati, B., Bruno, S., et al. (2007). Exogenous mesenchymal stem cells localize to the kidney by means of CD44 following acute tubular injury. *Kidney International, 72*(4), 430–441.

Hielscher, D., Kaebisch, C., Braun, B. J. V., et al. (2018). Stem cell sources and graft material for vascular tissue engineering. *Stem Cell Reviews, 14*(5), 642–667.

Horwitz, E. M., Le Blanc, K., Dominici, M., et al. (2005). Clarification of the nomenclature for MSC. The International Society for Cellular Therapy position statement. *Cytotherapy, 7*(5), 393–395.

Hourd, P., Chandra, A., Medcalf, N., et al. (2008). Regulatory challenges for the manufacture and scale-out of autologous cell therapies. StemBook, Harvard Stem Cell Institute, Cambridge, MA, USA.

Hu, J., Zhu, Y., Tong, H., et al. (2016). A detailed study of homogeneous agarose/hydroxyapatite nanocomposites for load-bearing bone tissue. *International Journal of Biological Macromolecules, 82*, 134–143.

Huang, B., Li, K., Yu, J., et al. (2013). Generation of human epidermis-derived mesenchymal stem cell-like pluripotent cells (hEMSCPCs). *Scientific Reports, 3*, 1933.

In't Anker, P. S., Scherjon, S. A., Kleijburg-van der Keur, C., et al. (2004). Isolation of mesenchymal stem cells of fetal or maternal origin from human placenta. *Stem Cells, 22*(7), 1338–1345.

Kaebisch, C., Schipper, D., Babczyk, P., et al. (2015). The role of purinergic receptors in stem cell differentiation. *Computational and Structural Biotechnology Journal, 13*, 75–84.

Lanzoni, G., Alviano, F., Marchionni, C., et al. (2009). Isolation of stem cell populations with trophic and immunoregulatory functions from human intestinal tissues. Potential for cell therapy in inflammatory bowel disease. *Cytotherapy, 11*(8), 1020–1031.

Lee, O. K., Kuo, T. K., Chen, W.-M., et al. (2004). Isolation of multipotent mesenchymal stem cells from umbilical cord blood. *Blood, 103*(5), 1669–1675.

Linero, I., & Chaparro, O. (2014). Paracrine effect of mesenchymal stem cells derived from human adipose tissue in bone regeneration. *PLoS One, 9*(9), e107001.

Liu, J. Y., Peng, H. F., Gopinath, S., et al. (2010). Derivation of functional smooth muscle cells from multipotent human hair follicle mesenchymal stem cells. *Tissue Engineering. Part A, 16*(8), 2553–2564.

Moraghebi, R., Kirkeby, A., Chaves, P., et al. (2017). Term amniotic fluid: An unexploited reserve of mesenchymal stromal cells for reprogramming and potential cell therapy applications. *Stem Cell Research & Therapy, 8*(1), 190.

Najima, Y., & Ohashi, K. (2017). Mesenchymal stem cells. The first approved stem cell drug in Japan. *Journal of Hematopoietic Stem Cell Transplantation, 6*(3), 125–132.

Nakagawa, M., Koyanagi, M., Tanabe, K., et al. (2008). Generation of induced pluripotent stem cells without Myc from mouse and human fibroblasts. *Nature Biotechnology, 26*(1), 101–106.

Ottensmeyer, P. F., Witzler, M., Schulze, M., et al. (2018). Small molecules enhance scaffold-based bone grafts via purinergic receptor signaling in stem cells. *International Journal of Molecular Sciences, 19*(11), 3601.

Pansky, A., Roitzheim, B., & Tobiasch, E. (2007). Differentiation potential of adult human mesenchymal stem cells. *Clinical Laboratory, 53*(1–2), 81–84.

Paris, J. L., Román, J., Manzano, M., et al. (2015). Tuning dual-drug release from composite scaffolds for bone regeneration. *International Journal of Pharmaceutics, 486*(1–2), 30–37.

Park, J. H., Kim, D. Y., Sung, I. Y., et al. (2012). Long-term results of spinal cord injury therapy using mesenchymal stem cells derived from bone marrow in humans. *Neurosurgery, 70*(5), 1238–1247.

Park, Y.-B., Ha, C.-W., Lee, C.-H., et al. (2017). Cartilage regeneration in osteoarthritic patients by a composite of allogeneic umbilical cord blood-derived mesenchymal stem cells and hyaluronate hydrogel. Results from a clinical trial for safety and proof-of-concept with 7 years of extended follow-up. *Stem Cells Translational Medicine, 6*(2), 613–621.

Patel, A. N., & Genovese, J. (2011). Potential clinical applications of adult human mesenchymal stem cell (Prochymal®) therapy. *Stem Cells Cloning., 4*, 61–72.

Pellegrini, G., Ardigò, D., Milazzo, G., et al. (2018). Navigating market authorization. The path Holoclar took to become the first stem cell product approved in the European Union. *Stem Cells Translational Medicine, 7*(1), 146–154.

Pittenger, M. F., Mackay, A. M., Beck, S. C., et al. (1999). Multilineage potential of adult human mesenchymal stem cells. *Science, 284*(5411), 143–147.

Poulos, J. (2018). The limited application of stem cells in medicine: A review. *Stem Cell Research & Therapy, 9*(1), 1.

Prasad, V. K., Lucas, K. G., Kleiner, G. I., et al. (2011). Efficacy and safety of ex vivo cultured adult human mesenchymal stem cells (Prochymal™) in pediatric patients with severe refractory acute graftversushost disease in a compassionate use study. *Biology of Blood and Marrow Transplantation, 17*(4), 534–541.

Prockop, D. J. (1997). Marrow stromal cells as stem cells for nonhematopoietic tissues. *Science, 276*(5309), 71–74.

Rajzer, I., Menaszek, E., & Castano, O. (2017). Electrospun polymer scaffolds modified with drugs for tissue engineering. *Materials Science & Engineering. C, Materials for Biological Applications, 77*, 493–499.

Rheinwald, J. G., & Green, H. (1975). Serial cultivation of strains of human epidermal keratinocytes. The formation of keratinizing colonies from single cells. *Cell, 6*(3), 331–343.

Ruiz-Navarro, F., & Kobinia, G. (2017). Point-of-care stem cell therapy (pocst). Multisite transplantation of autologous bone marrow-derived mononuclear cells in 85 patients with amyotrophic lateral sclerosis improves survival. *Journal of Stem Cell Research & Therapy, 2*(3), 98–110.

Sakaguchi, Y., Sekiya, I., Yagishita, K., et al. (2005). Comparison of human stem cells derived from various mesenchymal tissues. Superiority of synovium as a cell source. *Arthritis and Rheumatism, 52*(8), 2521–2529.

Salerno, S., Messina, A., Giordano, F., et al. (2017). Dermal-epidermal membrane systems by using human keratinocytes and mesenchymal stem cells isolated from dermis. *Materials Science & Engineering, C: Materials for Biological Applications, 71*, 943–953.

Schipper, D., Babczyk, P., El-Sayed, F., et al. (2017). The effect of nanostructured surfaces on stem cell fate. In A. M. Grumezescu & D. Ficai (Eds.), *Nanostructures for novel therapy* (Vol. 1, pp. 567–589). Amsterdam, Cambridge: Elsevier.

Schulze, M., & Tobiasch, E. (2012). Artificial scaffolds and mesenchymal stem cells for hard tissues. *Advances in Biochemical Engineering/Biotechnology, 126*, 153–194.

Secunda, R., Vennila, R., Mohanashankar, A. M., et al. (2015). Isolation, expansion and characterisation of mesenchymal stem cells from human bone marrow, adipose tissue, umbilical cord blood and matrix. A comparative study. *Cytotechnology, 67*(5), 793–807.

Simon, C. G. Jr. *3D scaffold fabrication approaches.* Available online: https://www.nist.gov/sites/default/files/images/mml/bbd/biomaterials/scaffoldomics.jpg. Accessed 29 Aug 2018.

Steigmann, S. A., & Fauza, D. O. (2007). Isolation of mesenchymal stem cells from amniotic fluid and placenta. *Current Protocols in Stem Cell Biology, 1*, 1E.2.1–1E.212.

Steinhoff, G., Nesteruk, J., Wolfien, M., et al. (2017). Cardiac function improvement and bone marrow response: Outcome analysis of the randomized PERFECT phase III clinical trial of intramyocardial CD133+ application after myocardial infarction. *eBioMedicine, 22*, 208–224.

Susa, M., Luong-Nguyen, N.-H., Cappellen, D., et al. (2004). Human primary osteoclasts: In vitro generation and applications as pharmacological and clinical assay. *Journal of Translational Medicine, 2*(1), 6.

Takagi, S., Mandai, M., Hirami, Y., et al. (2018). Frequencies of human leukocyte antigen alleles and haplotypes among Japanese patients with age-related macular degeneration. *Japanese Journal of Ophthalmology, 62*, 568–575.

Takahashi, K., & Yamanaka, S. (2006). Induction of pluripotent stem cells from mouse embryonic and adult fibroblast cultures by defined factors. *Cell, 126*(4), 663–676.

Takahashi, K., Tanabe, K., Ohnuki, M., et al. (2007). Induction of pluripotent stem cells from adult human fibroblasts by defined factors. *Cell, 131*(5), 861–872.

Thirumala, S., Zvonic, S., Floyd, E., et al. (2005). Effect of various freezing parameters on the immediate postthaw membrane integrity of adipose tissue derived adult stem cells. *Biotechnology Progress, 21*(5), 1511–1524.

Thomson, J. A., Itskovitz-Eldor, J., Shapiro, S. S., et al. (1998). Embryonic stem cell lines derived from human blastocysts. *Science, 282*(5391), 1145–1147.

Till, J. E., & McCulloch, E. A. (1961). A direct measurement of the radiation sensitivity of normal mouse bone marrow cells. *Radiation Research, 14*(2), 213.

Tobiasch, E. (2008). Adult human mesenchymal stem cells as source for future tissue engineering. In C. Zacharias, et al. (Eds.) *Forschungsspitzen und Spitzenforschung* (pp. 329–338). Physica-Verlag HD, Heidelberg, Germany.

Tsuji, O., Sugai, K., Yamaguchi, R., et al. (2018). Laying the groundwork for a first-in-human study of an induced pluripotent stem cell-based intervention for spinal cord injury. *Stem Cells, 37*, 6–13.

Ullah, I., Subbarao, R. B., & Rho, G. J. (2015). Human mesenchymal stem cells – Current trends and future prospective. *Bioscience Reports, 35*(2), e00191.

Velasco, M. A., Narváez-Tovar, C. A., & Garzón-Alvarado, D. A. (2015). Design, materials, and mechanobiology of biodegradable scaffolds for bone tissue engineering. *BioMed Research International, 2015*, 729076.

Wang, H.-S., Hung, S.-C., Peng, S.-T., et al. (2004). Mesenchymal stem cells in the Wharton's jelly of the human umbilical cord. *Stem Cells, 22*(7), 1330–1337.

White, I. A., Sanina, C., Balkan, W., et al. (2016). Mesenchymal stem cells in cardiology. *Methods in Molecular Biology, 1416*, 55–87.

Witzler, M., Alzagameem, A., Bergs, M., et al. (2018). Lignin-derived biomaterials for drug release and tissue engineering. *Molecules, 23*(8), E1885.

Wong, P. T., & Choi, S. K. (2015). Mechanisms of drug release in nanotherapeutic delivery systems. *Chemical Reviews, 115*(9), 3388–3432.

www.clinicaltrials.gov. Accessed 27 May 2019.

Yang, H. (2011). South Korea's stem cell approval. *Nature Biotechnology, 29*(10), 857.

Yao, W., Lay, Y.-A. E., Kot, A., et al. (2016). Improved mobilization of exogenous mesenchymal stem cells to bone for fracture healing and sex difference. *Stem Cells, 34*(10), 2587–2600.

Yu, J., Vodyanik, M. A., Smuga-Otto, K., et al. (2007). Induced pluripotent stem cell lines derived from human somatic cells. *Science, 318*(5858), 1917–1920.

Zhang, Y., Lau, P., Pansky, A., et al. (2014). The influence of simulated microgravity on purinergic signaling is different between individual culture and endothelial and smooth muscle cell coculture. *BioMed Research International, 2014*, 413708.

Zippel, N., Limbach, C. A., Ratajski, N., et al. (2012). Purinergic receptors influence the differentiation of human mesenchymal stem cells. *Stem Cells and Development, 21*(6), 884–900.

Zuk, P. A., Zhu, M., Ashjian, P., et al. (2002). Human adipose tissue is a source of multipotent stem cells. *Molecular Biology of the Cell, 13*(12), 4279–4295.

Cord Blood Stem Cells

Gesine Kogler

© Springer Nature Switzerland AG 2020
B. Brand-Saberi (ed.), *Essential Current Concepts in Stem Cell Biology*,
Learning Materials in Biosciences, https://doi.org/10.1007/978-3-030-33923-4_3

What You Can Learn in This Chapter
This chapter describes the research and clinical developments with cord blood stem and progenitor cells. You will learn about the development of hematopoietic cells including immune cells during fetal life and how the cells are utilized in clinical studies. Since the immaturity of the immune system and the biology of cord blood cells have an important impact on the transplantation situation with a mismatched donor, respective studies are highlighted. Due to biological advantages of the hematopoietic system, studies applying CD34+ expanded cells are discussed. You will learn in detail, why cord blood is the perfect raw material for reprogramming towards iPSC's products according to new GMP regulations.

Since cord blood also contains non-hematopoietic cells as endothelial cells and stromal cells, the characterization of the cells in vitro and in vivo is explained in detail. Infusion and application of cord blood cells for cerebral palsy, stroke and autism as of today is discussed.

3.1   Content of the Chapter

3.1.1   Biological Background of Hematopoietic Stem Cells During Fetal Development

As described by Waas and Maillard (2017), hematopoiesis first arises during early embryogenesis when passive diffusion of oxygen and nutrients becomes insufficient to support the developing organism. Early development is focused on the production of red blood cells and tightly linked to vascular development. Beyond these initial functions, the fetal hematopoietic system develops to generate mature elements as erythrocytes, platelets, macrophages, other myeloid cells as well as T and B lymphocytes. During fetal life, subsets of hematopoietic cells acquire long-term self- renewal potential as assessed after transplantation into lethally irradiated hosts, a function decades ago used to define hematopoietic stem cells (HSC) (Lorenz et al. 1951; Spangrude et al. 1988; Becker et al. 1963; Eaves 2015). Fetal cells are thought to seed the more quiescent adult HSC compartment that will sustain the adult hematopoietic system (Bowie et al. 2007, 2006; Jones et al. 2015). In a recent work Beaudin et al. (2017) identified and characterized a subset of fetal HSCs that are endowed with long term self-renewal potential in transplantation assays, but fail to persist into adulthood in physiological conditions, however they sustain long-term lymphoid based multi-lineage reconstitution after transplantation. These recent findings in mice are consistent with the coexistence (Waas and Maillard 2017) of at least two populations of fetal HSC: a population of "conventional" fetal HSC with transitions into adult HSC and seed to the adult hematopoietic system; and a population of HSC that is programmed for disappearance, unless introduced experimentally into irradiated adult recipients. In humans it is not clear yet whether these data could be also translated in observations after cord blood transplantation. Cord blood at birth (gestational age week 36–41) is characterized by a unique richness in hematopoietic stem and progenitor cells, particularly those 'early' cells which are detected in *in vitro* assays like the LTC-IC assay (Long Term Culture-Initiating Cell) and HPP-CFC assay (High Proliferative Potential-Colony Forming Cell) or *in vivo* due to their potential to repopulate NOD/SCID mice (SRC – SCID Repopulating Cells). As discussed above, hematopoietic stem cells (HSC) develop during embryogenesis and fetal life in a complex process involving multiple anatomic sites and niches (yolk sac, the aorta-gonad-mesonephrons region, placenta and fetal liver) (Cumano et al. 2001), before they colonize the bone marrow (BM). The precious ingredients in cord blood are the blood-forming hematopoietic stem and progenitor cells that replicate and diversify to replace a patient's entire blood and immune system. These cells are rare and are present with 0.1–1% in cord blood, but a typical collection contains millions of blood-forming cells and progenitor cells. As fetal and neonatal hematopoietic cells in cord blood are markedly different from adult HSC, it was conceivable that different mechanisms and/or niches control engraftment and self-renewal of HSC during fetal and adult life in humans. Since fetal blood is formed in close association with organs, the search for cell functions as niches similar to cell types present in adult bone marrow environment (osteoblasts, endothelial cells, fibroblasts, reticular cells) was a logical consequence. Although the HSC contained are currently the most relevant cells in cord blood with regard to clinical application, cord blood also contains non-hematopoietic cell types which bear interesting biological features in regenerative medicine as of today.

3.1.2 Current State of Cord Blood Stems Cell Banking and Transplantation

On September 2018, the 30th anniversary of the first hematopoietic stem cell (HSC) transplant using cord blood as a graft for a patient with Fanconi's anemia (Gluckman et al. 1989) was celebrated. The successful demonstration that cord blood was able to reconstitute a patient's blood and immune system in combination with the evidence that cord blood can be cryopreserved for later use, led to the establishment of cord blood banks in Europe (Düsseldorf, Barcelona, Milano) and the US (New York Blood Center) in the early 1990's (Ballen et al. 2015). Cord blood stem cell banking is performed in public cord blood banks, which store cord blood for future use as an allogeneic transplant or a sibling transplant for patients with indications. Private cord blood banking is performed for autologous use of the donor. As of January 2019 709,584 allogeneic donated cord blood units are stored in public banks (▶ https://wmda.info; https://www.uniklinik-duesseldorf.de/patienten-besucher/klinikeninstitutezentren/jose-carreras-stammzellbank); whereas it is estimated that over five million are stored in private cord blood banks (Kurtzberg 2017). The Düsseldorf José Carreras Cord Blood Bank (▶ www.stammzellbank.de) started the standardization of unrelated and related cord blood stem cell characterization and banking already in 1993. Hitherto more than 27.000 cord blood samples have been tested and cryopreserved (February 2019) including the storage of 464 family donations for directed indications of the patient. 1360 cord blood transplants were provided to transplant centers world-wide in 34 countries, mainly Europe and the US. 62% of the patients were adults, 38% children. 84% of the patients had malignant disease, 16% genetic, metabolic and immunological disorders. The majority of transplants were mismatched at one to two out of 6 HLA-antigens as defined by generic HLA-typing for HLA-class I and subtyping for HLA-class II for HLA-DRB1 and HLA-DQB1 alleles. The great majority of units had 5/6 (46.7%) or 4/6 degree (30.1%) of HLA matching (one or two differences) with the recipient. Only 20% were matched for 6 out of 6 HLA antigens (A, B antigenic level, DRB1 allelic level) in the single CBT. According to the defined criteria of EUROCORD the probability of overall survival at median follow-up (47.4 months) estimated by Kaplan-Meier method is 44% (±2), with significant better results in children than in adult patients and also in case of HLA match or only one difference than in two or more HLA discrepancies and in non-malignant diseases than in malignant ones.

This analysis in single CBT also shows better survival in case of increased number of nucleated or CD34+ cord blood cells infused. In the group of 490 patients transplanted with two cord blood units (double CBT) for malignant diseases the probability of disease-free survival estimated by Kaplan-Meier is 42% (±2) at median FU (36 months).

A recent National Marrow Donor Program (NMDP) study has demonstrated that whereas 75% of white European patients are likely to identify an 8/8 HLA-matched unrelated donor, the rate is much lower in minority patients or rare HLA-types (Gragert et al. 2014). 20% of all patients do not find a matched donor within 6 months in the world-wide inventor of more than 33 Million donors (▶ www.wmda.info). In the absence of an unrelated donor, cord blood or haploidentical related donor transplants are alternative options (Barker et al. 2017).

Multiple retrospective studies (Barker et al. 2017) have demonstrated that CBT performed in experienced centers can achieve disease-free survival rates comparable to the gold standard of HLA-matched unrelated bone marrow transplantation in patients with hematologic malignancies (Brunstein et al. 2012; Ponce et al. 2015; Warlick et al. 2015; Milano et al. 2016). These analyses are notable for the low relapse rate after CBT compared with unrelated PBSC/bone marrow transplantation with minimal residual disease (Milano et al. 2016). As discussed by Vago et al. 2018 at the ASH meeting on the observation of HLA-loss as a reason resulting in relapse after unrelated adult donor transplantation for either haploidentical or unrelated adult transplantation, CBT was not influenced by this mechanism. Relapse protection might have to do with the biology of cord blood stem and immune cells (NK, T-cells) at birth (see previous chapter).

3.1.3 New Developments in Cord Blood Transplantation to Improve Engraftment and Homing to the Bone Marrow

One of the limiting factors of the CBT is the delayed engraftment and immune reconstitution that may lead to infections, particularly uncommon viral infections. Over the last 10 years the data improved based on higher Total Nucleated cell count and CD34+ cells/kg bodyweight infused. The Düsseldorf data set reveal the median time to neutrophils engraftment as of 21 days. The date of neutrophils recovery is considered as the first of 3 consecutive days in which the laboratory results indicate that the neutrophils $\geq 0.5 \times 10e9/L$, without evidence of reconstitution with recipient bone marrow, nor graft rejection in the first 100 days.

Despite the improved results on engraftment due to the selection of the "largest" cord blood grafts, several strategies have been developed to improve this delayed immune recovery as homing to the bone marrow by intra bone injection, ex-vivo expansion of hematopoietic cells by cytokines in combination with Notch, Fucosylation and nicotinamide-based expansion with cytokines (◘ Table 3.1, Kurita et al. 2017; Horwitz et al. 2014; Popat et al. 2015; Delaney et al. 2010). The last approach resulted in engraftment in 13 days and is part of a randomized controlled trial. All studies are small and none has documented improved survival in randomized trials yet. Moreover one has to demonstrate that the manipulation of cord blood is not changing the relapse-preventing cell populations after CBT. If T-cell or NK cell function is impaired by expansion or manipulation, survival rates can be reduced by higher relapse rates or GvHD. The combination of haploidentical with either single or double CBT has definitively shown to improve engraftment (Bautista et al. 2009; Liu et al. 2011; Ponce et al. 2013). Since one

Table 3.1 Novel strategies to improve engraftment

Strategy	Mechanism	Investigators	Number	Days to absolute neutrophil count >500
Intrabone marrow injection	Homing	Kurita et al.	15	7
Nicotinamide	Expansion	Horwitz et al.	11	13
Fucosylation	Homining	Popat et al.	7	14
Notch	Expansion	Delaney et al.	10	16

Table 3.2 Selected active and recruiting regenerative medicine human cord blood trials

Disease	Agent	Investigator	Current trial
Autism	Auto or Allo UCB	Kurtzberg	NCT02847182
Cerebral palsy	Auto UCB	Carrol	NCT01072370
Cerebral palsy	Auto UCB and G-CSF	Lee	NCT02866331
Ischemic stroke	Allo UCB	Kurtzberg	NCT03004976
Chronic ischemic cardiomyopathy	UCB-derived mesenchymal stem cells	Dai	NCT02635464
Crohn's disease	UCB-derived stem cells	Lee	NCT02000362

Abbreviation: *G-CSF* granulocyte colony-stimulation factor, *UCB* umbilical cord blood

cord blood unit is not manipulated, the second graft can perform the bridging for better neutrophil, platelet recovery as well immune reconstitution.

Besides the hematopoietic stem cell, also T-cells and NK (Natural Killer) – cells from cord blood are candidates for off-the-shelf cells for immunotherapy. As described by the group of Rezvani 2018 (Liu et al. 2018) NK cells can be engineered to express CAR to redirect their specificity and a cytokine to enhance their in vivo proliferation and in vivo persistence. A first – in – human clinical trial to test the safety and efficacy of CAR19/IL15/iCas9 transduced cord blood NK cells is open at the MD Anderson Cancer Center in Houston, Texas.

3.1.4 Cord Blood Transplantation as of Today in Regenerative Medicine Applications

Completely new developments over the last 5 years are the use of either autologous or unrelated cord blood for application in neurological diseases, cardiology and endocrinology (Table 3.2). Cord blood has been used clinically to treat cerebral palsy, hypoxic ischemic encephalopathy, stroke and autism (Cotten et al. 2014; Dawson et al. 2017; Sun et al. 2017; Laskowitz et al. 2018). Cord blood derived mesenchymal stroma cells are in clinical trials for dilated cardiomyopathy and ischemic disease (Wu et al. 2007).

In cerebral palsy, intravenous autologous CB infusions were administered safely (Sun et al. 2017). Allogeneic cord blood infusions were given to 47 patients with severe cerebral palsy both intravenously and intrathecal (Feng et al. 2015) with improvement in motor function. Although a clinical effect is observed, the mechanism behind are still not defined yet. Although studies including our own have shown how subsets of cord blood cells differentiate under defined conditions into neurons, astrocytes and microglia *in vitro* by more or less artificial methods and substances, which do not reflect the *in vivo* situation, it is as of today common knowledge that cord blood stem cells secrete trophic factors that initiate and maintain the process of repair towards neurons *in vivo* (Schira et al. 2012). Therefore a hit and run mechanism towards injured tissue and secretion of cytokines seem to be the best explanation for the effects observed clinically. The first clinical trials in neurological disorders as cerebral palsy in children applying autologous or allogeneic matched cord blood with amelioration of the motor and cognitive dysfunction had no major side effects.

In 2001 Chen et al. (2001) were the first to demonstrate that the infusion of cord blood stem cells into rats that had been stroke-induced by occlusion of the middle cerebral artery was able to reduce have clearly shown that infusion as well as intracerebral transplantation of cord blood derived stem cells display beneficial effects. The mechanisms underlying the observed beneficial effects of these therapies have also not been elucidated. The most straight forward idea was at this time that stem cells differentiate into mature cell types and simply replace the lost tissue. However, that was never confirmed.

In 2019 there is a lot of evidence showing in preclinical models to achieve an IND in the US that transplanted cells may secret neurotrophic or neuroprotective factors as previously shown by our group (Trapp et al. 2008) that can counteract degeneration or promote regeneration. It was demonstrated that as an example stromal cells derived from human cord blood are strongly attracted by hepatocyte growth factor (HGF) that is secreted by ischemia-damaged brain tissue and by apoptotic neurons *in vitro* and *in vivo*. Necrotic neurons do not secret hepatocyte growth factor and have no potential to initiate migration of stromal cells. The secretion of HGF by neural target tissue and the expression of the HGF receptor c-MET in stromal cells directly correlated to the migration potential of MSC indicating that the HGF/c-MET axis is the driving force for migration towards neuronal injury. However this was just one factor. There might be others as many cytokines and their respective receptors acting in the same way but by different mechanisms.

Beside stroke and cerebral palsy the treatment of spinal cord injury with stem cell is in the focus of many researchers. The major problem in spinal cord injury is the breakdown of blood-spinal cord barrier associated with invasion of inflammatory cells, the activation of the glia and subsequently axonal degeneration (Schira et al. 2012). Schira et al. (2012) transplanted stroma cells from cord blood into a rodent model of acute spinal cord injury and investigated their survival, migration and neural differentiation potential as well their influence on axonal regrowth, lesion size and protection from spinal tissue loss. Moreover different locomotor tasks (open field Basso-Beattie-Bresnahan locomotion score, horizontal ladder walking test and CatWalk gait analysis) were applied. In the report, immune suppressed adult rats received a highly reproducible dorsal hemi section injury at thoracic level Th8. Immediately after hemi section cord blood cells were transplanted close to the site of the Injury. Two days after transplantation grafted cells were located at the injection site, 1 week after transplantation in the lesion center but without revealing immunoreactivity for the axon marker neurofilament, clearly showing that they were not able to be differentiated in vivo into neurons. Neurofilament positive host cells from the rat were present in the lesion center. Although the cord blood cells itself do not differentiated towards neurons or glia cells, they reduced the tissue loss significantly. Stroma cells release a wide amount of cytokines including stromal cell-derived factor 1 (SDF-1) (Kogler et al. 2005) which induces homing of hematopoietic and neural stem cells in ischemic and injured brain, and HGF, which is a known survival factor of neural development. The different growth factors or the combination of several as in other models of tissue regeneration are likely to participate in the positive regeneration effects observed pre-clinically but also as of today in the studies described above.

3.1.5 Endothelial Cells in Cord Blood

Endothelial progenitor cells (EPCs) have been investigated as a potential source of cells for vascular repair but also in tissue regeneration. First described in 1997 (Asahara et al. 1997), EPCs have been characterized by many investigators based on their morphology and surface antigen expression, but frequently without stringent in vivo analysis of function (Peichev 2000). By the group of Mervin Yoder (Ingram et al. 2004; Yoder 2018), ECFC in cord blood have been demonstrated to be the only circulating cells that possess all the characteristics of an endothelial cell progenitor, including distinct functions. To isolate ECFCs, cord blood derived mononuclear cells (MNC) or CD34$^+$/CD45$^-$ cells are plated on a collagen-coated surface and form adherent colonies with a cobblestone-like morphology between day 7 and 14 (Ingram et al. 2004). ECFCs are rare cells, found at a concentration of about 0.05–0.2 cells/ml in adult peripheral blood. They are enriched in human umbilical cord blood, being found at a concentration of about 2–5 cells/ml. ECFC can be enriched from each cord blood sample (fresh or cryopreserved) applying the isolated CD34$^+$-subpopulation as a basis. ECFC progeny express the cell surface antigens CD31, CD105, CD144, CD146, von-Willebrand-factor, and Kinase insert domain receptor (KDR) but do not express the hematopoietic or monocyte/macrophage cell surface antigens CD14, CD45, or CD115. Additionally, they are characterized by uptake of acetylated-low density lipoprotein. Functionally ECFC progeny form tubes when plated alone and form de novo functionally active human blood vessels in vivo. The highly proliferative ECFC from cord blood expressed higher levels of telomerase than ECFC cultured from adult peripheral blood cells. In 2014, human induced pluripotent stem cells (iPSC) –derived ECFC has been reported with properties that are similar to umbilical cord blood ECFC but with distinct differences in gene expression (Prasain et al. 2014). One potential clinical use of ECFCs is in the treatment of patients with ischemia and defective wound healing due to impaired neoangiogenesis (Shepherd et al. 2006). The authors state that the ability of implanted endothelial cells to form a vascular network when the host's angiogenic response is inhibited suggests this strategy could be useful in treating patients with impaired wound healing. These and other reports suggest that ECFCs represent an excellent cell source for vascular engineering strategies. While there are not so many data available of the use of ECFCs in human clinical trials, the results with pre-clinical rodent studies provide some hope for patients who suffer from poor vascular function. Moreover, based on their growth kinetic, they are interesting candidates for tissue engineering in combination with cells derived from tissue/tissue constructs. There are many preclinical models on the way to use the combination of endothelial cells, decellularized tissue and cells derived from a third party.

3.1.6 Stromal Cells in Cord Blood and Cord Tissue as Compared to Bone Marrow

The heterotopic transplantation of bone marrow results in the formation of ectopic bone and marrow (Tavassoli and Crosby 1968). This "osteogenic potential" is associated with non-hematopoietic stromal cells co-existing with hematopoietic stem cells in the bone marrow (Friedenstein et al. 1987). Friedenstein and colleges originally called these cells "osteogenic" or "stromal stem cells" (Friedenstein et al. 1987; Owen and Friedenstein 1988) in the following years the terms "mesenchymal stem cells", "mesenchymal stromal cells" or "skeletal stem cells" have been widely used in the literature (as summarized by Bianco and Robey 2015).

In 2004, our group detected cells in cord blood with a different proliferative potential, the so-called unrestricted somatic stromal cells (USSC) (Kogler et al. 2004) and in the following years these data were confirmed by other groups (Kim et al. 2005; Kern et al. 2006).

In the publication of 2004 (Kogler et al. 2004), USSC were described as a homogenous cell population with respect to their phenotype. During the last years, further detailed characterization *in vitro* and *in vivo* applying clonal cell population isolated and expanded from CB, clearly revealed distinct cell populations (Kluth et al. 2010; Kluth et al. 2013). The stroma cells were termed according to the revisited MSC concept (Bianco 2013) in unrestricted somatic stroma cells (USSC) and cord blood mesenchymal/multipotent stroma cells (CB MSC) (Liedtke et al. 2013). It could be also defined that clonal USSC and CB MSC lines differ in their developmental origin reflected by a distinct HOX gene expression and expression of the Delta-like 1 homologue (DLK-1), resulting in different differentiation capacities and regeneration *in vivo* focusing specific on skeletal regeneration. About 20 (out of 39) HOX genes are expressed in CB MSC (HOX positive) whereas native USSC (HOX negative) reveal no HOX gene expression (Liedtke et al. 2010, ◘ Fig. 3.1a. In addition, USSC display a lineage-specific lack of the adipogenic differentiation (◘ Fig. 3.1b) potential along with the expression of the adipogenic inhibitor Delta-like 1 homolog (DLK-1) (Kluth et al. 2012). Besides adult BM MSC, neonatal cord blood USSC and CB MSC are attractive cell sources for bone-regenerative approaches *in vivo* (Handschel et al. 2010; Klontzas et al. 2015). Like the "gold standard" BM MSC, neonatal cord blood-derived USSC and CB MSC can be differentiated *in vitro* into the chondrogenic and osteogenic lineages while showing a more immature osteogenic signature in comparison to adult BM MSC (Bosch et al. 2013) These cell type-associated signatures may be correlated to the specific expression of HOX genes. In the human system, 39 HOX genes located in four distinct clusters ABCD, are distributed among chromosomes 7 (11 HOXA genes), 17 (10 HOXB genes), 12 (9 HOXC genes) and 2 (9 HOXD genes). While the establishment of tightly regulated HOX expression patterns is important for developing limbs during embryonic and fetal development (Izpisua-Belmonte and Duboule 1992) specific HOX codes are maintained in adult cells, like fibroblasts (Chang et al. 2002) mesenchymal stromal cells (Ackema and Charite 2008) and osteoprogenitor cells (Leucht et al. 2008). Ackema et al. described characteristic topographic HOX codes in murine mesenchymal stromal cells from different anatomic sites (Ackema and Charite 2008). In line with this, our group determined the specific HOX code in human adult and neonatal cord blood stromal cell types revealing HOX expression in all four clusters in adult BM MSC similar to neonatal CB MSC, whereas USSC display absent or only marginal HOX

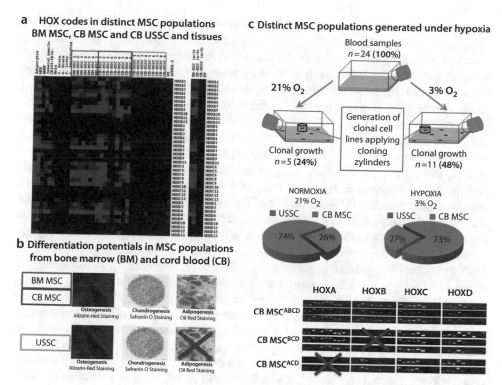

a HOX codes in distinct MSC populations BM MSC, CB MSC and CB USSC and tissues

b Differentiation potentials in MSC populations from bone marrow (BM) and cord blood (CB)

c Distinct MSC populations generated under hypoxia

◘ **Fig. 3.1** **a** HOX codes in distinct MSC populations BM MSC, CB MSC and CB USSC. On the left side RT-PCR (reverse transcription-polymerase chain reaction) expression results are presented as a heat map according to the respective expression level of HOX gene (bright red – high expression; darker shades of red – medium to marginal expression; black – no expression). $N = 7$ BM MSC (bone marrow mesenchymal stromal cells), $n = 9$ CB MSC (cord blood multipotent stromal cells) and $n = 9$ CB USSC (cord blood unrestricted somatic stromal cells) were tested for HOX gene expression as compared with the controls: H9 (embryonic stem cell line) and nTERA (embryonal teratocarcinoma cell line). On the right side the mean values of single RT-PCR experiments represent the specific HOX code of the distinct MSC populations. **b** Differentiation potentials in MSC populations from bone marrow (BM) and cord blood (CB). Representative pictures of typical differentiation assays for osteogenesis (bone), chondrogenesis (cartilage) and adipogenesis (fat) are presented. Bone is detected by Alizarin Red specifically staining calcium depots, cartilage is detected by a Safranin-O specifically staining proteoglycans and fat is detected by Oil Red specifically staining fat vesicles. BM MSC (bone marrow mesenchymal stromal cells) and CB MSC (cord blood multipotent stromal cells) reveal a tri-lineage potential into bone, cartilage and fat whereas USSC lack the potential to differentiate into fat. **c** Distinct MSC populations generated under hypoxia. $N = 24$ cord blood samples were divided and clonal cell lines generated under normoxia (21%) and hypoxia (3%). Under normoxia 74% of resulting clonal cell lines were HOX negative USSC and 26% HOX positive CB MSC. Under hypoxia (3%) this distribution changed to 73% of clonal CB MSC lines was generated and 27% of USSC lines. Within the group of CB MSC clonal cell lines, additional populations were generated under hypoxia (3%) missing a single HOX cluster (ABCD expression in HOX cluster A, B, C and D, BCD expression in HOX cluster B, C and D, ACD expression in cluster A, C and D. Exemplified RT PCR results after electrophoresis are given

expression. USSC and CB MSC derived from cord blood must be clearly distinguished from stromal cells derived from the umbilical cord (UC MSC) since UC MSC fail to differentiate *in vitro* and *in vivo* (Kaltz et al. 2008; Reinisch et al. 2015), differ in their typical *HOX* expression pattern, and have a different molecular chondro-osteogenic signature lacking relevant integrin-binding sialoprotein (*IBSP*) expression (Bosch 2012). *HOX* genes are known to be involved in cartilage formation and the transcriptional control of skeletogenesis (Goldring et al. 2006). With regard to bone regenerative approaches, it is therefore promising to characterize the individual inherent *HOX* code of potential cell sources linked to their inherent chondro-osteogenic potential. The heterogeneity among different or even the same cell sources is obvious but not easy to ressolve (McKenna et al. 2014). Not only the source-dependent and donor-dependent heterogeneity of individual cell types (Wegmeyer et al. 2013) but moreover the additional impact of low oxygen conditions on distinct cell types must be elucidated, as cells which are transplanted into injured tissues constantly encounter hypoxic stress. Effects of oxygen tension on the generation, expansion, proliferation, and differentiation of stromal cell types is widely described in the literature. However, data on the internal heterogeneity of applied cell populations at different O_2 levels and possible impacts on differentiation potentials are controversial. Liedtke et al. (2017) analyzed the expression of 39 human *HOX* genes applying hypoxic and normal (21% O_2 conditions. Whereas USSC lacking *HOX* gene expression and cord blood-derived multipotent stromal cells (CB MSC) expressing about 20 *HOX* genes are distinguished by their specific *HOX* code. Interestingly, 74% of generated clones at 21% O_2 were *HOX*-negative USSC, whereas 73% of upcoming clones at 3% O_2 were *HOX*-positive CB MSC. In order to better categorize distinct cell lines generated at 3% O_2, the expression of all 39 *HOX* genes within *HOX* cluster A, B, C and D were tested and new subtypes defined: Cells negative in all four *HOX* clusters (USSC), cells positive in all four clusters (CB MSC^{ABCD}) and subpopulations missing a single cluster (CB MSC^{ACD} and CB MSC^{BCD}). Extensive qPCR analyses of established chondro-osteomarkers revealed subtype-specific signatures verifiably associated with *in vitro* and *in vivo* differentiation capacity (◻ Fig. 3.1c).

3.1.7 Bone and Cartilage Forming Potential of Cord Blood Stroma Cells in vivo

For the purpose of bone and cartilage regeneration (lost due to trauma, surgical resection of tumors, skeletal disorders and aging), cell-based strategies are currently the standard of treatment. The use of freshly isolated CD146-positive bone marrow derived mesenchymal stromal cells (BM MSC), in contrast to the extensively expanded counterpart, provides an important therapeutic tool for bone regeneration although not for cartilage. Bone is a highly vascularized connective tissue undergoing continuous remodeling and regeneration processes. The intrinsic regeneration potential is initiated in response to injury, as well as during normal skeletal development reflected by continuous remodeling throughout adult life. Many bone and cartilage associated diseases require regeneration in large scale, e.g. large bone defects also known as "critical size" defects due to trauma, surgical resection of tumor, infection or skeletal disorders. Especially for treatment of these defects, a cell-based strategy is the most promising approach as long as a sufficient number of cells can be supplied. Clinically, stromal cells can be used as cell suspension expanded by culture or simply as bone marrow concentrate (Krampera et al. 2007). In this context, ex vivo expanded mesenchymal stromal cells (BMMSC) have demonstrated their ability to function as a tissue repair model in manifold therapeutic applications investigated in clinical trials. However, the outcome of tissue repair is strongly associated with the applied cell concentration, which is lower in bone marrow transplants as compared to cultured cells.

For de novo cartilage repair there are no established methods available, simply based on the fact, that adult bone marrow does not contain the early chondrogenic progenitors in sufficient amount to regenerate large areas of defects. For all clinical applications one should choose the best characterized cells for a directed and specific tissue repair. As summarized in Bosch et al. (2013; Liedtke et al. 2013) and many other publications *in vitro* and *in vivo*, fetal stroma cells (both USSC and CB MSC) have specific signatures for bone and cartilage formation.

In a landmark paper by Sacchetti, Bianco and co-workers (Sacchetti et al. 2016) have shown that cord blood derived stroma cells display the unique capacity to form cartilage in vivo spontaneously, in addition to a clear assayable osteogenic capacity. The data also supported the view that rather than a uniform class of "MSC", different mesoderm derivatives including distinct classes of tissue-specific committed progenitors exist, possibly of different developmental origin (Sacchetti et al. 2016).

3.1.8 Why Do We Have These Progenitors or Elusive Cells in Cord Blood?

The correct formation of the skeleton during embryogenesis and the fetal development and its preservation during the adult life is essential and maintained through the complementary activities of bone forming osteoblasts and bone-resorbing osteoclasts. The stability and strength of bones is accomplished by mineralization of the extracellular matrix, leading to a deposition of calcium hydroxyapatite. The osteogenesis can be split into two different processes: Intramembranous and endochondral ossification (Goldring et al. 2006). The intramembranous ossification is characterized by mesenchymal cells that condense and directly differentiate into osteoblasts and thereby deposit bone matrix. This process of bone formation is limited to certain parts of the skull as well as part of the clavicle. All other bones of the skeleton are formed by endochondral ossification. The formation of skeletal elements by endochondral ossification begins with the migration of undifferentiated mesenchymal cells to the zones that are destined to become bone. The undifferentiated cells condense, resulting in an increase in cell packing and forming of the cartilaginous anlagen. This process is regulated by mesenchymal-epithelial cell interactions. The next step, the aggregation of chondrogenic progenitor cells into precartilage condensations is dependent on cell-cell and cell-matrix interactions (Goldring et al. 2006). The following transition from a chondrogenic progenitor cell to a chondrocyte is marked by a change in the extracellular matrix composition. The chondrocytes hereby acquire a rounded morphology and undergo hypertrophy (substantially increase in size). This chondrocyte hypertrophy triggers the initial osteoblast differentiation from perichondrial cells. Blood vessels start to invade the cartilage from the perichondrium and thereby transport osteoclast cells into the bone to degrade the existing cartilage matrix producing marrow cavity. Additionally, the blood vessels transport perichondrial cells to nascent bone marrow, where they differentiate into osteoblasts. Many different transcription factors and regulatory signals are involved in the endochondral ossification, such as the transcription factor of the sex-determining region Y (SRY)-related high mobility group box, SOX9, the Runt domain-containing transcription factor 2 (RUNX2), Osterix (OSX) (Long 2011), bone sialoprotein (BSP) (Ogbureke et al. 2007) and parathyroid hormone-related protein (PTHLH). All of the transcription factors are regulated by a variety of developmental signals, including Hedgehog (HH) proteins, NOTCH signaling, WNT signaling,

BMP signaling and fibroblast growth factor (FGF)-signaling (Long 2011). It has been shown already that distinct populations can be defined in cord blood, each of them representing progenitors of skeletal development during fetal life (Buchheiser et al. 2012). The transcription factor analysis of subpopulation suggests that the stromal components in cord blood are elusive cells circulating from different stages of fetal development. For bone and cartilage forming cells in be concluded that CB contains natural progenitors, however with a different signature in vitro and a distinct in vivo regenerative capacity as compared to bone marrow MSC.

3.1.9 Reprogrammed Cells from Cord Blood

Since the first publication by the group of Yamanaka in 2007 (Takahashi et al. 2007) induced pluripotent stem cells (iPSC) including cord blood derived cells have received major attention by the scientific cord blood community (Okita et al. 2011, 2013). Besides the use of disease-specific iPSC for developing surrogate models of human diseases, platforms for drug discovery, the stem cell based cell replacement therapies are of major interest.

While original protocol described lentiviral insertion of genes for the transcription factors Oct4, Sox2, Klf4 and c-Myc, this method bears the potential risk of disrupting normal genes or of activating oncogenes in close proximity of the integration site. This can be avoided by using more sophisticated integration-free methods, such as appliance of episomal plasmids (Yu et al. 2009) or Sendai-virus (Fusaki et al. 2009). Although generation frequencies reported for different reprogramming systems vary, they are still low, ranging from 0.001% to 0.1%. Protooncogene c-Myc is seen critically, multiple different combinations factors including L-Myc are described. Although iPSC are already becoming a laboratory standard, it is very likely that it will still take some years until they are finally transferable into the clinic on a GMP-grade routine basis, however cord blood a source of raw material has huge advantages. Besides the inherent biological advantages of stem cells at birth as high proliferative potential, long telomeres (Buchheiser et al. 2012), good damage repair capacity (Liedtke 2015), the cells do not carry the risk of viral infections, a major prerequisite for the development of GMP-grade clinical trial samples. CB harbors a much lower risk for Cytomegalovirus (CMV), Herpes simplex virus (HSV), Epstein-Barr Virus (EBV) and many other viruses (Rubinstein 1993). CMV detected by PCR is only present in 0.25% of all CB, whereas adults have at least a frequency of 40%. HSV-6 is rarely detectable in CB; EBV and HSV are present in more than 90% of adult bone marrow donors and have a very low incidence in CB. As compared to skin biopsies the CB is not contaminated with a plenty of typical skin viruses as Herpes simplex Type 1, 6, 7, Herpes zoster, human papillomavirus (HPV 5, 8, 18, 32) and others. In the field of personalized medicine patient-specific iPSC presently are preferred candidates for cell-based autologous therapies (Lancaster and Knoblich 2014; Yamanaka 2009). Autologous therapies are not suitable for larger patient cohorts since time consuming and expensive (de Rham and Villard 2014). Therefore the use of HLAh (HLA- homozygous cells) allogenic CB units provides important novel aspects having impact on qualitative medicine and clinical practice as described also by Baghbaderani (Baghbaderani et al. 2015). Besides the immaturity of cord blood cells with all biological advantages (biological young cells), only few single nucleotide variations are found in human cord blood iPSC (Su et al. 2013) as compared to adult fibroblasts (Abyzov et al. 2012). As described in the first chapter allogeneic cord blood units are banked world-wide, can be selected for homozygous HLA-donors, thawed, isolated for CD34+ cells, expanded and reprogrammed. Based on the licensing/and or permission requirements and international accreditations existing word-wide, the cord blood units qualify already for all processes required for further clinical use.

3.2 Conclusion

Cord blood contains valuable hematopoietic and non-hematopoietic progenitor cells from different stages of fetal development circulating in the cord blood. Remarkable improvement in outcomes of adult recipients in more than 70 clinical indications has been observed over the last decade. Slower hematopoietic recovery can be solved by transplanting either two cord blood units or expand one product alone or in combination in increasing both the hematopoietic stem and progenitor cell expansion. Many studies are presently ongoing to prove the effect of cytokine driven expansion in combinations with other approaches. Clinical studies are also ongoing on the allogeneic umbilical cord blood infusions for adults with ischemic stroke and in children with autism. In order to characterize the other non-hematopoietic cells with regard to their true differentiation potential in cord blood (not cord!), for each cell population clonal cells and *in vivo* experimental design is required. The results summarized here clearly show that stroma cells from cord blood are different from bone marrow. Although MSC from cord blood have no clear "bone signature" they are able to differentiate towards bone and cartilage in vivo perhaps reflecting their natural origin during fetal life. Based on the developmental advantage of CB-subpopulations they might be ideal tools to analyze the fate of the distinct population in extensive pre-clinical models and define the mechanisms behind the improvements observed in the clinical trials. iPSC derived from homozygous allogeneic cord blood could have important impact for future differentiation and GMP-grade production for clinical trials also in the environment of Advanced Therapy Medicinal Products (ATMP) – regulations.

Take Home Message

1. Cord blood hematopoietic cells can be transplanted both in children and adults for more than 70 different indications
2. Cord blood contains stem cells from different fetal origins
3. Cord blood stem cells have long telomeres, are highly proliferative, harbor much lower risk for virus contamination and qualify as raw material for ATMP studies
4. Expansion of CD34+ cells for clinical application as well as for reprogramming towards iPSC is established
5. Cord blood derived stroma cells display the unique capacity to form cartilage in vivo spontaneously
6. Clinical studies for cerebral palsy, stroke and autism based on an IND are established

---

Acknowledgment   Special thanks to Dr. Stefanie Liedtke for assembling ◘ Fig. 3.1 and Sabine Többen for selecting references.

References

Abyzov, A., Mariani, J., Palejev, D., Zhang, Y., Haney, M. S., Tomasini, L., Ferrandino, A. F., Rosenberg Belmaker, L. A., Szekely, A., Wilson, M., Kocabas, A., Calixto, N. E., Grigorenko, E. L., Huttner, A., Chawarska, K., Weissman, S., Urban, A. E., Gerstein, M., & Vaccarino, F. M. (2012). Somatic copy number mosaicism in human skin revealed by induced pluripotent stem cells. *Nature, 492*, 438–442.

Ackema, K. B., & Charite, J. (2008). Mesenchymal stem cells from different organs are characterized by distinct topographic Hox codes. *Stem Cells and Development, 17*, 979–991.

Asahara, T., Murohara, T., Sullivan, A., Silver, M., van der Zee, R., Li, T., Witzenbichler, B., Schatteman, G., & Isner, J. M. (1997). Isolation of putative progenitor endothelial cells for angiogenesis. *Science, 275*, 964–967.

Baghbaderani, B. A., Tian, X., Neo, B. H., Burkall, A., Dimezzo, T., Sierra, G., Zeng, X., Warren, K., Kovarcik, D. P., Fellner, T., & Rao, M. S. (2015). cGMP-manufactured human induced pluripotent stem cells are available for pre-clinical and clinical applications. *Stem Cell Reports, 5*, 647–659.

Ballen, K. K., Verter, F., & Kurtzberg, J. (2015). Umbilical cord blood donation: Public or private? *Bone Marrow Transplantation, 50*, 1271–1278.

Barker, J. N., Kurtzberg, J., Ballen, K., Boo, M., Brunstein, C., Cutler, C., Horwitz, M., Milano, F., Olson, A., Spellman, S., Wagner, J. E., Delaney, C., & Shpall, E. (2017). Optimal practices in unrelated donor cord blood transplantation for hematologic malignancies. *Biology of Blood and Marrow Transplantation, 23*, 882–896.

Bautista, G., Cabrera, J. R., Regidor, C., Fores, R., Garcia-Marco, J. A., Ojeda, E., Sanjuan, I., Ruiz, E., Krsnik, I., Navarro, B., Gil, S., Magro, E., de Laiglesia, A., Gonzalo-Daganzo, R., Martin-Donaire, T., Rico, M., Millan, I., & Fernandez, M. N. (2009). Cord blood transplants supported by co-infusion of mobilized hematopoietic stem cells from a third-party donor. *Bone Marrow Transplantation, 43*, 365–373.

Beaudin, A. E., Boyer, S. W., Perez-Cunningham, J., Hernandez, G. E., Derderian, S. C., Jujjavarapu, C., Aaserude, E., MacKenzie, T., & Forsberg, E. C. (2017). A transient developmental hematopoietic stem cell gives rise to innate-like B and T cells. *Cell Stem Cell, 19*, 768–783.

Becker, A. J., Mc, C. E., & Till, J. E. (1963). Cytological demonstration of the clonal nature of spleen colonies derived from transplanted mouse marrow cells. *Nature, 197*, 452–454.

Bianco, P. (2013). Reply to MSCs: Science and trials. *Nature Medicine, 19*, 813–814.

Bianco, P., & Robey, P. G. (2015). Skeletal stem cells. *Development, 142*, 1023–1027.

Bosch, J., Houben, A. P., Radke, T. F., Stapelkamp, D., Bunemann, E., Balan, P., Buchheiser, A., Liedtke, S., & Kogler, G. (2012). Distinct differentiation potential of "MSC" derived from cord blood and umbilical cord: are cord-derived cells true mesenchymal stromal cells? *Stem Cells and Development, 21*, 1977–1988.

Bosch, J., Houben, A. P., Hennicke, T., Deenen, R., Kohrer, K., Liedtke, S., & Kogler, G. (2013). Comparing the gene expression profile of stromal cells from human cord blood and bone marrow: Lack of the typical "bone" signature in cord blood cells. *Stem Cells International, 2013*, 631984.

Bowie, M. B., McKnight, K. D., Kent, D. G., McCaffrey, L., Hoodless, P. A., & Eaves, C. J. (2006). Hematopoietic stem cells proliferate until after birth and show a reversible phase-specific engraftment defect. *Journal of Clinical Investigation*, 2808–16116.

Bowie, M. B., Kent, D. G., Dykstra, B., McKnight, K. D., McCaffrey, L., Hoodless, P. A., & Eaves, C. J. (2007). Identification of a new intrinsically timed developmental checkpoint that reprograms key hematopoietic stem cell properties. *Proceedings of the National Academy of Sciences of the United States of America, 104*, 5878–5882.

Brunstein, C. G., Eapen, M., Ahn, K. W., Appelbaum, F. R., Ballen, K. K., Champlin, R. E., Cutler, C., Kan, F., Laughlin, M. J., Soiffer, R. J., Weisdorf, D. J., Woolfrey, A., & Wagner, J. E. (2012). Reduced-intensity conditioning transplantation in acute leukemia: The effect of source of unrelated donor stem cells on outcomes. *Blood, 119*, 5591–5598.

Buchheiser, A., Houben, A. P., Bosch, J., Marbach, J., Liedtke, S., & Kogler, G. (2012). Oxygen tension modifies the 'stemness' of human cord blood-derived stem cells. *Cytotherapy, 14*, 967–982.

Chang, H. Y., Chi, J. T., Dudoit, S., Bondre, C., van de Rijn, M., Botstein, D., & Brown, P. O. (2002). Diversity, topographic differentiation, and positional memory in human fibroblasts. *Proceedings of the National Academy of Sciences of the United States of America, 99*, 12877–12882.

Chen, J., Sanberg, P. R., Li, Y., Wang, L., Lu, M., Willing, A. E., Sanchez-Ramos, J., & Chopp, M. (2001). Intravenous administration of human umbilical cord blood reduces behavioral deficits after stroke in rats. *Stroke, 32*, 2682–2688.

Cotten, C. M., Murtha, A. P., Goldberg, R. N., Grotegut, C. A., Smith, P. B., Goldstein, R. F., Fisher, K. A., Gustafson, K. E., Waters-Pick, B., Swamy, G. K., Rattray, B., Tan, S., & Kurtzberg, J. (2014). Feasibility of autologous cord blood cells for infants with hypoxic-ischemic encephalopathy. *The Journal of Pediatrics, 164*, 973–979.e971.

Cumano, A., Ferraz, J. C., Klaine, M., Di Santo, J. P., & Godin, I. (2001). Intraembryonic, but not yolk sac hematopoietic precursors, isolated before circulation, provide long-term multilineage reconstitution. *Immunity, 15*, 477–485.

Dawson, G., Sun, J. M., Davlantis, K. S., Murias, M., Franz, L., Troy, J., Simmons, R., Sabatos-DeVito, M., Durham, R., & Kurtzberg, J. (2017). Autologous cord blood infusions are safe and feasible in young children with autism Spectrum disorder: Results of a single-center phase I open-label trial. *Stem Cells Translational Medicine, 6*, 1332–1339.

de Rham, C., & Villard, J. (2014). Potential and limitation of HLA-based banking of human pluripotent stem cells for cell therapy. *Journal of Immunology Research, 2014*, 518135. https://doi.org/10.1155/2014/518135.

Delaney, C., Heimfeld, S., Brashem-Stein, C., Voorhies, H., Manger, R. L., & Bernstein, I. D. (2010). Notch-mediated expansion of human cord blood progenitor cells capable of rapid myeloid reconstitution. *Nature Medicine, 16*, 232–236.

Eaves, C. J. (2015). Hematopoietic stem cells: Concepts, definitions, and the new reality. *Blood, 125*, 2605–2613.

Feng, M., Lu, A., Gao, H., Qian, C., Zhang, J., Lin, T., & Zhao, Y. (2015). Safety of allogeneic umbilical cord blood stem cells therapy in patients with severe cerebral palsy: A retrospective study. *Stem Cells International, 2015*, 325652.

Friedenstein, A. J., Chailakhyan, R. K., & Gerasimov, U. V. (1987). Bone-marrow osteogenic stem-cells – Invitro cultivation and transplantation in diffusion-chambers. *Cell and Tissue Kinetics, 20*, 263–272.

Fusaki, N., Ban, H., Nishiyama, A., Saeki, K., & Hasegawa, M. (2009). Efficient induction of transgene-free human pluripotent stem cells using a vector based on Sendai virus, an RNA virus that does not integrate into the host genome. *Proceedings of the Japan Academy. Series B, Physical and Biological Sciences, 85*, 348–362.

Gluckman, E., Broxmeyer, H. A., Auerbach, A. D., et al. (1989). Hematopoietic reconstitution in a patient with Fanconi's anemia by means of umbilical-cord blood from an HLA-identical sibling. *The New England Journal of Medicine, 321*, 1174–1178.

Goldring, M. B., Tsuchimochi, K., & Ijiri, K. (2006). The control of chondrogenesis. *Journal of Cellular Biochemistry, 97*, 33–44.

Gragert, L., Eapen, M., Williams, E., Freeman, J., Spellman, S., Baitty, R., Hartzman, R., Rizzo, J. D., Horowitz, M., Confer, D., & Maiers, M. (2014). HLA match likelihoods for hematopoietic stem-cell grafts in the U.S. registry. *The New England Journal of Medicine, 371*, 339–348.

Handschel, J., Naujoks, C., Langenbach, F., Berr, K., Depprich, R. A., Ommerborn, M. A., Kubler, N. R., Brinkmann, M., Kogler, G., & Meyer, U. (2010). Comparison of ectopic bone formation of embryonic stem cells and cord blood stem cells in vivo. *Tissue Engineering. Part A, 16*, 2475–2483.

Horwitz, M. E., Chao, N. J., Rizzieri, D. A., Long, G. D., Sullivan, K. M., Gasparetto, C., Chute, J. P., Morris, A., McDonald, C., Waters-Pick, B., Stiff, P., Wease, S., Peled, A., Snyder, D., Cohen, E. G., Shoham, H., Landau, E., Friend, E., Peleg, I., Aschengrau, D., Yackoubov, D., Kurtzberg, J., & Peled, T. (2014). Umbilical cord blood expansion with nicotinamide provides long-term multilineage engraftment. *The Journal of Clinical Investigation, 124*, 3121–3128.

https://www.uniklinik-duesseldorf.de/patienten-besucher/klinikeninstitutezentren/jose-carreras-stammzellbank

https://www.wmda.info/

Ingram, D. A., Mead, L. E., Tanaka, H., Meade, V., Fenoglio, A., Mortell, K., Pollok, K., Ferkowicz, M. J., Gilley, D., & Yoder, M. C. (2004). Identification of a novel hierarchy of endothelial progenitor cells using human peripheral and umbilical cord blood. *Blood, 104*, 2752–2760.

Izpisua-Belmonte, J. C., & Duboule, D. (1992). Homeobox genes and pattern formation in the vertebrate limb. *Developmental Biology, 152*, 26–36.

Jones, M., Chase, J., Brinkmeier, M., Xu, J., Weinberg, D. N., Schira, J., Friedman, A., Malek, S., Grembecka, J., Cierpicki, T., Dou, Y., Camper, S. A., & Maillard, I. (2015). Ash1l controls quiescence and self-renewal potential in hematopoietic stem cells. *The Journal of Clinical Investigation, 125*, 2007–2020.

Kaltz, N., Funari, A., Hippauf, S., Delorme, B., Noel, D., Riminucci, M., Jacobs, V. R., Haupl, T., Jorgensen, C., Charbord, P., Peschel, C., Bianco, P., & Oostendorp, R. A. (2008). In vivo osteoprogenitor potency of human stromal cells from different tissues does not correlate with expression of POU5F1 or its pseudogenes. *Stem Cells, 26*, 2419–2424.

Kern, S., Eichler, H., Stoeve, J., Kluter, H., & Bieback, K. (2006). Comparative analysis of mesenchymal stem cells from bone marrow, umbilical cord blood, or adipose tissue. *Stem Cells, 24*, 1294–1301.

Kim, B. O., Tian, H., Prasongsukarn, K., Wu, J., Angoulvant, D., Wnendt, S., Muhs, A., Spitkovsky, D., & Li, R. K. (2005). Cell transplantation improves ventricular function after a myocardial infarction: A preclinical study of human unrestricted somatic stem cells in a porcine model. *Circulation, 112*, 196–104.

Klontzas, M. E., Kenanidis, E. I., Heliotis, M., Tsiridis, E., & Mantalaris, A. (2015). Bone and cartilage regeneration with the use of umbilical cord mesenchymal stem cells. *Expert Opinion on Biological Therapy, 15*, 1541–1552.

Kluth, S. M., Buchheiser, A., Houben, A. P., Geyh, S., Krenz, T., Radke, T. F., Wiek, C., Hanenberg, H., Reinecke, P., Wernet, P., & Kogler, G. (2010). DLK-1 as a marker to distinguish unrestricted somatic stem cells and mesenchymal stromal cells in cord blood. *Stem Cells and Development, 19*, 1471–1483.

Kluth, S. M., Radke, T. F., & Kogler, G. (2012). Potential application of cord blood-derived stromal cells in cellular therapy and regenerative medicine. *Journal of Blood Transfusion, 2012*, 365182.

Kluth, S. M., Radke, T. F., & Kogler, G. (2013). Increased haematopoietic supportive function of USSC from umbilical cord blood compared to CB MSC and possible role of DLK-1. *Stem Cells International, 2013*, 985285.

Kogler, G., Sensken, S., Airey, J. A., Trapp, T., Muschen, M., Feldhahn, N., Liedtke, S., Sorg, R. V., Fischer, J., Rosenbaum, C., Greschat, S., Knipper, A., Bender, J., Degistirici, O., Gao, J., Caplan, A. I., Colletti, E. J., Almeida-Porada, G., Muller, H. W., Zanjani, E., & Wernet, P. (2004). A new human somatic stem cell from placental cord blood with intrinsic pluripotent differentiation potential. *The Journal of Experimental Medicine, 200*, 123–135.

Kogler, G., Radke, T. F., Lefort, A., Sensken, S., Fischer, J., Sorg, R. V., & Wernet, P. (2005). Cytokine production and hematopoiesis supporting activity of cord blood-derived unrestricted somatic stem cells. *Experimental Hematology, 33*, 573–583.

Krampera, M., Franchini, M., Pizzolo, G., & Aprili, G. (2007). Mesenchymal stem cells: From biology to clinical use. *Blood Transfusion, 5*, 120–129.

Kurita, N., Gosho, M., Yokoyama, Y., Kato, T., Obara, N., Sakata-Yanagimoto, M., Hasegawa, Y., Uchida, N., Takahashi, S., Kouzai, Y., Atsuta, Y., Kurata, M., Ichinohe, T., & Chiba, S. (2017). A phase I/II trial of intrabone marrow cord blood transplantation and comparison of the hematological recovery with the Japanese nationwide database. *Bone Marrow Transplantation, 52*, 574–579.

Kurtzberg, J. (2017). A history of cord blood banking and transplantation. *Stem Cells Translational Medicine, 6*, 1309–1311.

Lancaster, M. A., & Knoblich, J. A. (2014). Organogenesis in a dish: Modeling development and disease using organoid technologies. *Science, 345*, 1247125.

Laskowitz, D. T., Bennett, E. R., Durham, R. J., Volpi, J. J., Wiese, J. R., Frankel, M., Shpall, E., Wilson, J. M., Troy, J., & Kurtzberg, J. (2018). Allogeneic umbilical cord blood infusion for adults with ischemic stroke: Clinical outcomes from a phase I safety study. *Stem Cells Translational Medicine, 7*, 521–529.

Leucht, P., Kim, J. B., Amasha, R., James, A. W., Girod, S., & Helms, J. A. (2008). Embryonic origin and Hox status determine progenitor cell fate during adult bone regeneration. *Development, 135*, 2845–2854.

Liedtke, S., Buchheiser, A., Bosch, J., Bosse, F., Kruse, F., Zhao, X., Santourlidis, S., & Kogler, G. (2010). The HOX code as a "biological fingerprint" to distinguish functionally distinct stem cell populations derived from cord blood. *Stem Cell Research, 5*, 40–50.

Liedtke, S., Freytag, E. M., Bosch, J., Houben, A. P., Radke, T. F., Deenen, R., Kohrer, K., & Kogler, G. (2013). Neonatal mesenchymal-like cells adapt to surrounding cells. *Stem Cell Research, 11*, 634–646.

Liedtke, S., Biebernick, S., Radke, T. F., Stapelkamp, D., Coenen, C., Zaehres, H., Fritz, G., & Kogler, G. (2015). DNA damage response in neonatal and adult stromal cells compared with induced pluripotent stem cells. *Stem Cells Translational Medicine, 4*, 576–589.

Liedtke, S., Sacchetti, B., Laitinen, A., Donsante, S., Klockers, R., Laitinen, S., Riminucci, M., & Kogler, G. (2017). Low oxygen tension reveals distinct HOX codes in human cord blood-derived stromal cells associated with specific endochondral ossification capacities in vitro and in vivo. *Journal of Tissue Engineering and Regenerative Medicine, 11*(10), 2725–2736.

Liu, H., Rich, E. S., Godley, L., Odenike, O., Joseph, L., Marino, S., Kline, J., Nguyen, V., Cunningham, J., Larson, R. A., del Cerro, P., Schroeder, L., Pape, L., Stock, W., Wickrema, A., Artz, A. S., & van Besien, K. (2011). Reduced-intensity conditioning with combined haploidentical and cord blood transplantation results in rapid engraftment, low GVHD, and durable remissions. *Blood, 118*, 6438–6445.

Liu, E., Tong, Y., Dotti, G., Shaim, H., Savoldo, B., Mukherjee, M., Orange, J., Wan, X., Lu, X., Reynolds, A., Gagea, M., Banerjee, P., Cai, R., Bdaiwi, M. H., Basar, R., Muftuoglu, M., Li, L., Marin, D., Wierda, W.,

Keating, M., Champlin, R., Shpall, E., & Rezvani, K. (2018). Cord blood NK cells engineered to express IL-15 and a CD19-targeted CAR show long-term persistence and potent antitumor activity. *Leukemia, 32*, 520–531.

Long, F. (2011). Building strong bones: Molecular regulation of the osteoblast lineage. *Nature Reviews. Molecular Cell Biology, 13*, 27–38.

Lorenz, E., Uphoff, D., Reid, T. R., & Shelton, E. (1951). Modification of irradiation injury in mice and Guinea pigs by bone marrow injections. *Journal of the National Cancer Institute, 12*, 197–201.

McKenna, D., Matthay, M. A., & Pati, S. (2014). Correspondence to: Soliciting strategies for developing cell-based reference materials to advance mesenchymal stem/stromal cell research and clinical translation. *Stem Cells and Development, 23*, 1717–1718.

Milano, F., Appelbaum, F. R., & Delaney, C. (2016). Cord-blood transplantation in patients with minimal residual disease. *The New England Journal of Medicine, 375*, 2204–2205.

Ogbureke, K. U., Nikitakis, N. G., Warburton, G., Ord, R. A., Sauk, J. J., Waller, J. L., & Fisher, L. W. (2007). Up-regulation of SIBLING proteins and correlation with cognate MMP expression in oral cancer. *Oral Oncology, 43*, 920–932.

Okita, K., Matsumura, Y., Sato, Y., Okada, A., Morizane, A., Okamoto, S., Hong, H., Nakagawa, M., Tanabe, K., Tezuka, K., Shibata, T., Kunisada, T., Takahashi, M., Takahashi, J., Saji, H., & Yamanaka, S. (2011). A more efficient method to generate integration-free human iPS cells. *Nature Methods, 8*, 409–412.

Okita, K., Yamakawa, T., Matsumura, Y., Sato, Y., Amano, N., Watanabe, A., Goshima, N., & Yamanaka, S. (2013). An efficient nonviral method to generate integration-free human-induced pluripotent stem cells from cord blood and peripheral blood cells. *Stem Cells, 31*, 458–466.

Owen, M., & Friedenstein, A. J. (1988). Stromal stem cells: Marrow-derived osteogenic precursors. *Ciba Foundation Symposium, 136*, 42–60.

Peichev, M., Naiyer, A. J., Pereira, D., Zhu, Z., Lane, W. J., Williams, M., Oz, M. C., Hicklin, D. J., Witte, L., Moore, M. A., & Rafii, S. (2000). Expression of VEGFR-2 and AC133 by circulating human CD34(+) cells identifies a population of functional endothelial precursors. *Blood, 92*, 952–958.

Ponce, D. M., Sauter, C., Devlin, S., Lubin, M., Gonzales, A. M., Kernan, N. A., Scaradavou, A., Giralt, S., Goldberg, J. D., Koehne, G., Perales, M. A., Young, J. W., Castro-Malaspina, H., Jakubowski, A., Papadopoulos, E. B., & Barker, J. N. (2013). A novel reduced-intensity conditioning regimen induces a high incidence of sustained donor-derived neutrophil and platelet engraftment after double-unit cord blood transplantation. *Biology of Blood and Marrow Transplantation, 19*, 799–803.

Ponce, D. M., Hilden, P., Devlin, S. M., Maloy, M., Lubin, M., Castro-Malaspina, H., Dahi, P., Hsu, K., Jakubowski, A. A., Kernan, N. A., Koehne, G., O'Reilly, R. J., Papadopoulos, E. B., Perales, M. A., Sauter, C., Scaradavou, A., Tamari, R., van den Brink, M. R., Young, J. W., Giralt, S., & Barker, J. N. (2015). High disease-free survival with enhanced protection against relapse after double-unit cord blood transplantation when compared with T cell-depleted unrelated donor Transplantation in patients with acute leukemia and chronic myelogenous leukemia. *Biology of Blood and Marrow Transplantation, 21*, 1985–1993.

Popat, U., Mehta, R. S., Rezvani, K., Fox, P., Kondo, K., Marin, D., McNiece, I., Oran, B., Hosing, C., Olson, A., Parmar, S., Shah, N., Andreeff, M., Kebriaei, P., Kaur, I., Yvon, E., de Lima, M., Cooper, L. J., Tewari, P., Champlin, R. E., Nieto, Y., Andersson, B. S., Alousi, A., Jones, R. B., Qazilbash, M. H., Bashir, Q., Ciurea, S., Ahmed, S., Anderlini, P., Bosque, D., Bollard, C., Molldrem, J. J., Chen, J., Rondon, G., Thomas, M., Miller, L., Wolpe, S., Simmons, P., Robinson, S., Zweidler-McKay, P. A., & Shpall, E. J. (2015). Enforced fucosylation of cord blood hematopoietic cells accelerates neutrophil and platelet engraftment after transplantation. *Blood, 125*, 2885–2892.

Prasain, N., Lee, M. R., Vemula, S., Meador, J. L., Yoshimoto, M., Ferkowicz, M. J., Fett, A., Gupta, M., Rapp, B. M., Saadatzadeh, M. R., Ginsberg, M., Elemento, O., Lee, Y., Voytik-Harbin, S. L., Chung, H. M., Hong, K. S., Reid, E., O'Neill, C. L., Medina, R. J., Stitt, A. W., Murphy, M. P., Rafii, S., Broxmeyer, H. E., & Yoder, M. C. (2014). Differentiation of human pluripotent stem cells to cells similar to cord-blood endothelial colony-forming cells. *Nature Biotechnology, 32*, 1151–1157.

Reinisch, A., Etchart, N., Thomas, D., Hofmann, N. A., Fruehwirth, M., Sinha, S., Chan, C. K., Senarath-Yapa, K., Seo, E. Y., Wearda, T., Hartwig, U. F., Beham-Schmid, C., Trajanoski, S., Lin, Q., Wagner, W., Dullin, C., Alves, F., Andreeff, M., Weissman, I. L., Longaker, M. T., Schallmoser, K., Majeti, R., & Strunk, D. (2015). Epigenetic and in vivo comparison of diverse MSC sources reveals an endochondral signature for human hematopoietic niche formation. *Blood, 125*, 249–260.

Rubinstein, P., Rosenfield, R. E., Adamson, J. W., & Stevens, C. E. (1993). Stored placental blood for unrelated bone marrow reconstitution. *Blood, 81*, 1679–1690.

Sacchetti, B., Funari, A., Remoli, C., Giannicola, G., Kogler, G., Liedtke, S., Cossu, G., Serafini, M., Sampaolesi, M., Tagliafico, E., Tenedini, E., Saggio, I., Robey, P. G., Riminucci, M., & Bianco, P. (2016). No identical "mesenchymal stem cells" at different times and sites: Human committed progenitors of distinct origin and differentiation potential are incorporated as adventitial cells in microvessels. *Stem Cell Reports, 6*, 897–913.

Schira, J., Gasis, M., Estrada, V., Hendricks, M., Schmitz, C., Trapp, T., Kruse, F., Kogler, G., Wernet, P., Hartung, H. P., & Muller, H. W. (2012). Significant clinical, neuropathological and behavioural recovery from acute spinal cord trauma by transplantation of a well-defined somatic stem cell from human umbilical cord blood. *Brain, 135*(Pt 2), 431–446.

Shepherd, B. R., Enis, D. R., Wang, F., Suarez, Y., Pober, J. S., & Schechner, J. S. (2006). Vascularization and engraftment of a human skin substitute using circulating progenitor cell-derived endothelial cells. *The FASEB Journal, 20*, 1739–1741.

Spangrude, G. J., Heimfeld, S., & Weissman, I. L. (1988). Purification and characterization of mouse hematopoietic stem cells. *Science, 241*, 58–62.

Su, R. J., Yang, Y., Neises, A., Payne, K. J., Wang, J., Viswanathan, K., Wakeland, E. K., Fang, X., & Zhang, X. B. (2013). Few single nucleotide variations in exomes of human cord blood induced pluripotent stem cells. *PLoS One, 8*, e59908.

Sun, J. M., Song, A. W., Case, L. E., Mikati, M. A., Gustafson, K. E., Simmons, R., Goldstein, R., Petry, J., McLaughlin, C., Waters-Pick, B., Chen, L. W., Wease, S., Blackwell, B., Worley, G., Troy, J., & Kurtzberg, J. (2017). Effect of autologous cord blood infusion on motor function and brain connectivity in young children with cerebral palsy: A randomized, Placebo-Controlled Trial. *Stem Cells Transl Med, 6*, 2071–2078.

Takahashi, K., Tanabe, K., Ohnuki, M., Narita, M., Ichisaka, T., Tomoda, K., & Yamanaka, S. (2007). Induction of pluripotent stem cells from adult human fibroblasts by defined factors. *Cell, 131*, 861–872.

Tavassoli, M., & Crosby, W. H. (1968). Transplantation of marrow to extramedullary sites. *Science, 161*, 54–56.

Trapp, T., Kogler, G., El-Khattouti, A., Sorg, R. V., Besselmann, M., Focking, M., Buhrle, C. P., Trompeter, I., Fischer, J. C., & Wernet, P. (2008). Hepatocyte growth factor/c-MET axis-mediated tropism of cord blood-derived unrestricted somatic stem cells for neuronal injury. *The Journal of Biological Chemistry, 283*, 32244–32253.

Vago, L., Toffalori, C., Ahci, M., Lange, V., Lang, K., Todaro, S., Lorentino, F., Stempelmann, K., Heinold, A., Stölzel, F., Waterhouse, M., Claus, R., Gendzekhadze, K., Onozawa, M., Devillier, R., Tang, R., Ulman, M., Kwon, M., Gojo, I., Ruggeri, L., Imovilli, A., Facchini, L., Lazarevic, D., Lupo Stanghellini, M. T., Peccatori, J., Steckel, N. K., Horn, P.A., Picardi, A., Manetta, S., Busca, A., Pinana, J. L., Sanz, J., Martínez-Laperche, C., Ciurea, S. O., Luznik, L., Velardi, A., Arcese, W., Sanz, G., Pini, M., Bruno, B., Kobbe, G., Al Malki, M., Teshima, T., Kröger, N., Finke, J., Nagler, A., Blaise, D., Mohty, M., Bornhäuser, M., Beelen, D.W., Schmidt, A.H., Ciceri, F., & Fleischhauer, K. *818 incidence of HLA loss in a global multicentric Cohort of post-transplantation relapses: Results from the Hlaloss Collaborative Study*. https://ash.confex.com/ash/2018/webprogram/Paper112142.html. 2018

Waas, B., & Maillard, I. (2017). Fetal hematopoietic stem cells are making waves. *Stem Cell Investigation, 4*, 25.

Warlick, E. D., Peffault de Latour, R., Shanley, R., Robin, M., Bejanyan, N., Xhaard, A., Brunstein, C., Sicre de Fontbrune, F., Ustun, C., Weisdorf, D. J., & Socie, G. (2015). Allogeneic hematopoietic cell transplantation outcomes in acute myeloid leukemia: Similar outcomes regardless of donor type. *Biology of Blood and Marrow Transplantation, 21*, 357–363.

Wegmeyer, H., Broske, A. M., Leddin, M., Kuentzer, K., Nisslbeck, A. K., Hupfeld, J., Wiechmann, K., Kuhlen, J., von Schwerin, C., Stein, C., Knothe, S., Funk, J., Huss, R., & Neubauer, M. (2013). Mesenchymal stromal cell characteristics vary depending on their origin. *Stem Cells and Development, 22*, 2606–2618.

Wu, K. H., Zhou, B., Mo, X. M., Cui, B., Yu, C. T., Lu, S. H., Han, Z. C., & Liu, Y. L. (2007). Therapeutic potential of human umbilical cord-derived stem cells in ischemic diseases. *Transplantation Proceedings, 39*, 1620–1622.

Yamanaka, S. (2009). A fresh look at iPS cells. *Cell, 137*, 13–17.

Yoder, M. C. (2018). Endothelial stem and progenitor cells (stem cells): (2017 Grover Conference Series). *Pulmonary Circulation, 8*, 2045893217743950.

Yu, J., Hu, K., Smuga-Otto, K., Tian, S., Stewart, R., Slukvin, I. I., & Thomson, J. A. (2009). Human induced pluripotent stem cells free of vector and transgene sequences. *Science, 324*, 797–801.

Neural Stem Cells and Their Niche

Jacqueline Reinhard, Lars Roll, Ursula Theocharidis, and Andreas Faissner

© Springer Nature Switzerland AG 2020
B. Brand-Saberi (ed.), *Essential Current Concepts in Stem Cell Biology*,
Learning Materials in Biosciences, https://doi.org/10.1007/978-3-030-33923-4_4

What You Will Learn in This Chapter

The central nervous system (CNS) consists of the brain and spinal cord. These structures arise from neural stem cells (NSCs), which undergo specific maturation steps, lineage decisions and commitment during development. They generate all major cell types of the CNS, neurons and glial cells, in a timely and spatially ordered fashion. NSCs show different morphologies and molecular characteristics in distinct phases of maturation that can be followed from embryogenesis over postnatal stages to adulthood. In the mature CNS only few NSCs are left in specific areas, the adult NSC niches. The characteristics and molecular composition of these niches as well as their interaction with the cells and their progeny are presented here. Finally, the last section focuses on the appearance of stem cells under pathological conditions in the adult, such as lesions or tumors.

4.1 Development of the Central Nervous System from Neural Stem Cells

The brain and spinal cord form the CNS, which is the most complex organ system in the body. Although it is made up of billions of cells with countless connections, the CNS only develops from a few cells during development that initially form a single layer.

During early embryonic development, three germ layers of the early organism are formed by cell divisions and migration of cells. The endoderm later gives rise to internal organs, such as lung, intestine and liver. Muscles, bones, blood vessels and the heart are formed from the mesoderm. External ectoderm forms the skin and from a special area, termed neuroectoderm, the nervous system arises. This neuroectodermal tissue, which is initially on the outside of the embryo, is folded inwards and then closes to form the neural tube. The caudal part of the neural tube will become the spinal cord whereas the rostral part develops into the brain. The neural tube, which pervades through the tissue, is not closed during further development, but forms the ventricles, a system of interconnected cavities filled with brain fluid. The CNS tissue is growing along this axis and thickened by cell proliferation and growth.

4.1.1 Neuroepithelial Cells and Early Neurogenesis

The neural tube initially only consists of a single cell layer, the so-called neuroepithelium. An epithelium is characterized by the tight connection of polarized cells that have contact to a tissue's surface and the underlying basal lamina. Neuroepithelial cells (NECs) have on their apical side contact to the cavity, i.e. the ventricle; on the basal side they face the basal lamina of the nervous system (◘ Fig. 4.1a). The nuclei of NECs lie in several layers because they migrate up and down during the cell cycle. Mitosis takes place at the most apical positions whereas the nucleus is relocated basally for DNA synthesis during S phase. This phenomenon is called interkinetic nuclear migration and leads to a pseudostratified appearance of the early neural tissue. NECs are NSCs, which are able to produce many descendants with the same characteristics by symmetrical cell divisions, but also generate daughter cells that can differentiate. Initially, neurons develop in this young tissue, which settle in more basal positions.

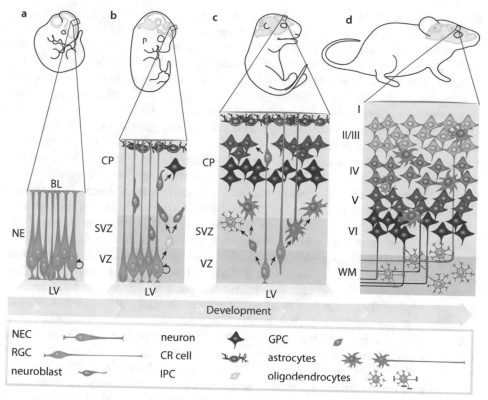

□ **Fig. 4.1** Development of the cerebral cortex in mice. The main cellular events and the structural organization are shown at different developmental stages (not to scale). **a** During early neural development NECs are the first NSCs in the NE. They divide predominantly symmetrically and enlarge the stem cell pool. Their cell bodies stretch from the LV to the BL. The BL is the attachment site for the endfeet of NECs (and later RGCs) that provides structural support and signaling cues to the cells. **b** During the second week of development RGCs overtake the stem cell functions. They proliferate and the daughter cells can be RGCs again or differentiate into neurons during neurogenesis. RGCs can also generate IPCs, which divide symmetrically in the SVZ to give rise to two more neurons. Neurons migrate basally through the SVZ and CP along the RGC processes to their final positions. Most basally located are CR cells secreting Reelin as guiding molecule. **c** In the perinatal phase, RGCs differentiate into oligodendrocytes and astrocytes during gliogenesis, partially via GPCs. Some RGCs directly transform to astrocytes, which reduces the NSC number in the VZ. **d** With ongoing maturation until adulthood, the CP is divided into several layers (layers I-VI) where neurons with subsequent birth dates are stacked upon each other. The oligodendrocytes enwrap the axons with their myelin sheaths, concentrating in the WM of the underlying corpus callosum. Astrocytes are intermingled in the tissue for structural and metabolic supply. BL basal lamina, CP cortical plate, CR cell Cajal-Retzius cell, GPC glial progenitor cell, IPC intermediate progenitor cell, LV lateral ventricle, NE neuroepithelium, NEC neuroepithelial cell, NSC neural stem cell, RGC radial glia cell, SVZ subventricular zone, VZ ventricular zone, WM white matter

4.1.2 Radial Glia Cells Generate Neurons and Glia

The NECs are replaced by cells that have two characteristic properties, which also give them their name: radial glial cells (RGCs). RGCs are elongated cells that are arranged radially to the tissue and have many properties of glial cells, such as the expression of marker proteins like the glutamate aspartate transporter (GLAST) and the brain-lipid binding protein (BLBP) (Doetsch 2003). The cell body of these cells lies in the apical region and a long thin extension stretches throughout the tissue towards the basal surface (◻ Fig. 4.1b). RGCs are NSCs of the CNS that can divide symmetrically or asymmetrically and thus produce either two identical or two different daughter cells. These offspring can be RGCs with NSC characteristics showing expression of the intermediate filament nestin and the transcription factor Sox2 (SRY (sex determining region Y)-box 2), or cells that are determined for differentiation (Cai et al. 2002). During early development the production of further stem cells by symmetric expansive divisions enlarges the stem cell pool. Later the division modes are preferentially asymmetric and lead to stem cell homeostasis with one daughter cell being a stem cell and the other differentiating. When the tissue matures the stem cells become less by exhausting in symmetric differentiative divisions.

The three important cell types of the CNS – neurons, astrocytes and oligodendrocytes – can arise from RGCs. Neurons are formed first in a process known as neurogenesis. Glia cells develop later in a subsequent process called gliogenesis. The switch in fate restriction is accompanied by the upregulation of molecules like the epidermal growth factor receptor (EGFR) and tenascin-C. Neurogenesis can either take place directly from RGCs or via intermediate progenitor cells (IPCs) that undergo further basal cell divisions in the tissue and form two neurons in a symmetrical manner. This division mode leads to a higher number of progeny because of the higher proliferative capacity of these transit-amplifying progenitors (TAPs). Theoretically, NECs and RGCs have an unlimited potential to self-renew by repeated cell divisions. Some progenitor cells have similar characteristics and show proliferative capacity as well as differentiation into neuronal or glial cells but their potential is restricted. The populations of NECs and RGCs are intermingled with more restricted progenitors and cannot be clearly distinguished from them. Therefore, it is reasonable to use the term neural stem/progenitor cells (NSPCs) for cells with variable fate restriction (Taverna et al. 2014).

Stem cells have been detected in diverse CNS regions, including the dorsal and ventral forebrain, the thalamus, the mid- and hindbrain as well as the spinal cord. The specification of cells is influenced by a set of morphogenic factors that are diffusible and show a gradual presence in the developing tissue. Bone morphogenic proteins (BMPs), Wnt (wingless/integrated) proteins, fibroblast growth factors (FGFs) and sonic hedgehog (Shh) interact with each other and with other regulating factors to induce signaling pathways in the cells determining their fate decisions at their distinct positions on the dorso-ventral and rostro-caudal axis (Altmann and Brivanlou 2001).

Newborn neurons use the long processes of RGCs as guiding substrates for their migration through the tissue. They move to outer layers of the developing brain where they leave their guiding fibers in their destination layers. Cajal-Retzius (CR) cells are the first neurons that develop and they produce the glycoprotein Reelin. New-born neurons leave the RGC fibers at their designated positions settling underneath this initial neuronal layer. Reelin is crucial for the proper localization of neurons and mutants show a destructed structural organization of the cerebral cortex. After the neuronal migration phase, the CR cells disappear when their mission is fulfilled (Kirischuk et al. 2014). Neurons are born in an inside-out fashion, which means that the first cells to occur are the ones that can later be found in the inner layers of the structured tissue whereas younger neurons settle in the outer layers.

Developmental processes have been intensively investigated in the cerebral cortex of mice and rats. Many of the processes described in these murine model systems can also be found in the human brain. However, there are also phenomena in human development that are not known to the same extent from rodents. This is mainly due to the fact that humans, but also sheep, ferrets and monkeys have a folded cerebral cortex, a so-called gyrencephalic cortex, in contrast to the smooth, lissencephalic cortical surfaces of mice and rats. In species with higher encephalization, the neocortex is folded in gyri. This larger outer cortical surface is caused by the proliferation of another cell type, which almost does not occur in species with a smooth cortex surface: the basal RGCs (bRGCs). In gyrencephalic species, these are located in the outer layer of a widened subventricular zone (oSVZ), and are therefore also called oSVZ precursors. These cells share many properties with apical RGCs, but they are located at more basal positions. They generate neurons directly or first via IPCs, which produce neurons in subsequent symmetrical cell divisions. Along the basal processes of the bRGCs, these newborn nerve cells migrate outwards until they reach their final positions and integrate into the resulting neuronal network.

4.1.3 Gliogenesis During Late Embryonic and Postnatal Development

Astrocytes are glia cells that are important for the structure and homeostasis in the brain and interact with the neuronal network. In the cerebral cortex they arise in the latest embryonic and first postnatal phase from RGCs via glial precursor cells and integrate in the network (◉ Fig. 4.1c). Additional astrocytes are formed by a direct transformation of RGCs that retract their processes and apical contacts to the ventricle before the cell bodies are translocated to basal positions (Tabata 2015). This would only lead to a limited number of astrocytes but indeed this cell type is highly proliferative and many descendants are generated by the proliferation of astrocytes in the cortical grey matter.

Neuronal signal transduction velocity is enhanced by the ensheathment with myelin. In the CNS oligodendrocytes are the cells providing the myelin membranes and these glia cells are also generated by RGCs. Three major waves of oligodendrocyte precursors arise in the brain at different positions and time points (Kessaris et al. 2006). The first two populations are generated in the ventral forebrain and invade the cortex by tangential migration. A third population is generated in the early postnatal cortex during gliogenesis from RGCs.

4.1.4 Architecture of the Mature Cortex

The timely organized generation of neurons and their radial migration along RGC processes leads to a cortical plate, which is structured into six layers (◉ Fig. 4.1d). Excitatory neurons, which are descendants of RGCs in the cortical VZ, are intermingled with inhibitory interneurons coming from progenitor cells in the ventral forebrain. Neurons in the distinct cortical layers build networks with defined communication partners within various brain regions. Their axons get enwrapped with oligodendrocytic myelin membranes and their signaling and homeostasis is supported and regulated by astrocytes contacting the nerve cell bodies and blood vessels.

4.2 Adult Neural Stem Cells

With ongoing maturation until adulthood, the number of multipotent NSPCs declines. Therefore, in the adult CNS, NSCs and neural progenitor cells (NPCs), which continuously proliferate and generate new neurons are rare. They reside in confined regions of the adult brain. These neurogenic regions or stem cell niches include the subependymal zone (SEZ) lining the lateral ventricle (LV) of the forebrain as well as the subgranular zone (SGZ) of the hippocampal dentate gyrus (◘ Fig. 4.2a–d).

Per definition, the NSC niche creates a privileged environment for the expansion, maintenance and neurogenic properties of NSCs. The NSC niche contains the NSCs themselves, called type B cells, TAPs, NSC-derived neuroblasts, blood vessel forming endothelial cells, pericytes and leptomeningeal cells. Blood vessels provide glucose and oxygen supply as well as many other factors that are important for the integrity of the NSC compartment. The cerebrospinal fluid transports signaling molecules that are sensed by small processes, called cilia, of the type B stem cells that reach into the ventricular space. Additionally, the extracellular matrix (ECM) environment of the NSCs that consists of basal lamina structures and a variety of glycoproteins and proteoglycans, is crucial. A detailed molecular composition of the NSC niche is described below (see ▸ Sect. 4.3).

4.2.1 Adult Neural Stem Cells in the Subependymal and Subgranular Zone

Adult NSCs (aNSCs) of the SEZ (type B cells) divide slowly and supposedly originated from the radial glia, the major neural stem/progenitor cell (NSPC) of the developing CNS. They generate neuroblasts via TAPs (◘ Fig. 4.2a, b). TAPs divide rapidly and generate type A neuroblasts, which proliferate in response to the molecule Shh. Type A cells expose the surface marker PSA-NCAM (polysialylated-neural cell adhesion molecule) and move to the olfactory bulb (OB) via a defined path - the rostral migratory stream (RMS). Finally, type A neuroblasts differentiate into interneurons in the cellular layer that surrounds the olfactory glomeruli, where they contribute to the processing of olfactory input.

aNSCs of the dentate gyrus follow a similar mode as aNSCs of the SEZ. Type B aNSCs generate early and late type D cells, which include precursors and neuroblasts (◘ Fig. 4.2c, d). Neuroblasts generate the granule cells (type G cells), the primary excitatory neurons of the dentate gyrus, which contribute to cognitive functions such as learning and memory. Neurogenesis in that niche is plastic, responds to physical exercise, to psychoactive drugs and enriched environments, revealing a remarkable plasticity of that stem cell pool. This plasticity, however, diminishes with increasing age. In view of its potential importance for regenerative medicine, considerable efforts are devoted to the clarification of regulatory mechanisms that control the NSC niches.

4.2.2 Characteristics of aNSCs

Type B cells of the SEZ display several properties of astrocytes. They express molecules such as the glial fibrillary acidic protein (GFAP), GLAST and BLBP (Doetsch 2003; Codega et al. 2014; Fuentealba et al. 2015; Llorens-Bobadilla et al. 2015). Also, on ultra-

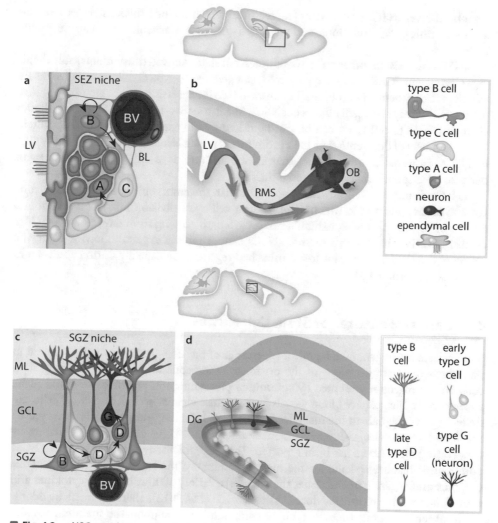

☐ **Fig. 4.2** aNSCs reside in confined neurogenic regions of the adult brain. The neurogenic regions or stem cell niches include the SEZ close to the LV of the forebrain **a** and **b** and the SGZ of the hippocampal dentate gyrus **c** and **d**. **a** The NSC niche of the SEZ contains the aNSCs called type B cells, which lie directly adjacent to the ependymal cells lining the LV, the NSC-derived NPCs termed type C cells/TAPs, neuroblasts (type A cells) and vessels formed by endothelial cells (red) and pericytes (brown). Signaling molecules represented by the interstitial matrix (light blue) and the BV-associated BL (dark blue) together with the different cell types contribute to the adult niche environment. **b** Neuroblasts of the LV migrate via the RMS to the OB and generate interneurons. **c** and **d** The SGZ of the hippocampus contains the type B cells, which give rise to early and late type D cells. Type D cells generate the excitatory granule neurons (type G cells) of the DG. aNSC adult neural stem cell, BL basal lamina, BV blood vessel, DG dentate gyrus, GCL granule cell layer, LV lateral ventricle, ML molecular layer, NPC neural progenitor cell, NSC neural stem cell, OB olfactory bulb, RMS rostral migratory stream, SEZ subependymal zone, SGZ subgranular zone, TAP transit-amplifying progenitor

structural level aNSCs share astrocytic features, which include thick intermediate filaments bundles, gap junctions and glycogen granules (Jackson and Alvarez-Buylla 2008).

aNSCs can exist in either an active or quiescent state. Activated, asymmetrically dividing, GFAP- and EGFR-positive type B cells, can give rise to type C cells, also called TAPs, which in turn generate neuroblasts. In contrast, GFAP-expressing cells, which are negative for the EGFR may represent quiescent aNSCs, but also astrocytes of the SEZ (Gengatharan et al. 2016). Both cell types can be distinguished by the expression of the glycoprotein CD133 (cluster of differentiation 133)/prominin-1, which is localized on the primary cilia of NSCs. However, due to the fact that aNSCs are rare, it is difficult to identify them because NSC specific markers are often missing.

aNSCs display a much slower division rate than the embryonic counterparts. In contrast to embryonic NSCs, which display a fast cell cycle division rate of several hours, aNSCs of the SEZ and SGZ, exhibit a cell cycle length ranging from days to weeks (Gotz et al. 2016). Perhaps for this reason, aNSCs cannot rapidly expand following injury or lesion of the CNS, which leads to a diminished regeneration capacity during adult stages (see ▶ Sects. 4.4 and 4.5).

4.3　General Features of Stem Cell Niches

Stem cell niches in general integrate a wide range of physiological stimuli and respond to pathophysiological alterations of the organism. They contain the slowly dividing stem cells, their progeny represented by committed progenitor and precursor cells and niche support cells. Niches have blood vessels in their close vicinity so that also endothelia can deliver specific signals to the niche territory, beyond the long-range signals conveyed by the blood stream (Rojas-Rios and Gonzalez-Reyes 2014). In some cases, signaling by innervating nerve fibers has been reported. The niche microenvironment is rich in ECM compounds that are specific for the niche. Presumably, the ECM constituents generate complex and dynamic interactomes that conceal a variety of morphogenes, cytokines and growth factors, due to specific docking sites (Brizzi et al. 2012). Thereby, the niche microenvironment functions as a reservoir of compounds that support the maintenance and persistence of the stem cell compartment.

4.3.1　The Extracellular Matrix Is Coded by a Complex Set of Genes

The ECM consists of glycoproteins, proteoglycans and complex glycans. As genomic sequences of a variety of species are decoded, a bioinformatics analysis has become possible. On that basis, the term "matrisome" has been introduced for the set of genes that can directly or indirectly be attributed to the ECM. A core matrisome of 300 genes has been defined that comprises 200 glycoprotein genes, 45 collagen and 35 proteoglycan genes. Matrix associated genes, regulatory genes and related molecules make up for additional 500 genes. Thereby, the ECM represents a complex set of molecules that exhibit tissue specific and temporally regulated expression patterns (Hynes and Naba 2012).

4.3.2 The Extracellular Matrix Occurs in Different Organization Forms

Glycoproteins and proteoglycans are complex molecules that comprise a large variety of domains and recognition sequences. These permit for interactions with other components of the ECM, with specific receptors and with a large variety of ligands. Thereby, constituents of the ECM form defined interactomes that can assemble to dynamic superstructures. One can distinguish different ECM organization forms. The interstitial matrix, which is composed of glycoproteins and proteoglycans, fills the intercellular space, is soluble in physiological buffers and is mainly distributed in a fluid phase. A second class of matrix structures is constituted by the basal laminae that are organized as two-dimensional layers. In numerous organs basal laminae underlie sheets of epithelia. Major constituents of the basal lamina are the glycoproteins laminin and nidogen as well as collagen IV and perlecan (◘ Fig. 4.3a). Basal laminae display some compositional variability depending on the organ and may comprise different types of laminin glycoproteins. A third organizational form of the ECM is the fibrillar matrix, prominent in the collagen fibers. Currently, more than 45 collagen genes are known, and collagens I, II and III are constituents of the fibrous matrix in connective tissues. Examples of the fibrillar matrix include collagen fibers and elastin fibrils. Overall, 28 homo- or heterotrimeric collagen types have been distinguished.

4.3.3 Glycoproteins

The glycoproteins of the ECM fall into subclasses that share sets of structural features. Laminin heterotrimers consist of α-, β- or γ-chains that are encoded by distinct genes. Five α-, four β- and three γ-proteins have been described so far. Based on their varying combinations of two or three individual chains, up to 14 different laminins have been identified. Laminins are key components of basal lamina structures and thereby intrinsic components of niches, for example in the intestine where epithelia are settling on basal lamina sheets, or as constitutive components of blood vessels.

The tenascins comprise a group of four genes in mammals, namely *tenascin-C* (*Tnc*), *tenascin-R* (*Tnr*), *tenascin-X* (*Tnx*), and *tenascin-W* (*Tnw*). On the structural level, tenascins display typical structural motifs, in particular a cysteine-rich amino terminus followed by characteristic EGF-type motifs, a sequence of fibronectin type three (FNIII) modules and homologies to fibrinogen β at the carboxy terminus (Tucker et al. 2006) (◘ Fig. 4.3b). The cysteine-rich amino terminus serves as an assembly domain for six monomers of Tnc to so-called hexabrachion structures. Considering the evolutionary origin, the first *tenascin* genes have emerged in the cephalochordate amphioxus and in urochordates. Tnc is among the first ECM glycoproteins that were identified as building blocks of the NSC niche (Faissner et al. 2017). There, Tnc is released by the type B astroglia-related slowly dividing stem cells of the niche. It became apparent that Tnc is also present in a variety of other niches in adult organisms, for example in the bone marrow, in the mesangium of kidney glomeruli and in the hair follicles (Chiquet-Ehrismann et al. 2014). Analysis of knock-out mouse lines has revealed subtle effects with regard to proliferation and differentiation behavior of stem and progenitor cells, in particular in the CNS (Faissner et al. 2017).

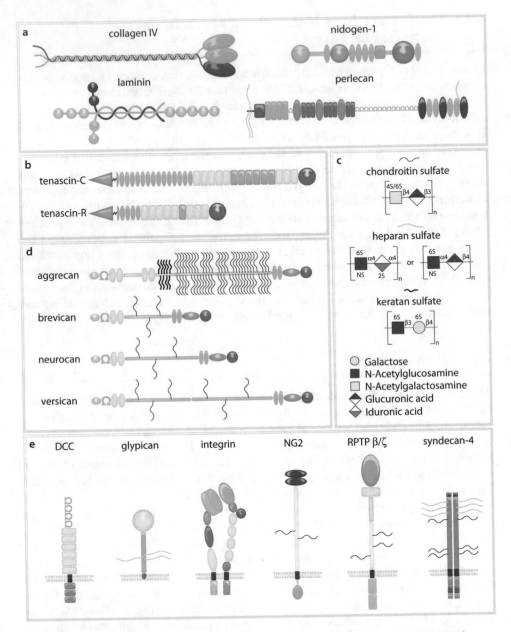

Fig. 4.3 Extracellular matrix molecules of the CNS. BL molecules, glycoproteins, proteoglycans and their complementary receptors are important functional components of stem cell niches. **a** Collagen IV, laminins, nidogen-1 and the HSPG perlecan are major constituents of the BL. **b** Tenascin-C and tenascin-R are glycoproteins of which tenascin-C is strongly associated with the neural stem cell niche. **c** The structural composition of the GAGs chondroitin sulfate, heparan sulfate and keratan sulfate is characterized by repeating disaccharide motifs (see legend in the figure). Numbers depict common positions of sulfate groups and the Greek letters refer to the link between the sugar units. **d** Aggrecan, brevican, neurocan and versican are lectican family CSPGs, which can be found in the interstitial matrix. **e** Important ECM receptors and membrane-associated molecules include DCC, glypicans, integrins, NG2, RPTP-β/ζ and syndecan-4. BL basal lamina, CNS central nervous system, CSPGs chondroitin sulfate proteoglycans, DCC deleted in colorectal cancer, GAG glycosaminoglycan, HSPG heparan sulfate proteoglycan, n number of repeats, NG2 neural/glial antigen 2, NS nitrogen-bound sulfation, RPTP-β/ζ receptor protein tyrosine phosphatase-β/ζ, S sulfation

4.3.4 **Proteoglycans**

Proteoglycans are formed by a glycoprotein core and at least one covalently attached glycosaminoglycan (GAG) chain (◻ Fig. 4.3c). GAG chains are generated by the repetitive alignment of carbohydrate dimer building blocks. Thus, dimers of glucuronic acid and N-Acetylgalactosamine are assembled to chondroitin sulfate. Depending on the GAG-complement proteoglycans are classified as chondroitin sulfate proteoglycans (CSPGs), heparan sulfate proteoglycans (HSPGs) and keratan sulfate proteoglycans (KSPGs). In general, proteoglycans embody particular properties due to the associated GAG chains, for example a high negative charge under physiological pH-conditions (Iozzo and Schaefer 2015). The GAG chains of proteoglycans are sulfated, with the exception of hyaluronic acid that is neither coupled to a core glycoprotein, nor sulfated. The sulfates are transferred to the GAG chains by sulfotransferases that attach the sulfate groups to the second, the fourth or the sixth carbon of the glycan composites of the building blocks of GAGs. The sulfate groups confer considerable charge to the GAGs. Beyond this biophysical implication, there is mounting evidence that the spatial patterns of sulfate groups on GAG chains create binding sites for selected ligands, for example for cytokines and growth factors (Purushothaman et al. 2012). Thereby, GAG chains may serve as reservoirs for the storage of distinct factors.

4.3.5 **Chondroitin Sulfate Proteoglycans**

The majority of CSPGs is distributed in the interstitial space. The lecticans represent a prominent family of CSPGs that encompass versican with its isoforms V_0, V_1, V_2 and V_3, neurocan that is specifically expressed in the CNS, brevican and aggrecan (◻ Fig. 4.3d). The latter is an abundant component of the synovial fluid. In the CNS, aggrecan is a fundamental constituent of the superstructure known as perineuronal net (Dzyubenko et al. 2016). Lecticans contain binding domains for hyaluronic acid, a lectin-like domain and an Ig-like motif. Different from the lecticans, the CSPG phosphacan is a soluble released isoform of the receptor protein tyrosine phosphatase (RPTP)-β/ζ, a large transmembrane receptor that intervenes in tyrosine kinase-dependent pathways (◻ Fig. 4.3e). RPTP-β/ζ and phosphacan have been observed in the stem cell niches of the CNS and also found associated with gliomas (Reinhard et al. 2016a). The RPTP-β/ζ receptor mediates pleiotrophin signaling, which plays a role in the proliferation of NSCs and glial progenitor cells (GPCs). Interestingly, RPTPs such as LAR (leukocyte antigen related tyrosine phosphatase) have been revealed as specific receptors for CSPGs, in particular for their GAG side chains. Increasing evidence highlights the importance of sulfation patterns of GAGs in that context.

NG2 (neural/glial antigen 2) is another prominent membrane-based CSPG of the CNS (◻ Fig. 4.3e). It is expressed by oligodendrocyte precursors and pericytes of the CNS. The NG2-expressing GPCs respond to lesion by extensive proliferation. Remarkably, the formation of excitatory synapses on the surface of NG2-positive GPCs has been described (Faissner and Reinhard 2015).

4.3.6 Heparan Sulfate Proteoglycans

Different from CSPGs, the majority of HSPGs are associated with the cell membrane. Two subfamilies can be distinguished, the syndecans and the glypicans (◘ Fig. 4.3e). Four syndecans have been described that are characterized by transmembrane domains. Due to their heparan sulfate GAG chains the syndecans can bind cytokines at specific binding sites in the carbohydrate polymer and present the molecules to their respective receptors. Thereby, syndecans play important roles as co-factors for signaling mechanisms, for example for FGF-2 (fibroblast growth factor)- or PDGF (platelet-derived growth factor)-dependent signaling mechanisms. Another function of syndecans is based on cis-interactions in the membrane that modulate the functions of cell adhesion receptors such as the integrins and cell adhesion molecules (CAMs) of the immunoglobulin (Ig) super-family (Sarrazin et al. 2011).

The glypicans are also associated with the cell membrane, but bind there via a small glycosyl-phosphatidyl-inositol (GPI)-anchor (◘ Fig. 4.3e). Glypicans are highly mobile in the membrane and similar to syndecans can interact with signaling proteins by their heparan sulfate side chains. As already mentioned above, the regulated distribution of sulfate groups along the carbohydrate polymers creates docking sites for distinct signaling molecules. On that basis, HSPGs are involved in critical signaling processes that are relevant for developmental mechanisms. For example, signaling by FGFs, by Wnt proteins and also by Shh depend on and are synergized by interactions with HSPGs (Kraushaar et al. 2013). All these factors have been found in niche microenvironments where they impact stem cell behavior. For example, Wnt-signaling drives the proliferation of intestinal stem cells and FGF-2 promotes the proliferation of NSCs.

4.3.7 Extracellular Matrix Receptors

Integrins are heterodimeric ECM receptors composed of α- and β-subunits (◘ Fig. 4.3e). A variety of α- and β-subunits have been discovered that assemble to 23 different integrin receptors that have been described in the human so far. Distinct integrins display different binding specificities for ECM glycoproteins. For example, integrin α1β1 is a collagen and integrin α6β4 is a laminin receptor (Hynes and Naba 2012). The group of β1-subunit carrying integrin receptors is particularly large and therefore it is not surprising that β1-integrin-containing receptors have been implicated in the proliferation of NSCs (Brizzi et al. 2012; Porcheri et al. 2014). Potential glycoprotein ligands of integrins in the NSC niche include laminin-1 and Tnc (Faissner and Reinhard 2015; Theocharidis et al. 2014; Tucker and Chiquet-Ehrismann 2015). As mentioned above, HSPGs of the syndecan family modulate integrin functions and thereby intervene in the ECM-dependent regulation of stem cell behavior.

Further receptors of the ECM include the CAMs of the Ig-superfamily. Ig-CAMs such as DCC (deleted in colorectal cancer; ◘ Fig. 4.3e) bind the laminin-related chemokine netrin-1 and the Ig-protein ROBO (roundabout) is the receptor for the repulsive molecule slit. Ig-superfamily CAMs also mediate calcium-independent intercellular adhesion. For example, the polysialylated form PSA-N CAM is expressed by neuronal progenitors derived from the slowly dividing stem cells of the granular layer in the adult hippocampal niche of rodents (Faissner and Reinhard 2015).

4.4 CNS Damage and Stem Cells

NSCs are not only present in the developing CNS and in adult niches, but are also associated with different pathological conditions. After CNS damage, cells with stem cell characteristics have been observed in regions that under healthy conditions are devoid of stem cells. Additionally, stem cells in the canonical niches are increased in number.

The appearance of potential stem/progenitor cells can be explained by two mechanisms: The first relies on the activation of local cells, whereas the second is based on the attraction of cells from the adult stem cell niches (☐ Fig. 4.4a, b). Both effects can be observed, depending on the type of injury, and do not exclude each other.

Activation of local astrocytes at the lesion site has been found after stab wound injury (Buffo et al. 2008). After demyelination in the rat spinal cord, NG2-positive glial progenitors respond by re-entering the cell cycle (Keirstead et al. 1998). Activation of the adult stem cell niche has been shown after stroke. Here, cells in the SEZ increased proliferation and migrated to the site of damage in the striatum (Arvidsson et al. 2002).

As cells can be attracted from the adult niches to the site of damage, a careful analysis is needed to verify by which mechanism stem cells are generated in a specific disease.

The fact that stem cells can be found in the diseased CNS leads to the question which signals induce stem cell properties. All cell types respond to stress induced by damage, among them are astrocytes, neurons and microglia, the immune cells of the CNS. These cells produce a plethora of signaling molecules that affect cell survival as well as proliferation, differentiation or migration. Shh has been identified as an important player in this context. It induces proliferation of type A cells, the neuroblasts formed in the adult SEZ stem cell niche. After damage, especially after invasive injury, Shh is expressed and responsible for the stem cell character of astrocytes (Sirko et al. 2013).

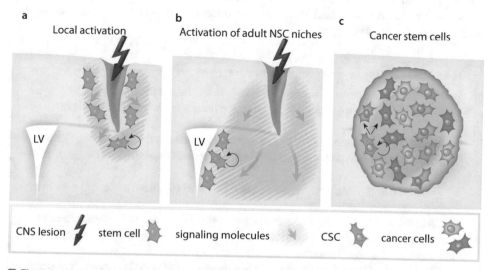

☐ **Fig. 4.4** Stem cells in the lesioned CNS. CNS damage can induce the activation of stem cells that originate from local cells like astrocytes **a** or can trigger cells in the adult NSC niches such as the SEZ of the LV **b**. Long-range and local environmental cues (blue, arrows) in the lesioned tissue influence the cellular response. **c** CSCs appear in tumor tissue and share similarities with NSCs. Also in the tumor, the cells react to stimuli of the surrounding. CNS central nervous system, CSC cancer stem cell, LV lateral ventricle, NSC neural stem cell, SEZ subependymal zone

With respect to the cells´ differentiation capacity, a gap is observed between cells that are isolated and cultivated in the dish (*in vitro*) on the one side and cells remaining on-site in the living organism (*in vivo*) on the other hand. *In vitro* the activated cells can differentiate into neurons, astrocytes and oligodendrocytes, thereby fulfilling the criterion of multipotency. In contrast, differentiation found *in vivo* is often restricted to the formation of new astrocytes. This discrepancy can be explained by the composition of signaling molecules present after lesion. For example, the surface receptor Notch, which is known to inhibit the neuronal cell fate, is present after different types of CNS damage. Therefore cells that are multipotent in principle, will not give rise to new neurons in this environment. Beside the question of differentiation, the functional integration of neurons, if formed at all, is a critical aspect. Increasing knowledge about the molecules involved in these processes can help to adjust therapy to support recovery of patients in the future.

4.5 Cancer Stem Cells in the Central Nervous System

A special case of lesions are tumors. Typically, the tumor mass is formed by a huge number of heterogeneous cell populations. In cancer tissue from different organs, including the CNS, cancer stem cells (CSCs) have been found (◘ Fig. 4.4c). CSCs can originate from stem cells, but can also derive from differentiated cell types. Accordingly, CSCs are defined as stem cells by their properties and not by their origin, namely self-renewal, differentiation capacity and formation of a new tumor with a similar cell composition. In addition to this potential, NSCs and CSCs share a number of similarities, for example in the expression profile of ECM (Reinhard et al. 2016b).

Several markers for CSCs have been identified so far, among them is the glycoprotein CD133/prominin-1. Due to the fact that CSCs can initiate new tumors and that they are often more resistant to standard therapies than other cells of a tumor mass, CSCs are the target of new therapeutic approaches.

> **Take Home Message**
> 1. Neuroepithelial cells are the first stem cells in the neural tube
> 2. Radial glia cells can generate neurons and glia cells in a timely and spatially ordered fashion
> 3. Neurons migrate along radial glia cell processes through the developing tissue
> 4. Early-born neurons settle in inner cortical layers and later-born ones in outer cortical layers
> 5. Astrocytes and oligodendrocytes can also be generated from radial glia cells in a process called gliogenesis
> 6. The subependymal zone lining the lateral ventricle and the subgranular zone of the hippocampal dentate gyrus are the two neurogenic regions (neural stem cell niches) of the adult forebrain
> 7. Type B cells of the subependymal zone generate transit-amplifying progenitors (type C cells), which give rise to neuroblasts (type A cells)
> 8. Subependymal zone-derived neuroblasts migrate through the rostral migratory stream into the olfactory bulb and give rise to olfactory input-processing interneurons

9. Neural stem cells of the subgranular zone (type B cells) generate type D cells, which give rise to excitatory granule cells (type G cells)
10. Neural stem cell niches create a privileged environment for the expansion, maintenance and neurogenic properties of neural stem cells
11. The adult neural stem cell niche contains the neural stem cells, transit-amplifying progenitors, neuroblasts, endothelial cells, leptomeningeal cells and pericytes
12. The extracellular matrix is a key constituent of the neural stem cell niche that consists of basal lamina structures, a variety of glycoproteins (laminins and tenascin-C), chondroitin and heparan sulfate proteoglycans as well as complex glycan structures
13. Damage of the central nervous system can induce stem cell properties in the adult, but the differentiation capacity of these cells is limited *in vivo*
14. Cancer stem cells are found in central nervous system tumors and share similarities with neural stem cells

❓ Questions

1. In which order do neural stem and progenitor cells appear?
2. What are the two neurogenic regions in the adult brain?
3. Which cell types can be found in the adult neural stem cell niche?
4. Which extracellular matrix glycoproteins and proteoglycans can be found in the neural stem cell niche?
5. Where do the stem cells that are found after central nervous system damage originate from? Explain the two underlying mechanisms.
6. How are cancer stem cells defined and from which cells do they derive?

✅ Answers

1. First, neuroepithelial cells are the stem cells, which increase cell numbers by symmetric divisions and generate first neurons. Then, radial glia cells appear, which give rise to neurons, astrocytes and oligodendrocytes during neuro- and gliogenesis. Radial glia cells can also generate intermediate neuronal or glial precursor cells, which proliferate but are lineage restricted.
2. The subependymal zone lining the lateral ventricle of the forebrain and the subgranular zone of the hippocampal dentate gyrus.
3. The adult neural stem cell niche contains the neural stem cells themselves, transit-amplifying progenitors, neuroblasts, endothelial cells, leptomeningeal cells as well as pericytes.
4. Laminins and tenascin-C glycoproteins, heparan sulfate proteoglycans (syndecans and glypicans) as well as chondroitin sulfate proteoglycans (lecticans, RPTP-β/ζ and phosphacan) can be found in the neural stem cell niche.
5. Central nervous system damage can activate cells locally at the site of damage or stimulate cells in the adult neural stem cell niches.
6. Cancer stem cells are defined by their ability to self-renew, their differentiation capacity and by the potential to form a new tumor with a similar cellular composition as the initial tumor. Cancer stem cells can derive from stem cells, but also from differentiated cells.

References

Altmann, C. R., & Brivanlou, A. H. (2001). Neural patterning in the vertebrate embryo. *International Review of Cytology, 203*, 447–482.

Arvidsson, A., Collin, T., Kirik, D., Kokaia, Z., & Lindvall, O. (2002). Neuronal replacement from endogenous precursors in the adult brain after stroke. *Nature Medicine, 8*(9), 963–970.

Brizzi, M. F., Tarone, G., & Defilippi, P. (2012). Extracellular matrix, integrins, and growth factors as tailors of the stem cell niche. *Current Opinion in Cell Biology, 24*(5), 645–651.

Buffo, A., Rite, I., Tripathi, P., Lepier, A., Colak, D., Horn, A. P., Mori, T., & Gotz, M. (2008). Origin and progeny of reactive gliosis: A source of multipotent cells in the injured brain. *Proceedings of the National Academy of Sciences of the United States of America, 105*(9), 3581–3586.

Cai, J., Wu, Y., Mirua, T., Pierce, J. L., Lucero, M. T., Albertine, K. H., Spangrude, G. J., & Rao, M. S. (2002). Properties of a fetal multipotent neural stem cell (NEP cell). *Developmental Biology, 251*(2), 221–240.

Chiquet-Ehrismann, R., Orend, G., Chiquet, M., Tucker, R. P., & Midwood, K. S. (2014). Tenascins in stem cell niches. *Matrix Biology: Journal of the International Society for Matrix Biology, 37*, 112–123.

Codega, P., Silva-Vargas, V., Paul, A., Maldonado-Soto, A. R., Deleo, A. M., Pastrana, E., & Doetsch, F. (2014). Prospective identification and purification of quiescent adult neural stem cells from their in vivo niche. *Neuron, 82*(3), 545–559.

Doetsch, F. (2003). The glial identity of neural stem cells. *Nature Neuroscience, 6*(11), 1127–1134.

Dzyubenko, E., Gottschling, C., & Faissner, A. (2016). Neuron-glia interactions in neural plasticity: Contributions of neural extracellular matrix and Perineuronal nets. *Neural Plasticity, 2016*, 5214961.

Faissner, A., & Reinhard, J. (2015). The extracellular matrix compartment of neural stem and glial progenitor cells. *Glia, 63*(8), 1330–1349.

Faissner, A., Roll, L., & Theocharidis, U. (2017). Tenascin-C in the matrisome of neural stem and progenitor cells. *Molecular and Cellular Neurosciences, 81*, 22–31.

Fuentealba, L. C., Rompani, S. B., Parraguez, J. I., Obernier, K., Romero, R., Cepko, C. L., & Alvarez-Buylla, A. (2015). Embryonic origin of postnatal neural stem cells. *Cell, 161*(7), 1644–1655.

Gengatharan, A., Bammann, R. R., & Saghatelyan, A. (2016). The role of astrocytes in the generation, migration, and integration of new neurons in the adult olfactory bulb. *Frontiers in Neuroscience, 10*, 149.

Gotz, M., Nakafuku, M., & Petrik, D. (2016). Neurogenesis in the developing and adult brain-similarities and key differences. *Cold Spring Harbor Perspectives in Biology, 8*(7), a018853.

Hynes, R. O., & Naba, A. (2012). Overview of the matrisome--an inventory of extracellular matrix constituents and functions. *Cold Spring Harbor Perspectives in Biology, 4*(1), a004903.

Iozzo, R. V., & Schaefer, L. (2015). Proteoglycan form and function: A comprehensive nomenclature of proteoglycans. *Matrix Biology, 42*, 11–55.

Jackson, E. L., & Alvarez-Buylla, A. (2008). Characterization of adult neural stem cells and their relation to brain tumors. *Cells, Tissues, Organs, 188*(1–2), 212–224.

Keirstead, H. S., Levine, J. M., & Blakemore, W. F. (1998). Response of the oligodendrocyte progenitor cell population (defined by NG2 labelling) to demyelination of the adult spinal cord. *Glia, 22*(2), 161–170.

Kessaris, N., Fogarty, M., Iannarelli, P., Grist, M., Wegner, M., & Richardson, W. D. (2006). Competing waves of oligodendrocytes in the forebrain and postnatal elimination of an embryonic lineage. *Nature Neuroscience, 9*(2), 173–179.

Kirischuk, S., Luhmann, H. J., & Kilb, W. (2014). Cajal-Retzius cells: Update on structural and functional properties of these mystic neurons that bridged the 20th century. *Neuroscience, 275*, 33–46.

Kraushaar, D. C., Dalton, S., & Wang, L. (2013). Heparan sulfate: A key regulator of embryonic stem cell fate. *Biological Chemistry, 394*(6), 741–751.

Llorens-Bobadilla, E., Zhao, S., Baser, A., Saiz-Castro, G., Zwadlo, K., & Martin-Villalba, A. (2015). Single-cell transcriptomics reveals a population of dormant neural stem cells that become activated upon brain injury. *Cell Stem Cell, 17*(3), 329–340.

Porcheri, C., Suter, U., & Jessberger, S. (2014). Dissecting integrin-dependent regulation of neural stem cell proliferation in the adult brain. *The Journal of Neuroscience, 34*(15), 5222–5232.

Purushothaman, A., Sugahara, K., & Faissner, A. (2012). Chondroitin sulfate "wobble motifs" modulate maintenance and differentiation of neural stem cells and their progeny. *The Journal of Biological Chemistry, 287*(5), 2935–2942.

Reinhard, J., Brosicke, N., Theocharidis, U., & Faissner, A. (2016a). The extracellular matrix niche microenvironment of neural and cancer stem cells in the brain. *The International Journal of Biochemistry & Cell Biology, 81*, 174–183.

Reinhard, J., Brosicke, N., Theocharidis, U., & Faissner, A. (2016b). The extracellular matrix niche microenvironment of neural and cancer stem cells in the brain. *The International Journal of Biochemistry & Cell Biology, 81*(Pt A), 174–183.

Rojas-Rios, P., & Gonzalez-Reyes, A. (2014). Concise review: The plasticity of stem cell niches: A general property behind tissue homeostasis and repair. *Stem Cells, 32*(4), 852–859.

Sarrazin, S., Lamanna, W. C., & Esko, J. D. (2011). Heparan sulfate proteoglycans. *Cold Spring Harbor Perspectives in Biology, 3*(7), a004952.

Sirko, S., Behrendt, G., Johansson, P. A., Tripathi, P., Costa, M., Bek, S., Heinrich, C., Tiedt, S., Colak, D., Dichgans, M., Fischer, I. R., Plesnila, N., Staufenbiel, M., Haass, C., Snapyan, M., Saghatelyan, A., Tsai, L. H., Fischer, A., Grobe, K., Dimou, L., & Gotz, M. (2013). Reactive glia in the injured brain acquire stem cell properties in response to sonic hedgehog. [corrected]. *Cell Stem Cell, 12*(4), 426–439.

Tabata, H. (2015). Diverse subtypes of astrocytes and their development during corticogenesis. *Frontiers in Neuroscience, 9*, 114.

Taverna, E., Gotz, M., & Huttner, W. B. (2014). The cell biology of neurogenesis: Toward an understanding of the development and evolution of the neocortex. *Annual Review of Cell and Developmental Biology, 30*, 465–502.

Theocharidis, U., Long, K., ffrench-Constant, C., & Faissner, A. (2014). Regulation of the neural stem cell compartment by extracellular matrix constituents. *Progress in Brain Research, 214*, 3–28.

Tucker, R. P., & Chiquet-Ehrismann, R. (2015). Tenascin-C: Its functions as an integrin ligand. *The International Journal of Biochemistry & Cell B, 65*, 165–168.

Tucker, R. P., Drabikowski, K., Hess, J. F., Ferralli, J., Chiquet-Ehrismann, R., & Adams, J. C. (2006). Phylogenetic analysis of the tenascin gene family: Evidence of origin early in the chordate lineage. *BMC Evolutionary Biology, 6*, 60.

Further Reading

Adams, K. V., & Morshead, C. M. (2018). Neural stem cell heterogeneity in the mammalian forebrain. *Progress in Neurobiology, 170*, 2–36.

Arai, Y., Pulvers, J. N., Haffner, C., Schilling, B., Nusslein, I., Calegari, F., & Huttner, W. B. (2011). Neural stem and progenitor cells shorten S-phase on commitment to neuron production. *Nature Communications, 2*, 154.

Dehay, C., & Kennedy, H. (2007). Cell-cycle control and cortical development. *Nature Reviews. Neuroscience, 8*(6), 438–450.

Gotz, M., & Huttner, W. B. (2005). The cell biology of neurogenesis. *Nature Reviews. Molecular Cell Biology, 6*(10), 777–788.

Lange, C., & Calegari, F. (2010). Cdks and cyclins link G1 length and differentiation of embryonic, neural and hematopoietic stem cells. *Cell Cycle, 9*(10), 1893–1900.

Ming, G. L., & Song, H. (2005). Adult neurogenesis in the mammalian central nervous system. *Annual Review of Neuroscience, 28*, 223–250.

Naba, A., Clauser, K. R., Ding, H., Whittaker, C. A., Carr, S. A., & Hynes, R. O. (2016). The extracellular matrix: Tools and insights for the "omics" era. *Matrix Biology, 49*, 10–24.

Ruddy, R. M., & Morshead, C. M. (2018). Home sweet home: The neural stem cell niche throughout development and after injury. *Cell and Tissue Research, 371*(1), 125–141.

Skeletal Muscle Stem Cells

Beate Brand-Saberi and Eric Bekoe Offei

© Springer Nature Switzerland AG 2020
B. Brand-Saberi (ed.), *Essential Current Concepts in Stem Cell Biology*,
Learning Materials in Biosciences, https://doi.org/10.1007/978-3-030-33923-4_5

What You Will Learn in This Chapter

To understand the biology of skeletal muscle stem cells involved in muscle regeneration, you have to understand myogenesis in the embryo. In this chapter, the steps and regulators of myogenesis are introduced. You will learn about the sources of muscle progenitors in the mesoderm and their distribution in the embryo. During this process, muscle stem cells are set aside which attach to the muscle fibres as undifferentiated quiescent satellite cells representing the main source for muscle growth and regeneration. In addition to satellite cells, pericytes, endothelial and interstitial cells, mesoangioblasts and side-population cells possess myogenic potential. You will learn about skeletal muscle specific transcription factors (MRFs) and their functions. Finally, you will learn about muscle wasting, which occurs during aging and muscle dystrophies. In this context, novel stem cell-based approaches involving reprogramming will be explained.

5.1 Myogenesis in the Embryo

Skeletal muscles, the most abundant tissue in the human body forms about 40% of the total body mass. It is involved in the control of movement, posture, breathing as well as control of whole-body metabolism (Frontera and Ochala 2015). The skeletal muscle tissue is made up of long terminally differentiated multinucleated cells (myofibres) that are ensheathed in several hierarchies of connective tissues containing blood vessels, nerves and stem cells. These myofibres contain specialized proteins; actin and myosin that enable the muscle to perform its contractile function to bring about various movements in the body as well as maintenance of posture of the body (Sambasivan and Tajbakhsh 2015). These muscle fibres are formed throughout the body and during the entire life of vertebrates. The progenitor cells of these elongated multinucleated cells in the vertebrate embryo originate from distinct mesoderm populations. The muscles of the trunk and its appendages are derived from the somites, bilateral paired blocks of paraxial mesoderm that form along both sides of the notochord and the neural tube (Yusuf and Brand-Saberi 2012).

The somites are initially epithelial spheres filled by losely-packed mesenchymal cells and are developed in a cranio-caudal sequence to form sequential portions of paraxial mesoderm (Christ and Ordahl 1995). With progress of the developmental processes, the somites are transformed into more complex structures: The dorsally located dermomyotome which yields epaxial and hypaxial skeletal muscle (Christ et al. 2000) together with other derivatives such as angioblasts, dermis and smooth muscle (Kalcheim et al. 1999; Ben-Yair and Kalcheim 2008), and the ventral sclerotome, which differentiates into axial cartilages of the vertebral column and ribs (Christ and Ordahl 1995).

The head and neck muscles are heterogeneous in origin. The head muscles such as the extraocular muscles, muscles of mastication and muscles of facial expression are derived from cells of the pre-otic paraxial head mesoderm and the prechordal mesoderm. The pre-otic paraxial mesoderm cells migrate into the first and second pharyngeal arches, respectively. The other muscles of the head such as the tongue muscles, hypobranchial muscles and the posterior pharyngeal muscles arise from the occipital somites that migrate into the third pharyngeal arch. The pharyngeal mesoderm forms the inner core of the pharyngeal arches and is made up of cells from the paraxial mesoderm and the splanchnic mesoderm which are almost not separable. The regulatory networks governing the development of the craniofacial muscles and the trunk muscles are distinct, both at signaling level and the level of the transcription factors (Tzahor 2015).

5.2 Molecular Regulation of Embryonic Myogenesis

Skeletal muscle development in the embryo is controlled both by intrinsic and extrinsic regulatory pathways. For example, Myf5; MyoD double knockout embryos in which MRF4 expression is not compromised fail to develop limb and craniofacial muscles whereas some trunk muscles are developed (Kassar-Duchossoy et al. 2004). On the other hand, mice lacking both Pax3/Myf5 (and MRF4) are unable to develop trunk muscles but are able to develop normal head muscles (Tajbakhsh et al. 1997), an indication that different molecular pathways are involved in the development of the craniofacial, trunk and limb muscles. Thus, Pax3 is required for the expression of MyoD in the trunk and not the head, which is consistent with the absence of expression of Pax3 in the muscle progenitors of the head (Hacker and Guthrie 1998; Harel et al. 2009).

Specification of skeletal myoblasts develops in the somites in response to signaling molecules from the neighbouring tissues such as the neural tube, notochord and dorsal ectoderm (Fan and Tessier-Lavigne 1994). These signaling molecules include the Wnt family, sonic hedgehog (SHH) and noggin as activators, and Bone morphogenic protein 4 (BMP4) as inhibitor (Hirsinger et al. 1997). In the trunk, expression of SHH and noggin by the notochord and the floor of the neural tube cause the ventral part of the somite to form the Pax1 and Pax9-positive sclerotome for vertebral column formation (Huang and Christ 2000). The ectoderm overlying the somite and the dorsal aspect of the neural tube express Wnts which in conjunction with the low levels of SHH causes the dorsal portion of the somite to form the dermomyotome; a sheet-like pseudostratified epithelium with ventrally curved lips. The cells of the dermomyotome express the paired- and homeodomain-containing transcription factors Pax3 (the first skeletal muscle-relevant myogenesis regulator) and Pax7 (Bober et al. 1994). Also, an interaction between the activating Wnts and inhibitory BMPs directs the dorsomedial portion of the dermomyotome to form the Myf5-positive muscle precursor cells of the primary myotome. The latter consists of elongated unit-length muscle pioneer cells spreading from the medial to the lateral extent throughout the myotome (Kahane et al. 2007). According to studies of mutant mice, the induction of Myf5 in the epaxial myotome relies on SHH (Borycki et al. 1999; ◘ Fig. 5.1).

The Wnt/β-catenin signaling pathway also regulates multiple steps of myogenesis by regulating step-specific targets (Suzuki et al. 2015). During the organization of the mesodermal epithelia to form somites, Wnt6 signaling from the overlying ectoderm maintains the epithelial structure of the dermomyotome of the somite. Transduction of Wnt6 signaling by its receptor molecule frizzled7 (Linker et al. 2005) is mediated by paraxis (bHLH transcription factor; Burgess et al. 1996) and leads to activation of β- catenin required for the maintenance of the epithelial structure of the somite (Linker et al. 2005). Indeed, mouse embryos deficient in Wnt/β-catenin signaling are embryonic lethal by E8.5 with increased cell death (Haegel et al. 1995; Girardi and Le Grand 2018), while those with conditional depletion of β-catenin in the muscle precursor Pax7+ cell lineage show reduced muscle mass and slow myofibres (Hutcheson et al. 2009). Moreover, upregulation of Dkk1/4, a Wnt/β-catenin antagonist (Zorn 2001; Hirata et al. 2011) inhibits muscle development. Hence Wnt/β-catenin signaling plays a crucial role in skeletal muscle development and homeostasis (Suzuki et al. 2015), because the proliferation of adult skeletal muscle stem/precursor cells is also regulated by Wnt/β-catenin signaling.

The myogenic regulatory factors (MRFs) were the first tissue-specific regulators of differentiation (Weintraub et al. 1991). They comprise four distinct muscle-specific tran-

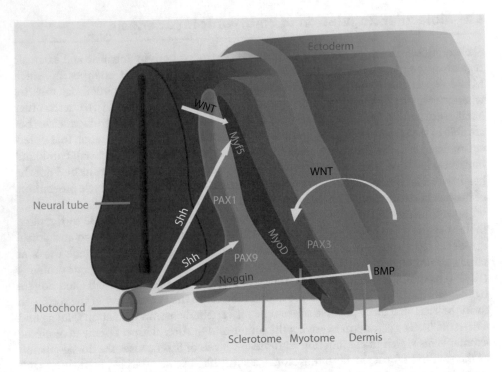

Fig. 5.1 Signals and genes controlling somite compartment formation. WNTs from the dorsal neural tube and ectoderm activate Pax3, which is initially expressed throughout the epithelial somite. It is subsequently maintained only in the dermomyotome, which contains muscle stem cells/progenitors. The myotome is formed from the dermomyotome in two waves: First, Myf5-positive pioneer cells arise from the dorsomedial lip of the dermomyotome. In a second wave, myotome cells are recruited from all four edges of the dermomyotome. The combined influence of activating WNT proteins and inhibitory BMP4 protein controls MyoD expression in the ventrolateral region to create the hypaxial muscle cell precursors. Noggin (BMP inhibitor), secreted by the notochord counteracts the BMPs from the lateral plate. SHH is produced by the notochord and also by the floor plate of the neural tube. It is essential for the expression of Myf5 in the DML (low levels), and also causes the ventral part of the somite to form the sclerotome (high levels). Pax1 and Pax9 are induced in the sclerotome, which control chondrogenesis and vertebra formation. (Adapted from Yusuf and Brand-Saberi (2012))

scription factors (MyoD, myogenin, Myf5 and MRF4) that are involved in the regulation of myogenesis in the embryo and in vitro. They belong to the basic-Helix-Loop-Helix superfamily that is involved in establishing as well as maintaining the myogenic lineage (Naidu et al. 1995). Traditionally, the MRFs were classified in those responsible for myogenic specification („early group": Myf5 and MyoD), myogenin and MRF4 were considered as control factors of muscle differentiation ("late group"). MRFs share many common features and it was later found that they exert overlapping functional activities, e.g. in the absence of Myf5, MRF4 carries out myogenic determination activity, although it was initially described to be involved in myotube differentiation (Summerbell et al. 2002; Kassar-Duchossoy 2004; Moncaut et al. 2013). Despite the overlapping functional activity of this gene family, the temporal and spatial expression patterns of individual members suggest that during normal myogenesis, each plays a unique role in controlling aspects of skeletal muscle myogenesis (Naidu et al. 1995). Indeed, gene ablation studies of the MRFs of this gene family showed their involvement in different aspects of myogenesis

(Hasty et al. 1993; Naidu et al. 1995). For instance, Myf-5 and MyoD act upstream of myogenin to specify myoblasts for terminal differentiation while myogenin and MRF4 are directly involved in the differentiation process and trigger the expression of myotube-specific genes (Bentzinger et al. 2012; Ganassi et al. 2018).

5.3 Skeletal Muscle Regeneration

Skeletal muscles have extensive metabolic and functional plasticity as well as a robust regenerative capacity (Tajbakhsh 2009), which enables them to generate new myofibres when they are damaged by injuries or diseases (Carlson 1973). This striking regenerative capacity of skeletal muscle makes it a good tool (Fry et al. 2015) for the study and application of regenerative medicine (Church et al. 1966; Zouraq et al. 2013). The satellite cells between the sarcolemma and the basal lamina of the skeletal muscle syncytium are the main players in the regeneration of skeletal muscles (Tedesco et al. 2010). The satellite cells also contribute to the postnatal growth of the myofibre, which is evident by the higher number (approximately 6–8 times) of nuclei in the adult myofibre as compared to that of the neonate (Mauro 1961). In addition to the satellite cells, other progenitor cells including pericytes, endothelial and interstitial cells located outside the basal lamina have shown some myogenic potential in vitro and after transplantation (Cossu and Biressi 2005). For some time now, there has been much interest in understanding the cellular and molecular mechanisms underlying the regeneration of skeletal muscles in different contexts as such knowledge might contribute to further development of therapies for diseases such as muscular dystrophy; this will be described in more detail in subsequent paragraphs.

Skeletal muscle regeneration employs essential aspects of embryonic myogenesis, and it is a very important homeostatic process in the adult muscle, which allows for repair of damaged muscle fibres. When a muscle fibre is damaged, the satellite cells respond to the injury by activation and re-entry into the cell cycle. The vast majority of the satellite cell-derived progenitors exits the cell cycle after one or more rounds of proliferation and enters a terminal (G0) phase that leads to differentiation, followed by either fusion to one another to generate new muscle fibres or to repair existing muscle fibres (Olguin and Olwin 2004; Olguín and Pisconti 2012; ◘ Fig. 5.2).

There are two concepts to understand the replenishment of the satellite cells in the regenerating myofibres; first, the activated satellite cells have been shown to divide asymmetrically giving rise to a daughter cell that has self-renewal capabilities and another daughter cell that becomes a myoblast (Kuang et al. 2007; Troy et al. 2012; Dumont et al. 2015). Simultaneously, the proliferating myoblasts are induced to upregulate Pax7, which inhibits myogenin expression and promotes the entry of the cell into a mitotically quiescent state (Olguin and Olwin 2004; Wen et al. 2012).

5.4 Stages of Skeletal Muscle Regeneration

Skeletal muscle regeneration proceeds through three sequential but overlapping stages: inflammatory reaction, satellite cell activation and formation of myofibres, and remodeling of the newly formed myofibres (Charge and Rudnicki 2004; Ciciliot and Schiaffino 2010).

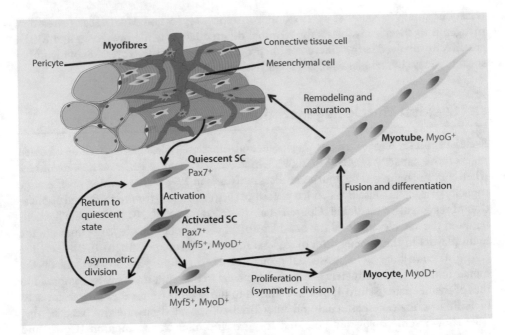

◘ Fig. 5.2 Skeletal muscle regeneration. During skeletal muscle injury, the satellite cell expressing Pax7 becomes activated (expressing Pax7, MyoD and Myf5) to enter into the cell cycle. It then divides asymmetrically giving rise to two daughter cells, one of which strongly expresses Pax7 and hence re-enters the quiescent state to replenish the satellite cell stock, while the other one expressing MyoD and Myf5 becomes a myoblast. The myoblast undergoes several cell divisions before cells fuse with each other and differentiate to form myotubes expressing myogenin. The myotube undergoes remodeling and maturation to either fuse with existing myofibres to repair the damaged fibre or form an entirely new myofibre to replace a completetly damaged fibre

During the inflammatory stage, there is an influx of calcium that leads to activation of calcium-dependent proteases such as calpains that disintegrate the myofibril and other cell constituents as a result of the damaged sarcolemma. This together with the entry of plasma proteins and activation of complement cascades induce chemotactic recruitment of neutrophils and macrophages (Tidball 2008). The early macrophages (called inflammatory macrophages) secrete pro-inflammatory cytokines such as tumor necrosis factor α (TNFα) and interleukin 1β (IL-1β) and are responsible for the removal of necrotic tissues from the damaged muscle. At about 24 hours after the onset of injury, these early invading macrophages expressing CD68 (a marker for late endosomes and lysosomes) reach their highest numbers and begin to be replaced by a second type of macrophages expressing CD163 (involved in the removal of proinflammatory ligands), possibly due to a phenotypic switch from CD68+/CD163- to CD68-/CD163+. These late macrophages, also called anti-inflammatory macrophages, secrete anti-inflammatory cytokines such as interleukin 10 (IL-10) that contribute to the termination of the inflammation and release factors that promote myogenic precursor proliferation, growth and differentiation (Cantini et al. 2002; Sonnet et al. 2006; Arnold et al. 2007). Thus, macrophages play a central role in skeletal muscle response to injury by removing necrotic tissues and promoting muscle regeneration (Ciciliot and Schiaffino 2010).

Towards the end of the inflammatory stage, the satellite cells undergo a finely orchestrated cellular and molecular response to regenerate well functional muscle fibres in two ways: By producing myocytes that either fuse with the existing functional fibres to repair them or fuse with each other to form new myofibres to replace the damaged ones (Charge and Rudnicki 2004). Nagata et al. (2006) indicated that multiple signals appear to trigger satellite cell activation, which include sphingosine-1-phosphate synthesized by the plasma membrane, that stimulate the entry of the satellite cells into the cell cycle. Abrogation of the synthesis of sphingosine-1-phosphate renders skeletal muscle regeneration defective. Nitric oxide (NO) has also been found to be necessary for the activation of satellite cells possibly through the activation of matrix metalloproteinases, which induce the production of hepatocyte growth factor (HGF) from the satellite cells. HGF contributes to satellite cell activation (Tatsumi et al. 1998, 2006) and at the same time inhibits satellite cell differentiation (Miller et al. 2000).

The switch of the myoblasts from the proliferation state to the differentiation state, just as in embryonic development, appears to be controlled by the Notch-Wnt signaling pathway with Notch signaling prevalent during the proliferation phase, while Wnt signaling is dominant during the differentiation phase (Conboy and Rando 2002; Brack et al. 2008). After injury, there is sustained Notch signaling, which ensures proper expansion of the satellite cell progeny, while the Wnt signaling drives the differentiation process (Brack et al. 2008).

The regenerated myofibres undergo a variety of remodeling and maturation processes usually based on conditions of the injury such as the type of muscle injury, involvement of blood vessels and re-establishment of neuromuscular and myotendinous connections (Ciciliot and Schiaffino 2010). One of the major factors of muscle regeneration is the successful establishment and maintenance of the basal lamina of the fibres within which satellite cells and myotubes can proliferate and fuse to form normal muscle fibres. In rodents and humans, freshly regenerated muscle fibres are characterized by the presence of their centrally located nuclei (Ciciliot and Schiaffino 2010; Fry et al. 2015). Regenerating muscle fibres may remodel to form different patterns, which include clusters of smaller muscle fibres as a result of non-fusion of myotubes within the same basal lamina or formation of fork fibres as a result of fusion at only one extremity (Schmalbruch 1976). After segmental necrosis, regenerative processes are concentrated at the level of the damaged stump and if the reconstitution of myofibre integrity is prevented by scar tissue that separates the two stumps, then a new myotendinous junctions will be formed (Järvinen et al. 2008) to repair the muscle tissue.

5.5 Types of Muscle Stem/Progenitor Cells

More than 50 years ago, the best described stem cells of skeletal muscle were discovered by Alexander Mauro in transmission electron micrographs: Satellite cells (Mauro 1961). Like most other skeletal muscle progenitors, they arise from the dermomyotomes of the somites within the population of Pax3 expressing cells that somewhat later also express Pax7 (Armand et al. 1983; Gros et al. 2005). In contrast to the postmitotic myonuclei, the satellite cells on muscle fibres are mitotically quiescent and can be activated due to injury or training-stimulated muscle growth. Satellite cells are specified and characterized by the expression of the paired box transcription factor Pax7 that protects them from apoptosis and is essential for the production of fetal myogenic progenitors and myofibres (Seale et al. 2000; Olguin and Olwin 2004; Kassar-Duchossoy et al. 2005; Relaix et al. 2005, 2006; Hutcheson et al. 2009).

Some of the satellite cells also express Myf5 and are thus committed to muscle formation. While the latter are considered as muscle progenitor cells, Pax7+ Myf5- cells are regarded as multipotent stem cells (Asakura et al. 2001). The expression of Myf5 depends on epigenetic changes of the Pax7 as well as the Myf5 locus. First of all, Pax7 has to be methylated at several arginine residues by an arginine methyltransferase called CARM1 (Kawabe et al. 2012). Subsequently, a histone methyltransferase complex is recruited to the Myf5 locus to allow for Myf5 transcription. However, Myf5 is not translated into protein immediately, but sequestered in mRNP granules (Crist et al. 2012). In this way, satellite cells are maintained in the quiescent state, but at the same time are poised for rapid entry into myogenic differentiation upon activation.

The aforementioned chromatin remodelling processes go along with the occurrence of asymmetrical divisions during which two kinds of daughter cells are being generated: the Myf5+ daughter cells poised for muscle differentiation and their Myf5- sisters that will remain quiescent stem cells. These asymmetrical divisions are also controlled by miRNA-489, which maintains stemness and quiescence in one of the daughter cells by inhibiting the translation of the oncogene DEK that leads to the proliferation of committed progenitor cells upon activation (Cheung et al. 2012). Furthermore, miRNA-31 is a component of the mRNP granules in poised daughter cells. It targets Myf5, thus preventing its translation (Crist et al. 2012).

In contrast to the situation in the trunk and limbs, satellite cells in the head and neck region are independent of the Pax3 pathway. Here, satellite cells are derived from the non-somitic cranial paraxial mesoderm and express Mesp1 and Isl1 (Harel et al. 2009). However, Pax7 is activated also in head muscle progenitors during the fetal period and retained in adult satellite cells (Sambasivan et al. 2009; Gnocchi et al. 2009). In addition to the transcription factor Pax7, surface markers have been established in satellite cells, among them c-met, M-cadherin, syndecan3 and 4, Vcam-1, NCAM-1, and CD34, E-cadherin, Vcam1, Icam1, Cldn5 (claudin 5), Esam (endothelial cell-specific adhesion molecule), and Pcdhb9 (Cornelison and Wold 1997; Irintchev et al. 1994; Beauchamp et al. 2000; Fukada et al. 2007). Interestingly, some of these cell surface molecules are shared with hematopoietic stem cells or with endothelial cells pointing to a common derivation within the mesoderm (Kardon et al. 2002), an issue to be kept in mind when reading the following paragraph.

The satellite cell niche is complex and comprises the extracellular matrix and neighbouring cells, which includes the cell contacts and secretome of the latter. The cellular neighbourhood of satellite cells comprises the myofibre, fibroblasts (interstitial cells) and endothelial cells of the capillaries (Christov et al. 2007). During inflammation, it also contains inflammatory cells. The endothelial cells can fuel the proliferation of satellite cells by IGF1, HGF, FGF2, VEGF and PDGFBB (Christov et al. 2007). Interestingly, myoblasts derived from satellite cells and cultivated in dispersed culture employ mechanisms of differentiation that differ from the ones in the presence of their niche, i.e. if they are cultivated together with their myofibre. A key regulator in support of myoblast quiescence in satellite cell–derived myoblast maintained in their niche is the oncogene p53, whereas the majority of dispersed myoblasts are undergoing differentiation upon downregulation of the ERK1 pathway after 3 days in culture (Flamini et al. 2018).

5.5.1 Interstitial Cells: Derived Muscle Stem Cells

Although satellite cells are the main source of muscle stem cells (reviewed by Relaix and Zammit (2012)), several other cells with myogenic potential have been described in the past. However, their topography is less well defined, because with one exception (for blood-derived myogenic stem cells, see below) all are residing outside the basement membrane of the myofibre within the surrounding connective tissue ("interstitium") in a stricter or broader sense. Some of them cling tightly to vessels. All of these myogenic progenitor cells differ in their potential to contribute to skeletal muscle regeneration by engrafting into preexisting myofibres to a certain extent, which may also partially depend on the assays used by the experimentors. The borders between the subpopulations are floating and the terminology in the literature partially depends on authors and their particular experimental approaches.

Based on their transplantation efficiency, the most important ones next to satellite cells are pericytes (Dellavalle et al. 2007). Fully mature pericytes are highly branched cells accompanying the microvessels in all body tissues. They are regarded as tissue-specific adult stem cells (Dellavalle et al. 2007, 2011) The ancestry of pericytes is diverse and consequently, several subtypes of pericytes can be distinguished on the basis of their markers and differentiation behaviour (Birbrair et al. 2013). Although their developmental origin remains elusive, muscle-derived pericytes share with brain-derived pericytes NG2 proteoglycan and with many others the more ubiquitous PDGFRs (Balabanov et al. 1996) and transitorily the intermediate filament Nestin, the decline of which can be used to distinguish pericytes with a myogenic potential from those with a neurogenic potential (Birbrair et al. 2013). The marker that distinguishes muscle pericytes from other pericytes as well as from satellite cells is Alkaline Phosphatase (AP). AP is neither found on satellite cells nor on myofibres (Dellavalle et al. 2007, 2011). Using AP, it was shown by genetic reporter expression (inducible Alkaline Phosphatase CreERT2) that AP-positive pericytes contributed to developing myofibres as well as to the satellite cell pool. Pericytes have been demonstrated to participate in muscle growth also in vivo (Dellavalle et al. 2011).

For our understanding of muscle dystrophies, it is important to note that pericytes also contribute to the production of adipose tissue and connective tissue within aging or diseased skeletal muscles (fibro-adipogenic progenitors, FAPs). In this context interestingly, two subtypes of muscle pericytes can be distinguished on the basis of PDGFRα: Type-1 muscle pericytes positive for both factors participate in adipogenesis and collagen production, whereas PDGFRα negative type-2 muscle pericytes are involved in myogenesis and angiogenesis (Birbrair et al. 2014; Lemos et al. 2015). Although pericytes have been shown to enter the satellite compartment, it has also been suggested that they exert additional interactive functions that have a positive impact on myogenesis. In this way the reinnervation of skeletal muscle via the recruitment of Schwann cells from pericytes is affected (Birbrair et al. 2013). Secondly, the angiogenic contribution to regenerating muscle is affected (Birbrair et al. 2014) and a third effect has been described via paracrine interactions from pericytes to satellite cells as well as endothelial cells which have been shown to be close neighbours (Christov et al. 2007).

Apart from pericytes, interstitial cells expressing PW1 (PW1+ interstitial cells; PICs) have been described to contribute to myonuclei in regenerating muscle (Relaix et al. 1996; Mitchell et al. 2010; Besson et al. 2011; Pannérec et al. 2013). PW1 is a zinc-finger transcription factor that interacts with the TNF receptor-2 and is also involved in the p53 axis, whereby it regulates the stress response, also in myoblasts (Schwarzkopf et al. 2006). PW1 is encoded by the paternally expressed gene 3 (PEG-3) and has been found to be a reliable marker for adult stem cells that can significantly contribute to the regeneration of tissues of ectodermal, mesodermal and endodermal origin (Besson et al. 2011). In invertebrates (planarians), pluripotent adult stem cells (neoblasts) capable of regenerating the whole body express Piwi, an orthologue of PW1. In vertebrates, PICs are multipotent, but have a strong preference to differentiate towards the mesodermal lineage, especially skeletal and smooth muscle tissue (Mitchell et al. 2010). Those involved in skeletal myogenesis start to express Pax7 in vivo only in response to local stimuli that recruit them to sites of muscle injury and the satellite cell compartment.

Mesenchymal-like cells closely abutting the endothelium of larger vessels have been detected and characterized by FLK1 expression (VEGFR2, KDR, CD309) that can participate in myogenesis under experimental conditions as well (Minasi et al. 2002; Sampaolesi et al. 2003; reviewed by Cossu and Bianco (2003)). These cells are called meso-angioblasts, because of their potential to adopt the angiogenic or the myogenic fate. When they were isolated from healthy dogs and grafted to the dystrophic Golden Retriever, muscle regeneration was significantly improved (Sampaolesi et al. 2006).

Such cell populations in the interstitium resemble mesenchymal (stromal) stem cells (MSCs) in morphology and topography resulting in ongoing discussions about whether or not mesenchymal stem cells can give rise to muscle cell types. First of all, care should be taken to distinguish between skeletal and smooth muscle cells, secondly MSC markers do not overlap with myogenic progenitor markers.

5.5.2 Side-Population

The so-called side-population (SP) represents another group of myogenic progenitors (Gussoni et al. 1999). The term "side population" which is not restricted to muscle stem/progenitor cells refers to the fact that this cell pool is not detected by the usual set of markers applied by flow cytometry for a particular cell population of interest, such as muscle stem/progenitor cells (Golebiewska et al. 2011).

SP cells express the membrane protein ABCG2 (ATP-binding cassette transporters) that is also found in hematopoietic stem cells and has been implied in therapy (multidrug) resistence in cancers. In the case of stem cells bearing this transporter, it is responsible for the exclusion of the dye Hoechst 33342. Thus, the myogenic SP cells are Hoechst-negative (Gussoni et al. 1999; Asakura et al. 2002).

The fact that some of the side population cells also express syndecan 3 and syndecan 4 (regarded as satellite cell markers) along with ABCG2 contributes to the difficulty or even impossibility to characterize muscle progenitor cell subpopulations without any overlap (Tanaka et al. 2009).

5.5.3 Blood-Derived Muscle Stem Cells

In contrast to the aforementioned groups of muscle stem/progenitor cells that reside within complex tissues (muscle resident), a fraction from the peripheral blood consisting of monocytes has been described to have myogenic potencies: The CD133-positive cells (Torrente et al. 2004; Péault et al. 2007; Negroni et al. 2009). CD133 is another name derived from the hematopoietic nomenclature for the multipass transmembrane protein Prominin-1, at first described in murine epithelial tissue (Weigmann et al. 1997) and also a well-known marker of hematopoietic stem and its AC133 epitope is also found on VEGFR2 of endothelial progenitor cells (Shmelkov et al. 2005).

From the clinical angle, it has enormous technical and patient-friendly advantages to obtain such stem cells from the patients' blood. Thus, the CD133+ cell has received much attention and has successfully been used in gene therapy (Torrente et al. 2007). The coexpression of CD34 allows predictions regarding the proliferative and differentiation behaviour of the CD133+ cells where CD133+/CD34+ cells had a higher myoblast/myotube fusion index after intramuscular injection and even yielded a better outcome in comparison to satellite cells due to a higher migration activity within the host muscle (Negroni et al. 2009).

5.6 Muscular Dystrophies

Muscular dystrophies are inherited diseases usually regarded as a distinct heterogeneous group of diseases that belong to the larger group of myopathies. They are characterized by progressive muscle wasting in an early life phase caused by defects or absence of structural proteins in the muscle tissue that can lead to severe mobility impairment and even a restricted lifespan. They have in common the histological aspect that is mainly characterized by variations in muscle fiber diameter due to different stages of regeneration attempts, and the invasion of macrophages and inflammatory cells. As the disease advances, the skeletal muscles accumulate fibrous and adipose tissue in replacement of the functional muscle tissue (Emery 1993).

We will restrict ourselves here to the brief introduction of three common congenital muscular dystrophies: Duchenne's muscular dystrophy (DMD), and its milder variant Becker's muscular dystrophy (BMD), and Limb girdle muscular dystrophy (LGMD) in which the underlying mutations have been identified. DMD is an X-linked disease affecting male newborns with an incidence of about 1 in 3600 per year (Greenberg et al. 1988). Like BMD, it is caused by different kinds of mutations in the largest human gene, encoding the sarcolemma protein dystrophin. Dystrophin is a major component of the dystrophin-dystroglycan complex that combines the cytoskeleton of the myofibre to the extracellular matrix of the endomysium. The gene spans more than two megabases and contains 79 exons and the described mutations comprise frameshifts, deletions and nonsense point mutations resulting in the absence of a functional dystrophin protein. As a consequence, the myofibres are destabilized during contraction and their myonuclei become apoptotic (Meryon 1851, 1852; Duchenne 1868).

Affected boys show weakness in the muscles of their shoulder and pelvic girdles and proximal leg muscles during the first 5 years of life. Patients suffer progressive muscle wasting, which results in the loss of ambulation at 12 or 13 years of age (reviewed by Emery (1993)). The life expectancy has recently slightly increased as a result of improvements in the medical care to about 30 years, however, secondary complications such as infections of the respiratory system and coagulative disorders in the context of surgical interventions may still cause a lethal outcome before the patients reach their twenties. Dystrophin is also an essential component in cardiac muscles; thus, dilatative cardiomyopathy is observed in patients after 10 years and heart failure is one of the inevitable causes of death in DMD patients (Emery 1993).

BMD is less abundant and dystrophin is usually not absent, but present at compromised quantities and qualities, such as shortened dystrophin variants. Thus, it shows typically a later onset at the age of about 12 years and is much less debilitating than DMB.

In contrast to DMD and BMD, Limb Girdle Muscular Dystrophy (LGMD) is an autosomal inherited disease and comprises a somewhat heterogeneous group of muscular dystrophies with 31 different underlying causes (reviewed by Nigro and Savarese (2014)). It can thus be divided into different types depending on the underlying genetic disorders. Thereof, eight mutations are autosomal dominant and 23 are autosomal recessive. The dominant forms (LGMD1) comprise for example myotilin (LGMD1A), lamin A/C (LGMD1B), caveolin 3 (LGMD1C), desmin (LGMD1E); the recessive ones comprise for example calpain (LGMD2A), dysferlin (LGMD2B), γ sarcoglycan (LGMD2C), α sarcoglycan (LGMD2D), β sarcoglycan (LGMD2E), δ sarco- glycan (LGMD2F), telethonin (LGMD2G), titin (LGMD2J), dystroglycan (LGMD2P), and again desmin (LG- MD2R), to name just a few.

In LGMD 1b, the LaminA/C (LMNA) gene is mutated, a feature shared with the Emery-Dreyfuss muscular dystrophy (Morris 2001; Bonne et al. 2000). The lamins are major constituents of the nuclear membrane and loss of function results in compromised cell function and survival. Thus, disruption of cell membrane components is not the only underlying disorder, which causes muscular dystrophies.

After reading the preceeding paragraphs of this chapter, you may wonder why satellite cells are unable to compensate for the loss of myofibres in muscular dystrophies, in particular in the case of defective dystrophin, which was not considered to be expressed in satellite cells (Miranda et al. 1988; Huard et al. 1991). In general, the failure of stem cell therapies was considered to be a result of stem cell exhaustion, similar to the situation in aging muscle (Sousa-Victor and Munoz-Canoves 2016).

However, it has been shown only recently that dystrophin is indeed present also prior to muscle differentiation and that it is critically involved in the asymmetrical divisions occuring during the activation of satellite cells in mdx-mice, a well established murine model for studying Duchenne muscular dystrophy (Dumont et al. 2015; Chang et al. 2016). Since asymmetrical divisions between Myf5+ and Myf5- satellite cells form the basis of generating daughter cells that are competent to enter the differentiation pathway (Myf5+) and to participate in the regeneration of myotubes after satellite cell activation (Kuang et al. 2007), loss of polarity results in a loss of asymmetrical divisions and a failure to produce myogenic satellite cells. Dystrophin associates with the asymmetry regulating proteins Mark2 (Par1b), a Ser/Thr kinase which enables the polarized distribution of Pard3 resulting in the asymmetric activation of p38/α/b and myogenic commitment of daughter cells. Loss of function in the Par complex results in abrogation of asymmetrical division and ensuing absence of myogenic differentiation (Troy et al. 2012). During this process, the dystrophin+ cell is maintained as satellite stem cell, whereas the other daughter cell is the satellite progenitor cell that enters into the myogenic program (◘ Fig. 5.3).

5.7 Stem Cell-Based Therapies

Until now, there has been no cure for the muscular dystrophies. Thus, therapy revolves around surgical interference for contractures, attention to respiratory care and cardiovascular complications. In the past decades, muscular dystrophies have been treated mainly to slow down the process of muscle loss, primarily by interfering with the secondary events such as immune response by administration of glucocorticoids to the patients.

Causative gene-therapies aiming at the restoration of a functional dystrophin by transgenic approaches have faced a number of problems. First of all, dystrophin is the largest gene in the human genome. High-capacity adenoviral vectors can accommodate and transfer full lengh dystrophin (Clemens et al. 1996), however the expression just persists transiently episomal and holds risks for human patients due to likelihood of an immune response or other complications resulting from the viral vectors. To overcome this difficulty, scientists have designed shortened versions of dystrophin comprising only the most essential exons, which could be incorporated into integrating systems like retro/lentiviral vectors or adeno-associated vector (AAV). Alternatively, another related smaller protein

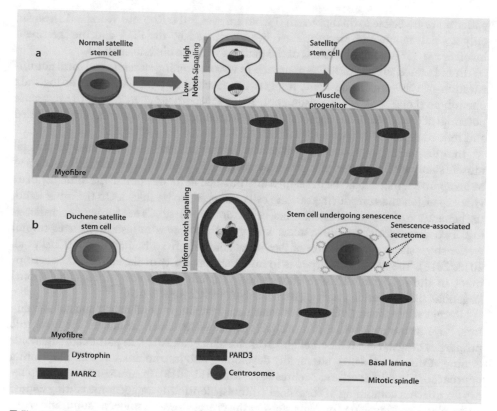

◻ Fig. 5.3 Cell polarity defect in dystrophin deficient satellite cells. **a** Normal satellite stem cells undergo asymmetric division upon dystrophin-dependent polarization of MARK2 and PARD3 to opposite sides along the apicobasal axis of the dividing cell. This results in asymmetrical distribution of cell fate determinants such as mediators of Notch signaling during mitosis to enforce different cell fates (stem cell self-renewal and myogenic commitment). **b** Dystrophin-deficiency leads to downregulation of MARK2 in the satellite cells resulting in equal distribution of PARD3 within the dividing cell. The absence of these polarity cues and the abnormal mitotic progression cause the satellite stem cells to undergo cell cycle arrest and they may enter senescence. (Graph redrawn and modified on the basis of Fig. 4 in Chang et al. (2016))

linking the cytoskeleton with the extracellular matrix in developing muscle, called utrophin ("dystrophin-related protein") has been explored to fulfill substitution of dystrophin function (Helliwell et al. 1992; Miura and Jasmin 2006). Adeno-associated vectors that can efficiently transduce satellite cells still have to be explored in more detail, because the commonly used AAV6 and AAV8 serotypes that transduce skeletal muscle successfully fail to transduce satellite cells or at very low efficiencies, rendering the delivery of functional dystrophin genes into satellite cells a challenge (Arnett et al. 2014; Chang et al. 2016).

To date, one of the most promising approaches is exon-skipping. The exon-skipping approach is offered by a few companies uses antisense-oligonucleotides to hide or mask particular defective exons in the gene sequence of the dystrophin gene, in order to avoid the truncation of the dystrophin protein during translation. Instead, a shorter but partially functional gene product will ensue. This will result in a milder manifestation of the disease, comparable to that of Becker's muscular dystrophy. Depending on the mutation, not all patients can benefit from the exon-skipping method (Fletcher et al. 2010; Aartsma-Rus et al. 2009). Especially frame-shift mutations constituting more than 80% of Duchenne patients are amenable to this approach (Pichavant et al. 2011; Koo and Wood 2013). So far, only a small number of products have been approved by the FDA and the treatment involves regular (weekly) infusions of the patients with the oligonucleotides.

In mdx mice, the CRISPR/Cas approach *in vivo* combining systemic and local administration of Cas9 and guide RNA, respectively, resulted in a partial rescue of the mdx phenotype (Long et al. 2016; Nelson et al. 2016; Tabebordbar et al. 2016). The immense potential of the CRISPR/Cas approach for genome editing of dystrophin is being exploited and refined successfully (Amoasii et al. 2017).

In order to achieve more lasting effects in human patients, genome editing in skeletal muscle stem cells or induced pluripotent stem cells (iPSC)-derived muscle cells appears to be an approach that should be developed and pursued in future efforts. The difficulty of *ex vivo* expansion and transducing satellite cells still needs to be solved. On the other hand, the generation of functional satellite-like cells from human iPSCs has turned out challenging. Two-dimensional culture systems with growth factors administrations over several weeks show success in provision of Pax7+ satellite cells from human iPSCs (Chal et al. 2015, 2016). The current challenge is to take the complexity of skeletal muscle development in the near-natural tissue context into account and develop three-dimensional 'organoid' differentiation systems (Brand-Saberi and Zaehres 2016).

Recent advances on tissue culture conditions reaching beyond media and soluble factors, have further revealed that a relatively elastic environment enhances the myogenic properties of stem cells (Gilbert et al. 2010; Hosseini et al. 2012). In particular, organoids aiming at disease modeling are rapidly developing also in combination with bioprinting approaches (Brand-Saberi and Zaehres 2016; Kim et al. 2018). In summary, interdisciplinary approaches combining 3D organoid cultures, bioprinting and genome editing appear particularly exciting and promising steps in the process towards understanding and treating skeletal muscle diseases.

Take Home Message

1. Skeletal muscle progenitors of the trunk and limbs arise from the dermomyotomes of the somites. Head and neck muscles are derived from the unsegmented preotic, the prechordal, the paraxial and splanchnic mesoderm.

2. The key regulatory factors of skeletal muscle specification and differentiation belong to the MyoD family of bHLH transcription factors: MyoD, Myf5, Mrf4 and Myogenin, summarized as the muscle regulatory factors (MRFs).

3. The best characterized and most abundant muscle stem cells are the satellite cells, but several other cell types in the interstitium and the blood have been shown to have myogenic competence; they are characterized by a plethora of different marker combinations.

4. Satellite cells depend on the paired box transcription factor Pax7 for their maintenance. They remain in a quiescent state and can undergo asymmetrical divisions upon activation.

5. Muscular dystrophies are inherited muscle wasting diseases still lacking cure; in some of them, cardiac muscle is also affected, for example Duchenne's muscular dystrophy, an X-linked lethal disease.

6. Stem cell-based approaches are being developed for disease modeling *in vitro* as well as for patient treatment combining organoid cultures, bioprinting and genome editing in the future.

Acknowledgments   At the beginning of 2019, more than 25.000 entries in Pubmed referred to muscle stem cells. The authors apologize to those authors who made significant contributions to this field, but were not mentioned in our textbook chapter due to space restrictions and the prominent teaching purpose of this book, which forced us to make a rigorous selection. The authors wish to thank PD Holm Zaehres, PhD, for discussions and Maximilian Mucke, M.Sc., for technical help. The financial support of the EU's sixth framework program MYORES (511978), DFG, FoRUM (F647-09, F732N-2011, F873-16), MERCUR (Pr-2012-0058), Deutsche Gesellschaft für Muskelkranke e.V. (DGM Foundation), Freiburg, and Deutsche Duchenne-Stiftung, action benni & co e. V., Bochum, is gratefully acknowledged.

References (and Further Reading)

Aartsma-Rus, A., Fokkema, I., Verschuuren, J., Ginjaar, I., Van Deutekom, J., van Ommen, G. J., & Den Dunnen, J. T. (2009). Theoretic applicability of antisense-mediated exon skipping for Duchenne muscular dystrophy mutations. *Human Mutation, 30*(3), 293–299.

Amoasii, L., Long, C., Li, H., Mireault, A. A., Shelton, J. M., Sanchez-Ortiz, E., & Hauschka, S. D. (2017). Single-cut genome editing restores dystrophin expression in a new mouse model of muscular dystrophy. *Science Translational Medicine, 9*(418), eaan8081.

Armand, O., Boutineau, A. M., Mauger, A., Pautou, M. P., & Kieny, M. (1983). Origin of satellite cells in avian skeletal muscles. *Archives d'Anatomie Microscopique et de Morphologie Expérimentale, 72*(2), 163–181.

Arnett, A. L., Konieczny, P., Ramos, J. N., Hall, J., Odom, G., Yablonka-Reuveni, Z., Chamberlain, J. R., & Chamberlain, J. S. (2014). Adeno-associated viral (AAV) vectors do not efficiently target muscle satellite cells. *Molecular Therapy: Methods & Clinical Development, 1*, 14038.

Arnold, L., Henry, A., Poron, F., Baba-Amer, Y., Van Rooijen, N., Plonquet, A., Gherardi, R. K., & Chazaud, B. (2007). Inflammatory monocytes recruited after skeletal muscle injury switch into antiinflammatory macrophages to support myogenesis. *The Journal of Experimental Medicine, 204*(5), 1057–1069.

Asakura, A., Rudnicki, M. A., & Komaki, M. (2001). Muscle satellite cells are multipotential stem cells that exhibit myogenic, osteogenic, and adipogenic differentiation. *Differentiation, 68*(4–5), 245–253.

Asakura, A., Seale, P., Girgis Gabardo, A., & Rudnicki, M. A. (2002). Myogenic specification of side population cells in skeletal muscle. *The Journal of Cell Biology, 159*(1), 123–134.

Balabanov, R., Washington, R., Wagnerova, J., & Dore-Duffy, P. (1996). CNS microvascular pericytes express macrophage-like function, cell surface integrin αM, and macrophage marker ED-2. *Microvascular Research, 52*(2), 127–142.

Beauchamp, J. R., Heslop, L., David, S. W., Tajbakhsh, S., Kelly, R. G., Wernig, A., & Zammit, P. S. (2000). Expression of CD34 and Myf5 defines the majority of quiescent adult skeletal muscle satellite cells. *The Journal of Cell Biology, 151*(6), 1221–1234.

Bentzinger, C. F., Wang, Y. X., & Rudnicki, M. A. (2012). Building muscle: Molecular regulation of myogenesis. *Cold Spring Harbor Perspectives in Biology, 4*(2), a008342.

Ben-Yair, R., & Kalcheim, C. (2008). Notch and bone morphogenetic protein differentially act on dermomyotome cells to generate endothelium, smooth, and striated muscle. *The Journal of Cell Biology, 180*(3), 607–618.

Besson, V., Smeriglio, P., Wegener, A., Relaix, F., Oumesmar, B. N., Sassoon, D. A., & Marazzi, G. (2011). PW1 gene/paternally expressed gene 3 (PW1/Peg3) identifies multiple adult stem and progenitor cell populations. *Proceedings of the National Academy of Sciences, 108*(28), 11470–11475.

Birbrair, A., Zhang, T., Wang, Z. M., Messi, M. L., Enikolopov, G. N., Mintz, A., & Delbono, O. (2013). Skeletal muscle pericyte subtypes differ in their differentiation potential. *Stem Cell Research, 10*(1), 67–84.

Birbrair, A., Zhang, T., Wang, Z. M., Messi, M. L., Mintz, A., Olson, J. D., & Delbono, O. (2014). Type-2 pericytes participate in normal and tumoral angiogenesis. *American Journal of Physiology. Cell Physiology, 307*, C25–C38, 2014.

Bober, E., Franz, T., Arnold, H. H., Gruss, P., & Tremblay, P. (1994). Pax-3 is required for the development of limb muscles: A possible role for the migration of dermomyotomal muscle progenitor cells. *Development, 120*(3), 603–612.

Bonne, G., Mercuri, E., Muchir, A., Urtizberea, A., Becane, H. M., Recan, D., Merlini, L., Wehnert, M., Boor, R., Reuner, U., & Vorgerd, M. (2000). Clinical and molecular genetic spectrum of autosomal dominant Emery-Dreifuss muscular dystrophy due to mutations of the lamin A/C gene. *Annals of Neurology, 48*(2), 170–180.

Borycki, A. G., Brunk, B., Tajbakhsh, S., Buckingham, M., Chiang, C., & Emerson, C. P. (1999). Sonic hedgehog controls epaxial muscle determination through Myf5 activation. *Development, 126*, 4053–4063.

Boutet, S. C., Cheung, T. H., Quach, N. L., Liu, L., Prescott, S. L., Edalati, A., & Rando, T. A. (2012). Alternative polyadenylation mediates microRNA regulation of muscle stem cell function. *Cell Stem Cell, 10*(3), 327–336.

Brack, A. S., Conboy, I. M., Conboy, M. J., Shen, J., & Rando, T. A. (2008). A temporal switch from notch to Wnt signaling in muscle stem cells is necessary for normal adult myogenesis. *Cell Stem Cell, 2*(1), 50–59.

Brand-Saberi, B., & Zaehres, H. (2016). The development of anatomy: From macroscopic body dissections to stem cell–derived organoids. *Histochemistry and Cell Biology, 146*(6), 647–650.

Burgess, R., Rawls, A., Brown, D., Bradley, A., & Olson, E. N. (1996). Requirement of the paraxis gene for somite formation and musculoskeletal patterning. *Nature, 384*(6609), 570.

Cantini, M., Giurisato, E., Radu, C., Tiozzo, S., Pampinella, F., Senigaglia, D., Zaniolo, G., Mazzoleni, F., & Vitiello, L. (2002). Macrophage-secreted myogenic factors: A promising tool for greatly enhancing the proliferative capacity of myoblasts in vitro and in vivo. *Neurological Sciences, 23*(4), 189–194.

Carlson, B. M. (1973). The regeneration of skeletal muscle—A review. *The American Journal of Anatomy, 137*(2), 119–149.

Chal, J., Oginuma, M., Al Tanoury, Z., Gobert, B., Sumara, O., Hick, A., & Tassy, O. (2015). Differentiation of pluripotent stem cells to muscle fiber to model Duchenne muscular dystrophy. *Nature Biotechnology, 33*(9), 962.

Chal, J., Al Tanoury, Z., Hestin, M., Gobert, B., Aivio, S., Hick, A., & Pourquié, O. (2016). Generation of human muscle fibers and satellite-like cells from human pluripotent stem cells in vitro. *Nature Protocols, 11*(10), 1833.

Chang, N. C., Chevalier, F. P., & Rudnicki, M. A. (2016). Satellite cells in muscular dystrophy–lost in polarity. *Trends in Molecular Medicine, 22*(6), 479–496.

Charge, S. B., & Rudnicki, M. A. (2004). Cellular and molecular regulation of muscle regeneration. *Physiological Reviews, 84*(1), 209–238.

Cheung, T. H., Quach, N. L., Charville, G. W., Liu, L., Park, L., Edalati, A., Yoo, B., Hoang, P., & Rando, T. A. (2012). Maintenance of muscle stem-cell quiescence by microRNA-489. *Nature, 482*(7386), 524.

Christ, B., & Ordahl, C. P. (1995). Early stages of chick somite development. *Anatomy and Embryology, 191*(5), 381–396.

Christ, B., Huang, R., & Wilting, J. (2000). The development of the avian vertebral column. *Anatomy and Embryology, 202*(3), 179–194.

Christov, C., Chrétien, F., Abou-Khalil, R., Bassez, G., Vallet, G., Authier, F. J., & Gherardi, R. K. (2007). Muscle satellite cells and endothelial cells: Close neighbors and privileged partners. *Molecular Biology of the Cell, 18*(4), 1397–1409.

Church, J. C., Noronha, R. F., & Allbrook, D. B. (1966). Satellite cells and skeletal muscle regeneration. *The British Journal of Surgery, 53*(7), 638–642.

Ciciliot, S., & Schiaffino, S. (2010). Regeneration of mammalian skeletal muscle: Basic mechanisms and clinical implications. *Current Pharmaceutical Design, 16*(8), 906–914.

Clemens, P. R., Kochanek, S., Sunada, Y., Chan, S., Chen, H. H., Campbell, K. P., & Caskey, C. T. (1996 Nov). In vivo muscle gene transfer of full-length dystrophin with an adenoviral vector that lacks all viral genes. *Gene Therapy, 3*(11), 965–972.

Conboy, I. M., & Rando, T. A. (2002). The regulation of notch signaling controls satellite cell activation and cell fate determination in postnatal myogenesis. *Developmental Cell, 3*(3), 397–340.

Cornelison, D. D., & Wold, B. J. (1997). Single-cell analysis of regulatory gene expression in quiescent and activated mouse skeletal muscle satellite cells. *Developmental Biology, 191*(2), 270–283.

Cossu, G., & Bianco, P. (2003). Mesoangioblasts - vascular progenitors for extravascular mesodermal tissues. *Current Opinion in Genetics & Development, 13*(5), 537–542.

Cossu, G., & Biressi, S. (2005). Satellite cells, myoblasts and other occasional myogenic progenitors: Possible origin, phenotypic features and role in muscle regeneration. *Seminars in Cell & Developmental Biology, 16*(4–5), 623–631.

Crist, C. G., Montarras, D., & Buckingham, M. (2012). Muscle satellite cells are primed for myogenesis but maintain quiescence with sequestration of Myf5 mRNA targeted by microRNA-31 in mRNP granules. *Cell Stem Cell, 11*(1), 118–126.

Dellavalle, A., Sampaolesi, M., Tonlorenzi, R., Tagliafico, E., Sacchetti, B., Perani, L., Innocenzi, I., Galvez, B. G., Messina, G., Morosetti, R., Belicchi, M., Peretti, G., Chamberlain SJ Wright, E. W., Torrente, Y., & Li, S. (2007). Pericytes of human skeletal muscle are myogenic precursors distinct from satellite cells. *Nature Cell Biology, 9*(3), 255.

Dellavalle, A., Maroli, G., Covarello, D., Azzoni, E., Innocenzi, A., Perani, L., & Cossu, G. (2011). Pericytes resident in postnatal skeletal muscle differentiate into muscle fibres and generate satellite cells. *Nature Communications, 2*, 499.

Duchenne, G. B. (1868). Recherches sur le paralysie musculaire pseudohypertrophique ou paralysie myoscle-rosique. I. Symptomatologie, marche, duree, terminaison. *Archives of General Internal Medicine, 11*, 179.

Dumont, N. A., Wang, Y. X., Von Maltzahn, J., Pasut, A., Bentzinger, C. F., Brun, C. E., & Rudnicki, M. A. (2015). Dystrophin expression in muscle stem cells regulates their polarity and asymmetric division. *Nature Medicine, 21*(12), 1455.

Emery, A. E. (1993). Duchenne muscular dystrophy—Meryon's disease. *Neuromuscular Disorders, 3*(4), 263–266.

Fan, C. M., & Tessier-Lavigne, M. (1994). Patterning of mammalian somites by surface ectoderm and noto-chord: Evidence for sclerotome induction by a hedgehog homolog. *Cell, 79*, 1175–1186.

Flamini, V., Ghadiali, R. S., Antczak, P., Rothwell, A., Turnbull, J. E., & Pisconti, A. (2018). The satellite cell niche regulates the balance between myoblast differentiation and self-renewal via p53. *Stem Cell Reports, 10*(3), 970–983.

Fletcher, S., Adams, A. M., Johnsen, R. D., Greer, K., Moulton, H. M., & Wilton, S. D. (2010). Dystrophin isoform induction in vivo by antisense-mediated alternative splicing. *Molecular Therapy, 18*(6), 1218–1223.

Frontera, W. R., & Ochala, J. (2015). Skeletal muscle: A brief review of structure and function. *Calcified Tissue International, 96*(3), 183–195.

Fry, C. S., Lee, J. D., Mula, J., Kirby, T. J., Jackson, J. R., Liu, F., Yang, L., Mendias, C. L., Dupont-Versteegden, E. E., McCarthy, J. J., & Peterson, C. A. (2015). Inducible depletion of satellite cells in adult, sedentary mice Impairs muscle regenerative capacity without affecting sarcopenia. *Nature Medicine, 21*(1), 76.

Fukada, S. I., Uezumi, A., Ikemoto, M., Masuda, S., Segawa, M., Tanimura, N., Yamamoto, H., Miyagoe-Suzuki, Y., & Takeda, S. I. (2007). Molecular signature of quiescent satellite cells in adult skeletal muscle. *Stem Cells, 25*(10), 2448–2459.

Ganassi, M., Badodi, S., Quiroga, H. P., Zammit, P. S., Hinits, Y., & Hughes, S. M. (2018). Myogenin promotes myocyte fusion to balance fibre number and size. *Nature Communications, 9*(1), 4232.

Gilbert, P. M., Havenstrite, K. L., Magnusson, K. E., Sacco, A., Leonardi, N. A., Kraft, P., & Blau, H. M. (2010). Substrate elasticity regulates skeletal muscle stem cell self-renewal in culture. *Science, 329*(5995), 1078–1081.

Girardi, F., & Le Grand, F. (2018). Wnt signaling in skeletal muscle development and regeneration. In J. Larraín & G. Olivares (Eds.), *Progress in molecular biology and translational science,* Amsterdam: Elsevier Academic Press, *153,* 157–179.

Gnocchi, V. F., White, R. B., Ono, Y., Ellis, J. A., & Zammit, P. S. (2009). Further characterisation of the molecu-lar signature of quiescent and activated mouse muscle satellite cells. *PLoS One, 4*(4), e5205.

Golebiewska, A., Brons, N. H., Bjerkvig, R., & Niclou, S. P. (2011). Critical appraisal of the side population assay in stem cell and cancer stem cell research. *Cell Stem Cell, 8*(2), 136–147.

Greenberg, C. R., Jacobs, H. K., Nylen, E., Rohringer, M., Averill, N., Van Ommen, G. J. B., & Wrogemann, K. (1988). Gene studies in newborn males with Duchenne muscular dystrophy detected by neonatal screening. *The Lancet, 332*(8608), 425–427.

Gros, J., Manceau, M., Thomé, V., & Marcelle, C. (2005). A common somitic origin for embryonic muscle progenitors and satellite cells. *Nature, 435*(7044), 954.

Gussoni, E., Soneoka, Y., Strickland, C. D., Buzney, E. A., Khan, M. K., Flint, A. F., & Mulligan, R. C. (1999). Dystrophin expression in the mdx mouse restored by stem cell transplantation. *Nature, 401*(6751), 390.

Hacker, A., & Guthrie, S. (1998). A distinct developmental programme for the cranial paraxial mesoderm in the chick embryo. *Development, 125*(17), 3461–3472.

Haegel, H., Larue, L., Ohsugi, M., Fedorov, L., Herrenknecht, K., & Kemler, R. (1995). Lack of beta-catenin affects mouse development at gastrulation. *Development, 121*(11), 3529–3537.

Harel, I., Nathan, E., Tirosh-Finkel, L., Zigdon, H., Guimarães-Camboa, N., Evans, S. M., & Tzahor, E. (2009). Distinct origins and genetic programs of head muscle satellite cells. *Developmental Cell, 16*(6), 822–832.

Hasty, P., Bradley, A., Morris, J. H., Edmondson, D. G., Venuti, J. M., Olson, E. N., & Klein, W. H. (1993). Muscle deficiency and neonatal death in mice with a targeted mutation in the myogenin gene. *Nature, 364*(6437), 501.

Helliwell, T. R., Man, N. T., Morris, G. E., & Davies, K. E. (1992). The dystrophin-related protein, utrophin, is expressed on the sarcolemma of regenerating human skeletal muscle fibres in dystrophies and inflammatory myopathies. *Neuromuscular Disorders, 2*, 177–184.

Hirata, H., Hinoda, Y., Nakajima, K., Kawamoto, K., Kikuno, N., Ueno, K., Yamamura, S., Zaman, M. S., Khatri, G., Chen, Y., & Saini, S. (2011). Wnt antagonist DKK1 acts as a tumor suppressor gene that induces apoptosis and inhibits proliferation in human renal cell carcinoma. *International Journal of Cancer, 128*(8), 1793–1803.

Hirsinger, E., Duprez, D., Jouve, C., Malapert, P., Cooke, J., & Pourquié, O. (1997). Noggin acts downstream of Wnt and Sonic Hedgehog to antagonize BMP4 in avian somite patterning. *Development, 124*(22), 4605–4614.

Hosseini, V., Ahadian, S., Ostrovidov, S., Camci-Unal, G., Chen, S., Kaji, H., & Khademhosseini, A. (2012). Engineered contractile skeletal muscle tissue on a microgrooved methacrylated gelatin substrate. *Tissue Engineering Part A, 18*(23–24), 2453–2465.

Huang, R., & Christ, B. (2000). Origin of the epaxial and hypaxial myotome in avian embryos. *Anatomy and Embryology, 202*(5), 369–374.

Huard, J., Labrecque, C., Dansereau, G., Robitaille, L., & Tremblay, J. P. (1991). Dystrophin expression in myotubes formed by the fusion of normal and dystrophic myoblasts. *Muscle & Nerve: Official Journal of the American Association of Electrodiagnostic Medicine, 14*(2), 178–182.

Hutcheson, D. A., Zhao, J., Merrell, A., Haldar, M., & Kardon, G. (2009). Embryonic and fetal limb myogenic cells are derived from developmentally distinct progenitors and have different requirements for β-catenin. *Genes & Development, 23*(8), 997–1013.

Irintchev, A., Zeschnigk, M., Starzinski-Powitz, A., & Wernig, A. (1994). Expression pattern of M-cadherin in normal, denervated, and regenerating mouse muscles. *Developmental Dynamics, 199*(4), 326–337.

Järvinen, T. A., Kääriäinen, M., Äärimaa, V., Järvinen, M., & Kalimo, H. (2008). Skeletal muscle repair after exercise-induced injury. In S. Schiaffino & T. Partridge (Eds.), *Skeletal muscle repair and regeneration* (Vol. 3, pp. 217–242). Dordrecht: Springer.

Kahane, N., Ben-Yair, R., & Kalcheim, C. (2007). Medial pioneer fibers pattern the morphogenesis of early myoblasts derived from the lateral somite. *Developmental Biology, 305*(2), 439–450.

Kalcheim, C., Cinnamon, Y., & Kahane, N. (1999). Myotome formation: A multistage process. *Cell and Tissue Research, 296*(1), 161–173.

Kardon, G., Campbell, J. K., & Tabin, C. J. (2002). Local extrinsic signals determine muscle and endothelial cell fate and patterning in the vertebrate limb. *Developmental Cell, 3*(4), 533–545.

Kassar-Duchossoy, L., Gayraud-Morel, B., Gomès, D., Rocancourt, D., Buckingham, M., Shinin, V., & Tajbakhsh, S. (2004). Mrf4 determines skeletal muscle identity in Myf5: Myod double-mutant mice. *Nature, 431*(7007), 466.

Kassar-Duchossoy, L., Giacone, E., Gayraud-Morel, B., Jory, A., Gomès, D., & Tajbakhsh, S. (2005). Pax3/Pax7 mark a novel population of primitive myogenic cells during development. *Genes & Development, 19*(12), 1426–1431.

Kawabe, Y. I., Wang, Y. X., McKinnell, I. W., Bedford, M. T., & Rudnicki, M. A. (2012). Carm1 regulates Pax7 transcriptional activity through MLL1/2 recruitment during asymmetric satellite stem cell divisions. *Cell Stem Cell, 11*(3), 333–345.

Kim, J. H., Seol, Y. J., Ko, I. K., Kang, H. W., Lee, Y. K., Yoo, J. J., & Lee, S. J. (2018). 3D bioprinted human skeletal muscle constructs for muscle function restoration. *Scientific Reports, 8*, 12307.

Koo, T., & Wood, M. J. (2013). Clinical trials using antisense oligonucleotides in Duchenne muscular dystrophy. *Human Gene Therapy, 24*(5), 479–488.

Kuang, S., Kuroda, K., Le Grand, F., & Rudnicki, M. A. (2007). Asymmetric self-renewal and commitment of satellite stem cells in muscle. *Cell, 129*(5), 999–1010.

Lemos, D. R., Babaeijandaghi, F., Low, M., Chang, C. K., Lee, S. T., Fiore, D., & Rossi, F. M. (2015). Nilotinib reduces muscle fibrosis in chronic muscle injury by promoting TNF-mediated apoptosis of fibro/adipogenic progenitors. *Nature Medicine, 21*(7), 786.

Linker, C., Lesbros, C., Gros, J., Burrus, L. W., Rawls, A., & Marcelle, C. (2005). β-Catenin-dependent Wnt signalling controls the epithelial organisation of somites through the activation of paraxis. *Development, 132*(17), 3895–3905.

Long, C., Amoasii, L., Mireault, A. A., McAnally, J. R., Li, H., Sanchez-Ortiz, E., & Olson, E. N. (2016). Postnatal genome editing partially restores dystrophin expression in a mouse model of muscular dystrophy. *Science, 351*(6271), 400–403.

Mauro, A. (1961). Satellite cell of skeletal muscle fibers. *The Journal of Biophysical and Biochemical Cytology, 9*(2), 493.

Meryon, E. (1851). On fatty degeneration of the voluntary muscles. *Lancet, 2*, 588–589.

Meryon, E. (1852). On granular and fatty degeneration of the voluntary muscles. *Medico-Chirurgical Transactions, 35*, 73.

Miller, K. J., Thaloor, D., Matteson, S., & Pavlath, G. K. (2000). Hepatocyte growth factor affects satellite cell activation and differentiation in regenerating skeletal muscle. *American Journal of Physiology. Cell Physiology, 278*(1), C174–C181.

Minasi, M. G., Riminucci, M., De Angelis, L., Borello, U., Berarducci, B., Innocenzi, A., & Boratto, R. (2002). The meso-angioblast: A multipotent, self-renewing cell that originates from the dorsal aorta and differentiates into most mesodermal tissues. *Development, 129*(11), 2773–2783.

Miranda, A. F., Bonilla, E., Martucci, G., Moraes, C. T., Hays, A. P., & Dimauro, S. (1988). Immunocytochemical study of dystrophin in muscle cultures from patients with Duchenne muscular dystrophy and unaffected control patients. *The American Journal of Pathology, 132*(3), 410.

Mitchell, K. J., Pannérec, A., Cadot, B., Parlakian, A., Besson, V., Gomes, E. R., Marazzi, G., & Sassoon, D. A. (2010). Identification and characterization of a non-satellite cell muscle resident progenitor during postnatal development. *Nature Cell Biology, 12*(3), 257.

Miura, P., & Jasmin, B. J. (2006). Utrophin upregulation for treating Duchenne or Becker muscular dystrophy: How close are we? *Trends in Molecular Medicine, 12*, 122–129.

Moncaut, N., Rigby, P. W., & Carvajal, J. J. (2013). Dial M (RF) for myogenesis. *The FEBS Journal, 280*(17), 3980–3990.

Morris, G. E. (2001). The role of the nuclear envelope in Emery–Dreifuss muscular dystrophy. *Trends in Molecular Medicine, 7*(12), 572–577.

Nagata, Y., Partridge, T. A., Matsuda, R., & Zammit, P. S. (2006). Entry of muscle satellite cells into the cell cycle requires sphingolipid signaling. *The Journal of Cell Biology, 174*(2), 245–253.

Naidu, P. S., Ludolph, D. C., To RQ, Hinterberger, T. J., & Konieczny, S. F. (1995). Myogenin and MEF2 function synergistically to activate the MRF4 promoter during myogenesis. *Molecular and Cellular Biology, 15*(5), 2707–2718.

Negroni, E., Riederer, I., Chaouch, S., Belicchi, M., Razini, P., Di Santo, J., Torrente, Y., Butler-Brown, S. G., & Mouly, V. (2009). In vivo myogenic potential of human CD133+ muscle-derived stem cells: A quantitative study. *Molecular Therapy, 17*(10), 1771–1778.

Nelson, C. E., Hakim, C. H., Ousterout, D. G., Thakore, P. I., Moreb, E. A., Rivera, R. M. C., & Asokan, A. (2016). In vivo genome editing improves muscle function in a mouse model of Duchenne muscular dystrophy. *Science, 351*(6271), 403–407.

Nigro, V., & Savarese, M. (2014). Genetic basis of limb-girdle muscular dystrophies: The 2014 update. *Acta Myologica, 33*(1), 1.

Olguin, H. C., & Olwin, B. B. (2004). Pax-7 up-regulation inhibits myogenesis and cell cycle progression in satellite cells: A potential mechanism for self-renewal. *Developmental Biology, 275*(2), 375–388.

Olguín, H. C., & Pisconti, A. (2012). Marking the tempo for myogenesis: Pax7 and the regulation of muscle stem cell fate decisions. *Journal of Cellular and Molecular Medicine, 16*(5), 1013–1025.

Pannérec, A., Formicola, L., Besson, V., Marazzi, G., & Sassoon, D. A. (2013). Defining skeletal muscle resident progenitors and their cell fate potentials. *Development, 140*(14), 2879–2891.

Péault, B., Rudnicki, M., Torrente, Y., Cossu, G., Tremblay, J. P., Partridge, T., & Huard, J. (2007). Stem and progenitor cells in skeletal muscle development, maintenance, and therapy. *Molecular Therapy, 15*(5), 867–877.

Pichavant, C., Aartsma-Rus, A., Clemens, P. R., Davies, K. E., Dickson, G., Takeda, S. I., & Tremblay, J. P. (2011). Current status of pharmaceutical and genetic therapeutic approaches to treat DMD. *Molecular Therapy, 19*(5), 830–840.

Relaix, F., & Zammit, P. S. (2012). Satellite cells are essential for skeletal muscle regeneration: The cell on the edge returns Centre stage. *Development, 139*(16), 2845–2856.

Relaix, F., Weng, X., Marazzi, G., Yang, E., Copeland, N., Jenkins, N., & Sassoon, D. (1996). Pw1, a novel zinc finger gene implicated in the myogenic and neuronal lineages. *Developmental Biology, 177*(2), 383–396.

Relaix, F., Rocancourt, D., Mansouri, A., & Buckingham, M. (2005). A Pax3/Pax7-dependent population of skeletal muscle progenitor cells. *Nature, 435*(7044), 948.

Relaix, F., Montarras, D., Zaffran, S., Gayraud-Morel, B., Rocancourt, D., Tajbakhsh, S., Mansouri, A., Cumano, A. and Buckingham, M. (2006). Pax3 and Pax7 have distinct and overlapping functions in adult muscle progenitor cells. *Journal of Cell Biology 172*, 91–102.

Sambasivan, R., & Tajbakhsh, S. (2015). Adult skeletal muscle stem cells. In B. Brand-Saberi (Ed.), *Vertebrate myogenesis. Results and problems in cell differentiation 56* (pp. 191–213). Berlin, Heidelberg: Springer.

Sambasivan, R., Gayraud-Morel, B., Dumas, G., Cimper, C., Paisant, S., Kelly, R. G., & Tajbakhsh, S. (2009). Distinct regulatory cascades govern extraocular and pharyngeal arch muscle progenitor cell fates. *Developmental Cell, 16*(6), 810–821.

Sampaolesi, M., Torrente, Y., Innocenzi, A., Tonlorenzi, R., D'Antona, G., Pellegrino, M. A., Barresi, R., Bresolin, N., De Angelis, M. G., Campbell, K. P., Bottinelli, R., & Cossu, G. (2003). Cell therapy of α-sarcoglycan null dystrophic mice through intra-arterial delivery of mesoangioblasts. *Science, 301*(5632), 487–492.

Sampaolesi, M., Blot, S., D'Antona, G., Granger, N., Tonlorenzi, R., Innocenzi, A., Mognol, P., Thibaud, J. L., Galvez, B. G., Barthelemy, I., Perani, L., Mantero, S., Guttinger, M., Pansarasa, O., Rinaldi, C., Cusella De Angelis, M. G., Torrente, Y., Bordignon, C., Bottinelli, R., & Cossu, G. (2006). Mesoangioblast stem cells ameliorate muscle function in dystrophic dogs. *Nature, 444*(7119), 574.

Schmalbruch, H. (1976). The morphology of regeneration of skeletal muscles in the rat. *Tissue & Cell, 8*(4), 673–692.

Schwarzkopf, M., Coletti, D., Sassoon, D., & Marazzi, G. (2006). Muscle cachexia is regulated by a p53–PW1/Peg3-dependent pathway. *Genes & Development, 20*(24), 3440–3452.

Seale, P., Sabourin, L. A., Girgis-Gabardo, A., Mansouri, A., Gruss, P., & Rudnicki, M. A. (2000). Pax7 is required for the specification of myogenic satellite cells. *Cell, 102*(6), 777–786.

Shmelkov, S. V., Clair, R. S., Lyden, D., & Rafii, S. (2005). AC133/CD133/Prominin-1. *The International Journal of Biochemistry & Cell Biology, 37*(4), 715–719.

Sonnet, C., Lafuste, P., Arnold, L., Brigitte, M., Poron, F., Authier, F., Chrétien, F., Gherardi, R. K., & Chazaud, B. (2006). Human macrophages rescue myoblasts and myotubes from apoptosis through a set of adhesion molecular systems. *Journal of Cell Science, 119*(12), 2497–2507.

Sousa-Victor, P., & Munoz-Canoves, P. (2016). Regenerative decline of stem cells in sarcopenia. *Molecular Aspects of Medicine, 50*, 109–117.

Summerbell, D., Halai, C., & Rigby, P. W. (2002). Expression of the myogenic regulatory factor Mrf4 precedes or is contemporaneous with that of Myf5 in the somitic bud. *Mechanisms of Development, 117*(1–2), 331–335.

Suzuki, A., Pelikan, R. C., & Iwata, J. (2015). WNT/β-catenin signaling regulates multiple steps of myogenesis by regulating step-specific targets. *Molecular and Cellular Biology, 35*(10), 1763–1776. https://doi.org/10.1128/MCB.01180-14.

Tabebordbar, M., Zhu, K., Cheng, J. K., Chew, W. L., Widrick, J. J., Yan, W. X., & Cong, L. (2016). In vivo gene editing in dystrophic mouse muscle and muscle stem cells. *Science, 351*(6271), 407–411.

Tajbakhsh, S. (2009). Skeletal muscle stem cells in developmental versus regenerative myogenesis. *Journal of Internal Medicine, 266*(4), 372–389.

Tajbakhsh, S., Rocancourt, D., Cossu, G., & Buckingham, M. (1997). Redefining the genetic hierarchies controlling skeletal myogenesis: Pax-3 and Myf-5 act upstream of MyoD. *Cell, 89*(1), 127–138.

Tanaka, K. K., Hall, J. K., Troy, A. A., Cornelison, D. D. W., Majka, S. M., & Olwin, B. B. (2009). Syndecan-4-expressing muscle progenitor cells in the SP engraft as satellite cells during muscle regeneration. *Cell Stem Cell, 4*(3), 217–225.

Tatsumi, R., Anderson, J. E., Nevoret, C. J., Halevy, O., & Allen, R. E. (1998). HGF/SF is present in normal adult skeletal muscle and is capable of activating satellite cells. *Developmental Biology, 194*(1), 114–128.

Tatsumi, R., Liu, X., Pulido, A., Morales, M., Sakata, T., Dial, S., Hattori, A., Ikeuchi, Y., & Allen, R. E. (2006). Satellite cell activation in stretched skeletal muscle and the role of nitric oxide and hepatocyte growth factor. *American Journal of Physiology-Cell Physiology, 290*(6), C1487–C1494.

Tedesco, F. S., Dellavalle, A., Diaz-Manera, J., Messina, G., & Cossu, G. (2010). Repairing skeletal muscle: Regenerative potential of skeletal muscle stem cells. *The Journal of Clinical Investigation, 120*(1), 11–19.

Tidball, J. G. (2008). Inflammation in skeletal muscle regeneration. In S. Schiaffino & T. Partridge (Eds.), *Skeletal muscle repair and regeneration Vol. 3* (pp. 243–268). Dordrecht: Springer.

Torrente, Y., Belicchi, M., Sampaolesi, M., Pisati, F., Meregalli, M., D'Antona, G., & Pellegrino, M. A. (2004). Human circulating AC133+ stem cells restore dystrophin expression and ameliorate function in dystrophic skeletal muscle. *The Journal of Clinical Investigation, 114*(2), 182–195.

Torrente, Y., Belicchi, M., Marchesi, C., D'antona, G., Cogiamanian, F., Pisati, F., & Lamperti, C. (2007). Autologous transplantation of muscle-derived CD133+ stem cells in Duchenne muscle patients. *Cell Transplantation, 16*(6), 563–577.

Troy, A., Cadwallader, A. B., Fedorov, Y., Tyner, K., Tanaka, K. K., & Olwin, B. B. (2012). Coordination of satellite cell activation and self-renewal by par-complex-dependent asymmetric activation of p38α/β MAPK. *Cell Stem Cell, 11*(4), 541–553.

Tzahor, E. (2015). Head muscle development. In B. Brand-Saberi (Ed.), *Vertebrate Myogenesis. Results and problems in cell differentiation 56* (pp. 123–142). Berlin, Heidelberg: Springer.

Weigmann, A., Corbeil, D., Hellwig, A., & Huttner, W. B. (1997). Prominin, a novel microvilli-specific polytopic membrane protein of the apical surface of epithelial cells, is targeted to plasmalemmal protrusions of non-epithelial cells. *Proceedings of the National Academy of Sciences of the United States of America, 94*(23), 12425–12430.

Weintraub, H., Davis, R., Tapscott, S., Thayer, M., Krause, M., Benezra, R., Blackwell, T. K., Turner, D., Rupp, R., Hollenberg, S., Zhuang, Y., & Lassar, A. (1991). The myoD gene family: Nodal point during specification of the muscle cell lineage. *Science, 251*, 761–766.

Wen, Y., Bi, P., Liu, W., Asakura, A., Keller, C., & Kuang, S. (2012). Constitutive notch activation upregulates Pax7 and promotes the self-renewal of skeletal muscle satellite cells. *Molecular and Cellular Biology, 32*(12), 2300–2311. https://doi.org/10.1128/MCB.06753-11.

Yusuf, F., & Brand-Saberi, B. (2012). Myogenesis and muscle regeneration. *Histochemistry and Cell Biology, 138*(2), 187–199.

Zhou, B. P., Deng, J., Xia, W., Xu, J., Li, Y. M., Gunduz, M., & Hung, M. C. (2004). Dual regulation of snail by GSK-3β-mediated phosphorylation in control of epithelial–mesenchymal transition. *Nature Cell Biology, 6*(10), 931.

Zorn, A. M. (2001). Wnt signalling: Antagonistic Dickkopfs. *Current Biology, 11*(15), R592–R595.

Zouraq, F. A., Stölting, M., & Eberli, D. (2013). Skeletal muscle regeneration for clinical application. In *Regenerative medicine and tissue engineering*, London: InTech Open, 680–712. https://doi.org/10.5772/55739.

Heart Muscle Tissue Engineering

Advantages and Challenges in Cardiac Microtissues

Michelle Coffee, Santoshi Biswanath, Emiliano Bolesani, and Robert Zweigerdt

© Springer Nature Switzerland AG 2020
B. Brand-Saberi (ed.), *Essential Current Concepts in Stem Cell Biology*,
Learning Materials in Biosciences, https://doi.org/10.1007/978-3-030-33923-4_6

What You Will Learn in This Chapter

This chapter is focused on highlighting the efforts that researchers have made to better elucidate the complexities of the human heart. It will explore what we know about the cellular composition of the heart and interactions *in vivo* and how this knowledge has subsequently been used to model the heart *in vitro*, including using stem cells to obtain cardiomyocytes. Building on this, you will learn about current cardiac tissue engineering approaches and how *in vitro* models are constantly being adapted to advance cardiac research and the advantages and challenges associated with it.

6.1 Introduction

The most critical function of the human heart is to pump blood and transport nutrients and oxygen to the entire body – a process requiring approximately 3 billion contractions and resulting in nearly 200 million litres of blood delivery in the average lifespan (Pomeroy et al. 2019). This process is controlled by the four heart chambers and the accompanying valves. It requires that the ventricular cardiac muscle produces forces strong enough to pump blood throughout the entire body down to the distal tips of extremities – and subsequently back to the heart against gravitational force. Within the cardiac tissue structure, cardiomyocytes (CMs) account for the largest volume, but are surrounded by a number of supporting, yet vital non-myocytes. These mainly include blood vessel- and capillary- lining endothelial cells (ECs), blood vessel-supporting smooth muscle cells (SMCs), connective tissue-forming fibroblasts (FBs) as well as tissue-perfusing blood cells and specific neurons. Together, these cells are embedded into an extracellular matrix (ECM), which is vital to ensure the required tissue stability and rigidity of the heart (Kurokawa and George 2016; Kofron and Mende 2017). The orchestrated coordination of the complex contraction pattern of the heart is controlled by intercellular communication and an underlying network of tightly regulated biochemical, electrical and biomechanical cues. These mechanisms must particularly ensure the correct contractility of the pump-force-generating CMs, which form a functional syncytium within the individual heart chambers (Wanjare and Huang 2017).

6.2 Composition, Structure and Function of Native Human Heart Tissue

Communication between the CMs and non-myocytes is critical for cardiac development and function (Kofron and Mende 2017). Furthermore, the non-myocytes are of utmost importance following cardiac injury such as myocardial infarction (MI) which may occur as a consequence of coronary heart vessel occlusion (Zweigerdt 2007). CMs have essentially no proliferation/regeneration potential and hardly respond to apoptosis and necrosis resulting from MI-induced tissue ischemia. In contrast, non-myocytes – and in particular cardiac fibroblasts and blood-born immune cells – alter their phenotype towards proliferation and secretion of enzymes and chemical mediators to reorganise the extracellular matrix. Consequently, in a process known as tissue remodelling, terminally depleted CMs are "replaced" by the formation of an akinetic, stiff fibrotic scar (Gray et al. 2018). This process prevents heart rupture but cannot compensate for the loss of contractile function and may eventually lead to heart failure (Andrée and Zweigerdt 2016).

ECs form a barrier between the myocardium and the perfusing blood while being part of a high-density capillary network penetrating the oxygen- and nutrition-demanding heart tissue. These cells communicate with CMs by having roles in cardiac development, vascular homeostasis and in supporting CM organisation and survival (Gray et al. 2018; Aird 2007). This communication is typically based on the release of specific paracrine or autocrine factors such as nitric oxide (NO), endothelin-1 (ET-1) and Neuregulin (NRG-1). NO is known to be involved in vascular relaxation and has been shown to have cardio-protective effects (Paulus et al. 1994; Jones et al. 2003) whereas ET-1 is an important regulator of cardiac pathophysiology and plays a substantial role in cardiac development (Leucker and Jones 2014). Finally, NRG-1 has been shown to have a vital role in the maintenance of normal cardiac structure and function, due to NRG-1/ErB4 signalling (Slamon et al. 2001; Falls 2003).

FBs are currently considered to be the most abundant cell type within the heart (Souders et al. 2009). Besides being crucial for cardiac development and function, these cells have vital roles in ECM remodelling, electrical coupling and paracrine signalling (Rother et al. 2015). Cardiac FBs are the main producers of collagens, fibronectin, elastin, glycoproteins, proteoglycans and matrix metalloproteinases (MMPs), all of which are found in the ECM (Fan et al. 2012). When directly coupled to CMs, cardiac FBs allow propagation of electrical signals via connexins and can have substantial effects on the electrophysiology of CMs (Camelliti et al. 2004; Yue et al. 2011). They also crosstalk with ECs and macrophages to promote matrix synthesis and angiogenesis (Gray et al. 2018).

6.2.1 Quantitative Assessment of the Cellular Composition of Heart Tissue

Based on the function and importance of non-myocytes, the cardiac cellular composition and interplay can therefore be considered as an essential focus for numerous studies, the most relevant of which have been reviewed by Zhou and Pu (2016). The first studies aiming to identify the cellular composition of the heart were in the late 1970's – early 1980's and analysed samples from both human and rat hearts – specifically papillary muscles and ventricles. These studies found that in the human heart ECs were the most abundant in volume fraction, being nearly two-fold higher than CMs; and FBs being two–threefold less abundant than the CMs (Anversa et al. 1978; Zhou and Pu 2016). Later studies in rats suggested that FBs were in fact more prevalent than previously shown and that all three cell types were present in a nearly 1:1:1 ratio (Anversa et al. 1980). However, most recent studies again suggest that ECs are the most abundant cell type – at least in mouse hearts – which was confirmed by stereology and flow cytometry (Pinto et al. 2016). Finally, when considering the cell composition within the human heart, the most comprehensive study to date was by Bergmann and colleagues (Bergmann et al. 2015), where left ventricular (LV) tissue samples were analysed by means of stereology and flow cytometry, using PCM1, Ulex Europaeus Agglutinin I (UEA1) and the remaining unmarked cells as indicators of CMs, ECs and mesenchymal cell populations respectively. Stereological methods indicated 18% CMs, 24% ECs and 58% mesenchymal cells (used as a more general term for fibroblast-like cells), while flow cytometry suggested 33% CMs, 24% ECs, and 43% mesenchymal cells. Although the two methods yielded different numbers for each cell population, the consensus was with the FB population, which was the most abundant cell population within the human heart if LV tissue samples were to be considered as an adequate representative of the whole organ.

However, to date, the cell composition within the cardiac environment remains an estimate due to the major variations in studies performed so far, coupled with the lack of further studies, appropriate techniques and suitable markers to identify individual cell populations.

6.3 The Unlimited Availability of Human Pluripotent Stem Cell (hPSC)-Derived Cardiomyocytes (hPSC-CMs) Supports Research in Drug Development, Disease Modelling, Heart Regeneration and Heart Development

The damaged human heart has very limited potential for renewal and organ transplantation is therefore currently the standard treatment for acute heart failure. However, shortage of organ supply and side effects of accompanying immunosuppression, which is required to avoid transplant rejection, highlights the strong need for alternative treatments. This has led to the advent of cardiac regenerative therapies, which has centred on human pluripotent stem cells (hPSCs) in recent years. The application of hPSCs, including embryonic and induced pluripotent stem cells (hESCs and hiPSCs), provides promising options for regenerative medicine. At the pluripotent stage, these cells have – in principle – an unlimited expansion and differentiation potential *in vitro*, including the generation of bona fide CMs (Xiu et al. 2009; Weber et al. 2016), ECs (Olmer et al. 2018), blood cell progenies (Ackermann et al. 2018; Eicke et al. 2018) and other functional lineages specific to the heart or any other organ.

With respect to the generation of hPSC-derived cardiomyocytes (hPSC-CM), targeting the WNT and/or bone morphogenetic protein (BMP) signalling pathways (to mimic their role at specific stages of heart development), has been effectively applied to direct hPSCs' cardiomyogenic differentiation (Kempf and Zweigerdt 2017; Kempf et al. 2016a; Kehat et al. 2001; Chen et al. 2012; Lian et al. 2013; Zhang et al. 2012; Burridge et al. 2014). Recent research has also focused on delivering CMs in large quantities, at relatively high purities (Kropp et al. 2016; Kempf et al. 2015) and with a more mature phenotype (Jiang et al. 2018). Finally, if derived from patient-specific hiPSC lines, CMs could be patient-compatible, thereby minimizing their immunogenicity (Batalov and Feinberg 2015; Moran et al. 2014). Together, hPSC-CMs provide a continuous source of cells, which has recently enabled both their routine use for *in vitro* experimentation and substantial progress towards their future application for heart repair *in vivo* (Batalov and Feinberg 2015).

In terms of *in vitro* applications, hPSC-CMs have been used for modelling of cardiac disorders, drug discovery / safety pharmacology and developmental studies. By utilising hiPSCs from specific patients, both monogenic and complex multifactorial disorders have been modelled (Giacomelli et al. 2017a), including cardiac arrhythmia and channelopathies (Egashira et al. 2012; Liang et al. 2013; Ma et al. 2015), cardiomyopathies (Carvajal-Vergara et al. 2010; Caspi et al. 2013; Ma et al. 2013), cardiometabolic diseases (Kawagoe et al. 2013; Hashem et al. 2015; Raval et al. 2015) as well as non-cardiovascular diseases displaying cardiac traits (Lin et al. 2015).

Moreover, it is thought that by using hPSC-CMs for drug screening and safety pharmacology, the limitations of current *in vitro* assays for drug development could be reduced (Matsa and Denning 2012).

At present, these tests still largely rely on primary CMs from animal sources, which have inadequate physiological properties. While healthy human hearts have a beating rate of ~60–70 bpm during a resting state, rodent hearts beat at 250–500 bpm. Additionally, the human Ether-à-go-go-Related Gene (HERG) channel, a crucial K⁺-channel involved in repolarization of the human heart, is non-existent in mouse hearts. This is a considerable limitation since the HERG channel is not only involved in the human long QT syndromes (due to mutations in the HERG gene), but is often inadvertently affected by diverse drugs, making it a point of concern during pharmacological drug development (Sanguinetti and Tristani-Firouzi 2006; Mummery 2018). However, hPSC-CMs have been readily used in a number studies which aimed to model human cardiac arrhythmias and channelopathies, revealing not only expected effects of established drugs, but also aiding in understanding of the underlying disease mechanism (Egashira et al. 2012; Liang et al. 2013; Moretti et al. 2010; Wang et al. 2014).

Importantly, by studying the differentiation process from hPSCs to CMs, mechanisms controlling early human cardiogenesis can be better investigated and deciphered (Gaspari et al. 2018; Konze et al. 2017; Rajala et al. 2011). This knowledge, in parallel to the optimization of cardiogenic differentiation protocols, has supported the identification of pathways playing the most critical roles in cardiac specification, namely the WNT, FGF, BMP and TGFß/activin/NODAL signalling pathways (Matsa and Denning 2012).

These examples highlight that hPSC-CMs are a valuable tool for the understanding of cardiac development, diseases and drug response and, if used in the right environment, can provide unmatched insights into these aspects.

6.4 2D Versus 3D Cell Culture Models

While two dimensional (2D) models have proven to be useful for initial *in vitro* hPSC-CM studies in terms of cell characterisation, developmental studies and disease modelling or drug screening (Liang et al. 2013; Caspi et al. 2009; Matsa et al. 2011; Navarrete et al. 2013), it has become increasingly more evident that cells need to be cultured in a 3D microenvironment in order to more accurately mimic *in vivo* conditions (Griffith and Swartz 2006). Pitfalls of 2D research with regards to hPSC-CMs include, but are not limited to, the immature "fetal-like" phenotype of hPSC-CMs and structural/cellular alignment of CMs (Robertson et al. 2013). While 3D cell culture cannot completely overcome these challenges, studies have shown that it can be an improvement. Multicellular tumour spheroids have been used for more than three decades (Sutherland 1988) and have been shown to demonstrate enhanced sensitivity to chemotherapeutic agents due to altered gene expression in comparison to 2D culture models (Dubessy et al. 2000).

Notably, the first evidence of enhanced function in 3D was as early as 1959, when Moscona demonstrated that 3D aggregations of embryonic chicken CMs showed improved functionality compared to standard 2D culture of the same CMs (Moscona 1959; Hirt et al. 2014).

Generally, the working hypothesis is that the more closely *in vitro* models can recapitulate the *in vivo* (tissue/organ) environment in terms of structure, function and responses, the more the use of animal models could be replaced, while also closing the gap between current *in vitro* models and ultimate *in vivo* applications (Fennema et al. 2013). Contrary to 2D monolayer cell cultures, CMs cultured in 3D are able to retain contractile properties and remain viable for longer (Polonchuk et al. 2017). It has been shown that neonatal rat CMs demonstrated contractile function longer in 3D than in 2D (Chan et al. 2015) and CMs were able to maintain their signature sarcomeric structure for as long as 2 weeks when cultured in a 3D alginate matrix as opposed to a few days when cultured in 2D (Decker et al. 1991). Comparative studies of functionality of hPSC-CMs in 2D and 3D have shown maturation of CMs in 3D, based on the downregulation of the fetal genes natriuretic peptide precursor A & B (NPPA, NPPB) and myosin heavy chain 6 (MYH6), along with enhanced contractile function, including a more negative resting membrane potential and higher action potential (AP) upstrokes velocities (Nunes et al. 2013). However, not many studies have yet directly compared drug screening or disease modelling using hPSC-CMs in 2D and 3D, due to obvious challenges such as the technical limitations between analysis techniques suitable for both formats (Zuppinger 2016).

Building on the advantages and potential of 3D cell culture, the following sections will elaborate on cardiac tissue engineering and microtissues in cardiac research.

6.5 Cardiac Tissue Engineering

Cardiac tissue engineering aims to mimic cardiac tissue by incorporating most of the native environment as discussed above. The principle of this field is to use cells with regenerative capacity in biological or synthetic scaffolding materials, with the purpose of *in vitro* studies or *in vivo* applications (Gálvez-Montón et al. 2013; Ronaldson-Bouchard et al. 2018).

Current methods include hydrogel-based techniques, the use of prefabricated matrices, decellularizing heart tissue or forming cell sheets (Hirt et al. 2014) and studies have shown that constructs rely on strain (Fink et al. 2000; Zimmermann et al. 2006), electrical stimulation (Nunes et al. 2013), incorporation of non-myocytes (Naito 2006; Banerjee et al. 2007) and have displayed spontaneous vascularization (Stevens et al. 2009a) and an enhanced maturation phenotype (Tiburcy et al. 2011). Pioneering work from Eschenhagen and colleagues (Eschenhagen et al. 1997; Zimmermann et al. 2002; Zimmermann and Eschenhagen 2003) cast neonatal rat CMs in a mixture of collagen I and matrix factors into circular molds which subsequently underwent phasic mechanical stretching and finally formed ring-shaped engineered heart tissue (EHT). The EHT exhibited characteristics of differentiated myocardium, comparable to adult rat native heart tissue in terms of highly organised sarcomeres, adherence and gap junctions, a well-developed T-tubule system as well as contractile function and response to inotropic compounds (Zimmermann et al. 2002). Since then these constructs have grown in importance and relevance for cardiac research.

The promise of using engineered heart tissue as advanced *in vitro* models lies within the 3D, native-like structure and function of these constructs and the incorporation of hPSCs into these models have rendered it even more promising.

Engineered vascularized cardiac muscle generated from hESCs were demonstrated by Caspi and colleagues (Caspi et al. 2007) by combining hESC-CMs, hESC-ECs/HUVECS and embryonic FBs in biodegradable scaffolds of porous sponges composed of 50% polylactic-glycolic acid (PLGA) and 50% poly-L-lactic acid (PLLA). The engineered structures were spontaneously contracting and contained endothelial vessel networks, and characterization proved cardiac-specific molecular and functional properties (Caspi et al. 2007). Similarly, enhanced engineered cardiac constructs were generated by embedding hPSC-CMs in a 3D collagen matrix (along with ECs and stromal cells), while subject-ing the constructs to uniaxial mechanical stress. These constructs exhibited increased CM alignment and myofibrillogenesis as well as improved CM proliferation, survival and formation of vessel-like structures (Tulloch et al. 2011). Finally, in a model that aimed to build on the potential of hiPSCs, Masumoto and colleagues (Masumoto et al. 2016) dif-ferentiated and combined CMs, ECs and vascular mural cells (MCs) from hiPSCs and showed engineered cardiac tissue (ECT) with preferential electromechanical properties and improved CM maturity and sarcomere alignment. This supported the engraftment of the ECTs into immune tolerant rat hearts, exhibited graft- and host-derived vasculature and ultimately improved cardiac dysfunction (Masumoto et al. 2016).

Notably, Kensah et al. revealed that bioartificial cardiac tissues (BCTs) can be gener-ated directly from suspension culture-derived cardiac cell aggregates supporting the effi-cient construction of force-developing heart muscle-like structures (Kensah et al. 2013; Kempf et al. 2014).

In an effort to circumvent matrix dependency central to tissue engineering studies, Stevens and colleagues (Stevens et al. 2009a; Stevens et al. 2009b) aimed to generate scaffold-free cardiac patches suitable for implantation into rat hearts. Improvements yielded functionally enhanced vascularised "second-generation" patches consisting of hESC-CMs and -ECs along with FBs, which, once implanted, functionally anastomosed with the host rat heart (Stevens et al. 2009a).

However, due to the size and complexity of these constructs, oxygen and nutrient dif-fusion is limited and a considerable amount of research has been devoted to developing the *in vitro* tissue vascularization aspects (Vunjak-Novakovic et al. 2010) but with limited progress to date (Andrée et al. 2019). Furthermore, imaging and non-invasive analysis techniques of engineered cardiac tissues in real time and for extended periods of time need to be improved (Ly et al. 2008; Haraguchi et al. 2017). In addition to these challenges, engineered constructs are only compatible with drug screening at lower throughput (Eder et al. 2016; Sala et al. 2018) but cannot be used for high throughput screenings (HTS) established in pharmacological companies, thus requesting the application of smaller tis-sue formats (Hirt et al. 2014). Together, this highlights the need for developing "microtissue"-based approaches in the field of cardiac research.

6.6 Microtissues and Their Upcoming Role in Cardiac Research

At the beginning of 2015, a PubMed literature search for "cardiac microtissue" by us deliv-ered 15 results, of which only 9 were in fact about cardiac microtissues, which further-more included various versions of what a "cardiac microtissue" can be defined as. However, an overview of published literature would support the general description of cardiac microtissues to be miniature constructs (smaller than 500 μm) mimicking cardiac tissue in terms of cell composition, cardiomyocyte alignment or a 3D environment. While microtissues (often also referred to as "spheroids" or "aggregates") were not an entirely new concept, there was a sudden surge in interest regarding this field. As mentioned above, 3D cell culture was proving to be a more accurate system for *in vitro* applications, in terms of resembling the *in vivo* conditions. Furthermore, combining the two fields of 3D cell culture and tissue engineering, and incorporating multiple cell types within a con-trolled 3D microenvironment, microtissues were thought to become the pinnacle of *in vitro* cardiac research. Four years later, the same PubMed search delivered 64 results (the

◻ **Table 6.1**   Summary of the most relevant studies regarding cardiac microtissues

Publication (Reference)	Platform	Type of cardiac micro-tissues	Cells used	Duration of culture	Relevance/outcome of study
Kelm et al. (2004)	Scaffold-free (hanging drop)	Mono- & multicel-lular	Neonatal rat & mouse CMs & FBs	3 weeks	Pioneering study for 3D cardiac microtissues
Garzoni et al. (2009)	Scaffold-free (96-well agarose-coated plates)	Mono- & multicel-lular	Murine embryonic CMs, HUVECs, BM-MSCs	<24 days	Angiogenesis & vasculogenesis in cardiac microtissues due to ECs and MSCs
Desrosches et al. (2012)	Scaffold-free (micromolded nonadhesive agarose hydrogels with microcavities)	Mono- & multicel-lular	NRV CMs & CFs	7 days	Advanced 3D CM-CF microtissues
Thavandi-ran et al. (2013)	Scaffold-based (microfabri-cated con-straints with collagen)	Mono- & multicel-lular	hPSC-CMs, hPSC-CD90+ cells	7 days	3D matrix, directed mechanical stress & co-culture improves CM function & maturation
Beauchamp et al. (2015)	Scaffold-free (hanging drop)	Monocel-lular	hiPSC-CMs	4 weeks	Well-developed & functioning hiPSC-CM 3D microtissues
Huebsch et al. (2016)	Scaffold-based (micro-heart muscle (μHM) arrays)	Multicel-lular	hiPSC-CMs and stromal cells	2 weeks	iPSC-CM μHM with uniaxial contractility and alignment
Noguchi et al. (2016)	Scaffold-free (ULA 96-U well plates)	Multicel-lular	RNVCMs, HCMECs, HNDFBs	< 3 days	Generated vascularized cardiac patches for transplanta-tion into rats
Ravenscroft et al. (2016)	Scaffold-free (ULA 96-U well plates)	Mono- & multicel-lular	hiPSC-CMs, hCMECs, hCFs, hDMECs, NhDFs	14–28 days	Multicellular microtissues were more accurate in inotropic drug prediction
Giacomelli et al. (2017b)	Scaffold-free (96-well V-bottom plates)	Mono- & multicel-lular	hPSC-CMs & -ECs	7–20 days	3D cardiac microtissues consisting of hPSC-cells

▣ **Table 6.1** (continued)

Publication (Reference)	Platform	Type of cardiac micro-tissues	Cells used	Duration of culture	Relevance/outcome of study
Pointon et al. (2017); Archer et al. (2018)	Scaffold-free (ULA 384-U well plates)	Multicellular	hiPSC-CMs, hCMECs & hCFs	14–28 days	Improved prediction accuracy of drug-induced changes in CM structure and contractility
Lee et al. (2019)	Scaffold-free (ULA 96-U well plates)	Multicellular	hESC-CMs and -MSCs	2 weeks	Cardiac microtissues for the study of drug-induced cardiac fibrosis

The studies listed below are the most relevant in terms of 3D cardiac microtissues and highlight the differences in platforms, type of cardiac microtissues, duration and relevance/outcome of the study

Abbreviations: *hPSC-CMs* human pluripotent stem cell derived cardiomyocytes, *hPSC-ECs* human pluripotent stem cell derived endothelial cells, *hAMSCs* human amniotic mesenchymal stem cells, *HUVECS* human umbilical vein ECs, *EmFs* embryonic fibroblasts, *BM-MSCs* bone marrow MSCs, *NRV* neonatal rat ventricular, *RNVCMs* rat neonatal ventricular CMs, *HCMECs* human coronary microartery ECs, *HNDFBs* human normal dermal FBs, *hCMECs* human cardiac ECs, *hCFs* human cardiac FBs, *hDMECs* human dermal microvascular ECs, *NhDFs* normal human dermal FBs

most relevant of which are summarised in ▣ Table 6.1), showing a fourfold increase in publications. While this is a substantial number, there have been various limitations hindering the output of publications in this field and the sections below will aim to explore the current state of cardiac microtissues in terms of generation, characterisation and application.

6.6.1 Platforms to Generate Cardiac Microtissues

Platforms that have been used to generate cardiac microtissues can broadly be divided into two categories, namely scaffold-based and scaffold-free. The former is mainly matrix/hydrogel-based (▣ Fig. 6.1d), or uses microcarrier beads (▣ Fig. 6.1e) or micropatterned plates (▣ Fig. 6.1f), while the latter type of platform includes the hanging drop technique (▣ Fig. 6.1a), formation in low attachment plates (▣ Fig. 6.1b), spinner flasks (▣ Fig. 6.1c), magnetic levitation (▣ Fig. 6.1g), magnetic bioprinting (▣ Fig. 6.1h) and microfluidics (▣ Fig. 6.1i) (Duval et al. 2017, Debbie King 2019) (Summarized in ▣ Fig. 6.1). The advantage of scaffold-based approaches is that scaffolds provide an ECM for the cells. In contrast, scaffold-free approaches rely on cells' own (potentially more native-like) ECM formation (Fennema et al. 2013; Zuppinger 2016; Alépée et al. 2014). Furthermore, hydro-

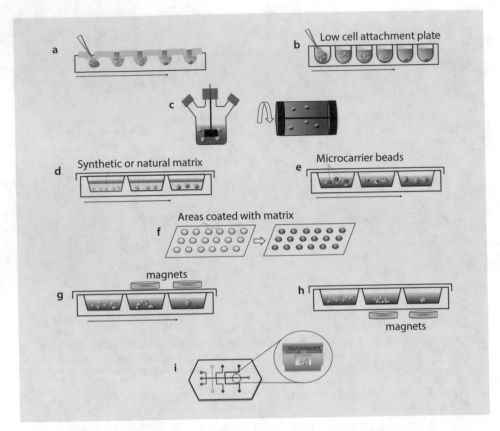

□ **Fig. 6.1** Platforms for microtissue production. Summary of the different platforms for microtissue production, showing scaffold-based and scaffold-free techniques in the form of **a** hanging drop technique, **b** low attachment plates, **c** spinner flasks, **d** matrices, **e** microcarrier beads, **f** micropatterned plates, **g** magnetic levitation, **h** magnetic bioprinting and **i** microfluidic devices. (Figure taken from online publication by The Dish, 2019 (Debbie King 2019))

gel-based scaffolds do not have spatial constraints, thereby delivering heterogenous microtissues of varying sizes, while scaffold-free approaches tend to be highly controlled in this respect (Asthana and Kisaalita 2012). The use of bioreactors for the formation of 3D spheroids or aggregates is beyond the scope of this article but the concept has recently been reviewed elsewhere (Kempf et al. 2016b).

Based on the number of microtissues produced and ease of production, there are "more conventional methods" available in most laboratories versus specialized higher throughput methods (ranging from high to ultra-high throughput). However, it should be mentioned that the term "high throughput" is a broad term used varyingly throughout literature (specifically with regards to microtissues) and it is expected that "true high throughput" production platforms will advance in the near future, due to growing demands. Currently, there are limited methods for high throughput production of CM microtissues and attempts to scale up conventional methods have been to use 384 well plates (Pointon et al. 2017; Archer et al. 2018), micromolded hydrogels with microcavities (Desroches et al. 2012) and 3D high throughput printing (Boyer et al. 2018) allowing for up to 384 and 822 simultaneously produced microtissues per platform, respectively, for the former two methods. Importantly, the latter method was demonstrated by using 96 well plates (with one microtissue per well), however, considering the speed, ease and automation of 3D printing, qualifies this as a high throughput technique. However, researchers are increasingly realising the need for high throughput production of well-developed cardiac microtissues. For instance, the work from Huebsch and colleagues (2016) aimed to combine cardiac spheroids with advance tissue engineering techniques to generate Micro-Heart Muscles (µHM) in microarrays, which was suggested as a suitable foundation for mass production of advanced µHM for future studies due to the small size and ease of fabrication.

While microcavity-based platforms such as Statarrays© MCA (microcavity array) Low Attachment Plates (300MICRONS) and AggreWell™ 400 plates (Stemcell Technologies™) are attractive options for pragmatic higher throughput microtissue production (specifically cardiac microtissues), these platforms have been limited to use for embryoid body (EB) formation of hPSC in general or before cardiac differentiation (to ensure homogenous hPSC embryoid bodies of specific size and therefore a more efficient cardiac differentiation process) or to enrich hPSC-CM populations after cardiac differentiation (Ungrin et al. 2008; Nguyen et al. 2014; Zhang et al. 2015; Budash et al. 2016; Vrij et al. 2016). In fact, the long-term use of such platforms in various fields has not been extensively explored, with microtissues or spheroids being transferred to alternative platforms within 24–48 h (Bratt-Leal et al. 2013; Fey and Wrzesinski 2012; Lim et al. 2011; Lei et al. 2014; Kabiri et al. 2012; Wallace and Reichelt 2013).

Larger scale production is an absolute necessity for thorough *in vitro* modelling or drug screening, in terms of the required repeats and having enough cell material for characterization. Furthermore, substantial amounts of microtissues are required to generate larger constructs as shown in previous studies using up to 14,000 spheroids (consisting of 1000 cells each) to generate scaffold-free cardiac patches (Noguchi et al. 2016). The potential of high throughput production platforms is massively understated, which was evident in a market survey of 3D spheroid culture technology (done by HTStec in 2016 (Comley 2017)), where surveyed participants showed the most awareness toward low-mid throughput production platforms such as hanging drop plates (92%) and 96-well ULA plates (90%) and by far the least awareness of such high throughput production microcavity array platforms (54%), thereby highlighting limited use of these platforms.

6.6.2 The Cardiac Microenvironment – Relevance for *in vitro* Studies

In addition to capturing the 3D state of *in vivo* conditions, the human heart is a complex multicellular organ that cannot simply be represented by CMs only *in vitro* – albeit differentiated from hPSCs. Understandably, all the intricacies of the heart cannot possibly be accounted for in an *in vitro* model, since it would be practically unmanageable and potentially unnecessarily complicated (Kurokawa and George 2016). Regardless, major factors that could influence CMs/cells' biology should be controlled. ☐ Figure 6.2 demonstrates the impact and importance of incorporating multiple cell types within an *in vitro* 3D cardiac model. By incorporating additional relevant cell types, such as ECs and FBs into *in vitro* models, this could enhance paracrine signalling, ECM remodelling and electrophysiological coupling. Furthermore, the CMs could respond with changes in conduction

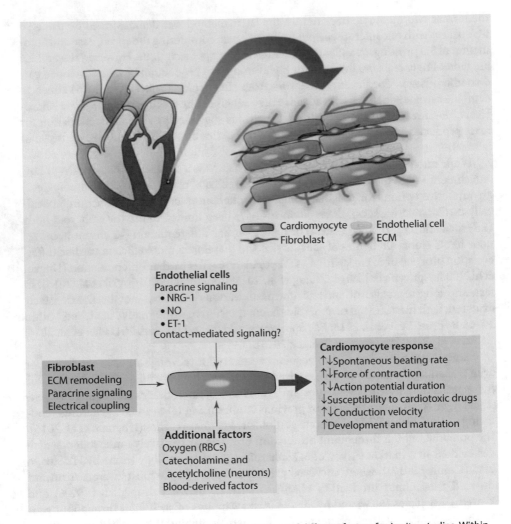

Fig. 6.2 The cardiac microenvironment and relevance of different factors for *in vitro* studies. Within the cardiac microenvironment, CMs are surrounded by ECs, FBs and an ECM. The presence of EC, FBs and additional factors in CM cultures could aid paracrine signalling, ECM remodelling and electrical coupling, thereby causing a number of functional changes in CM behaviour. (Figure taken from Kurokawa and George (2016))

velocity, spontaneous beating rate and force of contractions, with decreased sensitivity to cardiotoxic drugs and increased development and maturation of the CMs.

Considering the crucial role in regulating CM metabolism, survival and contractile function (Narmoneva et al. 2004; Hsieh et al. 2006) ECs might be vital for monitoring CMs' response to drugs *in vitro* or modelling cardiac disease states (Brutsaert 2003; Leucker et al. 2013; Parodi and Kuhn 2014) and is a major component of generating vascularized cardiac tissue constructs. Furthermore, FBs/MSCs have been proven to provide the required ECM for scaffold-free 3D cardiac microtissue models, thereby improving the structural function and survival of the CMs (Desroches et al. 2012; Furtado et al. 2016).

As summarised in ◘ Table 6.1, a number of studies have used multicellular cardiac microtissues to analyse the impact of microtissue formation, behaviour and response to pharmaceutical compounds. Studies incorporating only one additional cell type with CMs, found that FBs caused an increase in collagen production (Kelm et al. 2004) as well as increased expression of integrins and adhesion proteins, (Thavandiran et al. 2013) all of which contribute to the ECM and have been shown to increase cell-cell contacts. These contributions can in turn further the maturation of CMs, as was shown by gene expression analysis of key cardiac maturation markers (Thavandiran et al. 2013). Finally, Desroches and colleagues (Desroches et al. 2012) were able to show that the presence of FBs (specifically cardiac FBs) resulted in a twofold AP prolongation of cardiac microtissues. Multicellular microtissues consisting of CMs and ECs or MSCs demonstrated enhanced vascular structures and angiogenic sprouting within the structure, which supported survival and contractility for at least 24 days (Garzoni et al. 2009).

With the advent of hPSC-based *in vitro* models, it is not surprising that many of the studies have also incorporated hPSC-CMs, -ECs or -CD90+ cells (which could be considered FB-like). Beauchamp and colleagues (Beauchamp et al. 2015) used hiPSC-CMs only, in a study that compared these cells in 2D vs 3D models. They effectively showed that hiPSC-CMs in the form of 3D microtissues exhibited well-developed with homogeneously distributed myofibrils and showed spontaneous contractions, which were temperature sensitive and responsive to electrical pacing and pharmacological compounds. Microtissues consisting of CMs and ECs co-differentiated from hPSCs, resulted in an increase in the shift from myosin light chain 2 (MYL2) to myosin light chain 7 (MYL7) and myosin heavy chain (MYH7) to MYH6, both of which are gene expression patterns associated with maturity of hPSC-CMs (Giacomelli et al. 2017b). Finally, hPSC-based 3D microtissue models comprised of all three cell types showed enhanced contractile function (including spontaneous beating rate and Ca^{2+} transient amplitudes), possibly as a result of enhanced maturity of the CMs. Additionally, the multicellular cardiac microtissues showed more accurate responses to pharmaceutical compounds as opposed to CMs-only microtissues (Ravenscroft et al. 2016; Pointon et al. 2017; Archer et al. 2018). Caspi and colleagues (2007) generated a vascularised engineered cardiac structure, however in this study, they demonstrated that hPSC-EC/HUVEC survival was improved by including embryonic FBs in the cardiac environment, thereby demonstrating the advantage of all three cell types.

As previously discussed, the cell populations' ratios within the heart are still largely debated. Therefore, in an effort to model *in vivo* cell ratios, different proportions of each cell type would also have to be considered for *in vitro* studies. In addition to performing the aforementioned studies in 3D and using multiple cell types, the cell ratios were also shown to impact microtissue performance. Thavandiran and colleagues (2013) showed that 75% CMs and 25% FBs exhibited optimal tissue remodelling dynamics with enhanced structural and functional properties, as opposed to 0%, 50% or 75% FBs in miniaturised cardiac tissue. The ratio of ECs to CMs within microtissues was shown to be best at 15% ECs and 85% CMs. This ratio showed a better EC distribution and organisation in comparison to microtissues with 40% ECs (Giacomelli et al. 2017b). Finally, Noguchi and colleagues (Noguchi et al. 2016) demonstrated that cardiac microtissues consisting of 70% CMs, 15% ECs and 15% FBs were structurally most suitable for larger cardiac patch formation as opposed to 100% CMs or 70% CMs with 30% ECs or FBs. While cell ratios would most likely depend on cell source, platform and application, it highlights the need for further studies to elucidate cell stoichiometry within the human heart, in order to have a relevant comparison and basis for *in vitro* studies.

6.6.3 **Applications of Cardiac Microtissues**

Since the surge in interest in cardiac microtissues, the applications have ranged from *in vitro* drug testing and disease modelling to *in vivo* therapeutic approaches.

It was previously shown that by 2009, the number of post-approval drug withdrawals from the market, due to cardiotoxicity, was a staggering 45% (Ferri et al. 2013). This could be considered among the many reasons why there was a growing interest in alternative *in vitro* models for cardiac research. Indeed, in recent years, studies were able to demonstrate the advance predictive power of 3D cardiac microtissues as compared to conventional monocellular or 2D models, when evaluating drug-induced responses.

By forming scaffold free 3D cardiac microtissues consisting of hiPSC-CMs, human cardiac ECs and human cardiac FBs, studies have been able to prove that such microtissues were capable of predicting the inotropic and non-inotropic effects of drugs better than CM-only microtissues. The convenience of using these microtissues was due to the fact that they were uniform in size, demonstrated contractile functionality and most likely expressed markers of all three cell types. Additionally, due to the small size (~200 μm in diameter) they showed no hypoxic core and were compatible with high throughput screening technologies. Exposure to various inotropic and non-inotropic compounds revealed an 80% increased sensitivity and 91% increased specificity to predict inotropic response and assess cardiac contractility risk (Ravenscroft et al. 2016; Pointon et al. 2017). Furthermore, these microtissues were also proven to be useful in detecting structural changes due to cardiotoxicity after being subjected to various structural and non-structural cardiotoxins (Archer et al. 2018).

In theory, highly complex 3D models would be able to provide valuable insights into diseased states of the heart. In addition to predictive compound testing, 3D cardiac microtissues have also proven to be of use for disease modelling. By combining hESC-CMs with MSCs and focusing on the most common pathway involved in cardiac diseases, namely cardiac fibrosis, Lee and colleagues (2019) successfully established a model to understand cardiac fibrosis and to evaluate the effects of pro-fibrotic agents. MSCs within the cardiac microtissues were able to form FBs and upon exposure to transforming growth factor beta 1 (TGFß1), a primary inducer of fibrosis, underwent transition to myofibroblasts, thereby causing increased collagen deposition and the presence of necrotic CMs – known characteristics of cardiac fibrosis (2019). This study provided a valuable tool for a feature characteristic of most myocardial pathologies and highlighted the promise of 3D cardiac microtissues for *in vitro* disease modelling.

Finally, one key application of microtissues would be as building blocks for larger tissue constructs – macrotissues. The main purpose of this would be to utilize the macrotissues for *in vivo* applications, as was done by Noguchi et al. (2016). Cardiac patches/macrotissues were formed by the self-assembly of multicellular cardiac microtissues consisting of neonatal rat ventricular (NRV) CMs and human ECs and FBs. Upon formation, the resulting vascularised cardiac patches were implanted into rat hearts, and, after 5 days, were shown to be viable with functioning microvascular structures. This study was essential in demonstrating the regenerative capacity of cardiac microtissues.

Moreover, it is well established that the injection of dissociated single cells to the heart results into an extensive, almost immediate loss of the applied cells from the tissue (Feyen et al. 2016; Rojas et al. 2017). Thus, the injection of microtissue, which seems to have a better retention and survival rate in the heart, represent a viable intermediate between single cells on the one hand and more complex engineered tissues on the other hand, thus forming an innovative concept for heart repair which is currently under development (Templin et al. 2012; Technologies for Breakthrough in Heart Therapies: TECHNOBEAT).

6.6.4 Tools & Technologies for the Analysis of Microtissues

2D monolayer cell culture has the advantage of providing unobstructed access to cellular features, for both handling and analysis. Media changes, cell manipulation and the inspection of cells are routine steps in monolayer cell culture, however when adapted to a 3D environment these trivial steps become serious considerations which are challenged due to cells being easily disturbed in suspension or optically hidden in thick layers of cells or matrices (Zuppinger 2016). While techniques developed for larger tissue samples could be applied to 3D cell culture to a certain extent, these techniques are often laborious, time consuming and not suitable for small-to-mid throughput analysis.

Despite the growing trend in microtissue production and application, methods for functional assessment have been limited and are still lagging. Methods must provide a full assessment of tissue response in terms of number of replicates and unique conditions (Chen et al. 2010). Current techniques lean more towards structural characterisation, with most of the efforts dedicated to the development of microscopy techniques (including fluorescence based imaging or histological sections) or low throughput assessment such as RNA extraction (Pampaloni et al. 2007). In fact, in a recent survey of 3D cell culture technologies, consumers indicated that microscopy was the most frequently used tool / method of assessment (listed by 83% of participants), followed by plate-readers and PCR cyclers (Comley 2017). Methods and technologies used the least were next generation sequencers and specialised spheroid imaging devices (gaining 2–4% interest from participants) (Comley 2017), thereby highlighting the need for pragmatic, reliable and routine techniques suitable for a general laboratory setting. Notably, the tools and technologies highlighted by respondents of the survey appeared to have an impact on the type of assessment that they were interested in, with features such as spheroid morphology and viability gaining the most attention, while more advanced assessments such as protein interaction was of least interest (Comley 2017). This is a major hurdle for 3D cell culture, specifically microtissues, since the immense potential is actively being limited due to a lack of suitable tools and technologies.

Many assays cannot be adapted from 2D to 3D culture. For instance, when performing a conventional Lactate Dehydrogenase (LDH) assay, fluorescent signals can linger in a 3D environment even after the chemical reaction has stopped, which makes it difficult to assess cell viability – an assessment which works seamlessly in 2D cultures (Friedrich et al. 2007). Furthermore, confocal microscopy – one of the most relied-on techniques in microtissue analysis – is only effective with structures smaller than 320 μm in diameter (le Roux et al. 2008). Few specialised 3D formats of routine assays have been developed in recent years, such as the CellTiter-Glo® 3D Cell Viability Assay (Promega) as a 3D version of the conventional CellTiter-Glo® Luminescent Cell Viability Assay (Promega) and the Seahorse XFe96 Spheroid Microplates as a 3D compliment to the standard 96 well culture plates. However, these alternatives often require extensive optimization.

Currently, due to the limitations in analytical tools, there is largely a "set" of methods that have been used for characterization and analysis of cardiac microtissues that are in most of the relevant studies to date. These methods can be divided according to structural or functional assessment. Structural analysis techniques include (i) brightfield or fluorescent time-lapse imaging to monitor cell aggregation, (ii) fluorescence based confocal imaging to analyse individual cell populations and microstructures, (iii) live/dead staining followed by fluorescence microscopic assessment of dead and viable cells and (iv) microscopic assessment of stained cryosections or whole microtissues. Functional assessment includes (i) gene expression analysis by means of quantitative RT-PCR, (ii) contractile analysis via video-based edge detection or calcium imaging (which notable also relies on microscopy) and (iii) electrophysiological analysis via patch-clamp and multielectrode arrays.

These limitations in techniques pose a serious problem and highlight one of many challenges related to the microtissue technology in general and in particular in cardiac research.

6.6.5 Challenges and Future Perspective of Current Microtissue-Based Systems

While one of the most promising approaches to *in vitro* studies, there are numerous current shortcomings of microtissue-based systems, including microtissue generation and analysis as readily indicated above. In terms of generation, as previously mentioned, larger quantities of microtissues are required for thorough analysis and application. However, high throughput production in this sense has been limited.

Furthermore, the lack of suitable techniques has created a bottleneck in the advancement of 3D microtissue generation and application. In addition to the limited readouts and resulting data, many researchers are also discouraged to implement 3D culture, due to the challenging nature of their production (requiring the coordinated, large scale availability of respective cell types), thereby further slowing down the advancement of such systems in general laboratory settings (Pampaloni et al. 2007; Celli et al. 2015). By reviewing the studies previously highlighted, the majority of characterization of cardiac microtissues was done by means of conventional microscopy. In the event of a more advanced technique being used, it was often costly or not feasible for routine analysis. This emphasized the void with regards to analytical tools available, thereby prompting the establishment of pragmatic techniques which could routinely be used for analysis of 3D cardiac microtissues. Furthermore, due to the size of microtissues and the cell numbers which are required to generate them, large cell quantities of several different cell types at the same time are inevitably necessary in order to perform thorough tissue production and characterisation. This, in turn, highlights the need for feasible mass production of respective lineages as well as higher throughput production platforms for functional 3D cardiac microtissues.

There is currently a need to improve reproducibility of standardised and validated 3D models, capacity for high throughput analysis and the compatibility of analysis techniques (Edmondson et al. 2014). However, much of the efforts that have been devoted to developing suitable tools and technologies are highly specialised and/or costly (Chen et al. 2010). This not only emphasizes the challenges of microtissue-based systems, but also highlights the topics that require further attention.

6.7 Conclusion

In conclusion, heart muscle tissue engineering, whether in the form of conventional EHT or microtissues, is a crucial yet continuously evolving field that has provided valuable insights into the complexities of the human heart by essentially aiming to mimic it. Novel tools and technologies have incorporated stem cell research and 3D cell culture to produce miniaturised cardiac tissues compatible with high throughput screenings, thereby providing a higher yield of results, which is essential for such a fast-growing area of research. Recent studies have leapt from understanding the fundamentals of cardiac microtissue generation and function to utilizing these constructs for disease modelling and drug screening. It is expected that this trend will increase even faster in the near future, with a strong focus on the translational aspects of these microtissues and unlocking the potential for ultimate cardiac therapy or repair.

Take Home Message

Based on the topics discussed in this chapter, you should now have a comprehensive understanding of the current state of cardiac tissue engineering, with a focus on cardiac microtissues. You have been introduced to the basic concepts which are essential for this field, a brief history of what has been achieved as well as the approaches of current research, with the most important messages being:

1. The human heart is a complex organ consisting of multiple cell types and subsequent interactions, which should be taken into account when aiming to model it.
2. Cardiac tissue engineering is based on capturing and understanding the structural and functional aspects of the human heart in order to elucidate the biology thereof.
3. 3D cell culture aims to imitate the 3D microenvironment reminiscent of the *in vivo* conditions.
4. Cardiac microtissues are miniature constructs (typically <500 μm), which combines approaches from cardiac tissue engineering and 3D cell culture in order to mimic the human heart.
5. Cardiac tissue constructs (EHTs or microtissues) can be produced in a number of platforms, with different cell compositions and has been used to study heart development, diseases and drug effects.
6. This field requires a substantial amount of development; however it holds a great deal of potential as is evident in its fast evolving nature and progress made in recent years. The interdisciplinary nature of the challenges posed by this area of research will stimulate the development of novel tools and technologies.

Acknowledgements RZ received funding from: German Research Foundation (DFG; Cluster of Excellence REBIRTH EXC 62/2, ZW64/4-1, KFO311/ZW64/7-1), German Ministry for Education and Science (BMBF, grants: 13N14086, 01EK1601A, 01EK1602A), the European Union H2020 program to the project TECHNOBEAT (grant 66724).

MC received funding from the Deutscher Akademischer Austauschdienst (DAAD).

References

Ackermann, M., et al. (2018). Bioreactor-based mass production of human iPSC-derived macrophages enables immunotherapies against bacterial airway infections. *Nature Communications, 9*, 5088.

Aird, W. C. (2007). Phenotypic heterogeneity of the endothelium. *Circulation Research, 100*, 174–190.

Alépée, N., et al. (2014). State-of-the-art of 3D cultures (organs-on-a-chip) in safety testing and pathophysiology. *ALTEX, 31*, 441–477.

Andrée, B., & Zweigerdt, R. (2016). Directing cardiomyogenic differentiation and rransdifferentiation by ectopic gene expression – Direct transition or reprogramming detour? *Current Gene Therapy, 16*, 14–20.

Andrée, B., et al. (2019). Formation of three-dimensional tubular endothelial cell networks under defined serum-free cell culture conditions in human collagen hydrogels. *Scientific Reports, 9*, 5437.

Anversa, P., Loud, A. V., Giacomelli, F., & Wiener, J. (1978). Absolute morphometric study of myocardial hypertrophy in experimental hypertension. II. Ultrastructure of myocytes and interstitium. *Laboratory Investigation, 38*, 597–609.

Anversa, P., Olivetti, G., Melissari, M., & Loud, A. V. (1980). Stereological measurement of cellular and subcellular hypertrophy and hyperplasia in the papillary muscle of adult rat. *Journal of Molecular and Cellular Cardiology, 12*, 781–795.

Archer, C. R., et al. (2018). Characterization and validation of a human 3D cardiac microtissue for the assessment of changes in cardiac pathology. *Scientific Reports, 8*, 10160.

Asthana, A., & Kisaalita, W. S. (2012). Microtissue size and hypoxia in HTS with 3D cultures. *Drug Discovery Today, 17*, 810–817.

Banerjee, I., Fuseler, J. W., Price, R. L., Borg, T. K., & Baudino, T. A. (2007). Determination of cell types and numbers during cardiac development in the neonatal and adult rat and mouse. *American Journal of Physiology-Heart and Circulatory Physiology, 293*, H1883–H1891.

Batalov, I., & Feinberg, A. W. (2015). Differentiation of cardiomyocytes from human pluripotent stem cells using monolayer culture. *Biomarker Insights, 10*, 71–76.

Beauchamp, P., et al. (2015). Development and characterization of a scaffold-free 3D spheroid model of induced pluripotent stem cell-derived human cardiomyocytes. *Tissue Engineering. Part C, Methods, 21*, 852–861.

Bergmann, O., et al. (2015). Dynamics of cell generation and turnover in the human heart. *Cell, 161*, 1566–1575.

Boyer, C. J., et al. (2018). High-throughput scaffold-free microtissues through 3D printing. *3D Printing in Medicine, 4*, 9.

Bratt-Leal, A. M., Nguyen, A. H., Hammersmith, K. A., Singh, A., & McDevitt, T. C. (2013). A microparticle approach to morphogen delivery within pluripotent stem cell aggregates. *Biomaterials, 34*, 7227–7235.

Brutsaert, D. L. (2003). Cardiac endothelial-myocardial signaling: Its role in cardiac growth, contractile performance, and rhythmicity. *Physiological Reviews, 83*, 59–115.

Budash, G. V., Bilko, D. I., & Bilko, N. M. (2016). Differentiation of pluripotent stem cells into cardyomyocytes is influenced by size of embryoid bodies. *Biopolymers & Cell, 32*(2), 118–125. https://doi.org/10.7124/bc.000914.

Burridge, P. W., et al. (2014). Chemically defined generation of human cardiomyocytes. *Nature Methods, 11*, 855–860.

Camelliti, P., Green, C. R., LeGrice, I., & Kohl, P. (2004). Fibroblast network in rabbit sinoatrial node. *Circulation Research, 94*, 828–835.

Carvajal-Vergara, X., et al. (2010). Patient-specific induced pluripotent stem-cell-derived models of LEOPARD syndrome. *Nature, 465*, 808–812.

Caspi, O., et al. (2007). Tissue engineering of vascularized cardiac muscle from human embryonic stem cells. *Circulation Research, 100*, 263–272.

Caspi, O., et al. (2009). In vitro electrophysiological drug testing using human embryonic stem cell derived cardiomyocytes. *Stem Cells and Development, 18*, 161–172.

Caspi, O., et al. (2013). Modeling of Arrhythmogenic right ventricular cardiomyopathy with human induced pluripotent stem cells. *Circulation. Cardiovascular Genetics, 6*, 557–568.

Celli, J. P., et al. (2015). An imaging-based platform for high-content, quantitative evaluation of therapeutic response in 3D tumour models. *Scientific Reports, 4*, 3751.

Chan, V., et al. (2015). Fabrication and characterization of optogenetic, multi-strip cardiac muscles. *Lab on a Chip, 15*, 2258–2268.

Chen, A., Underhill, G., & Bhatia, S. (2010). Populational analysis of suspended microtissues for high-throughput, multiplexed 3D tissue engineering. *Integrative Biology (Cambridge), 2*, 517–527.

Chen, V. C., et al. (2012). Scalable GMP compliant suspension culture system for human ES cells. *Stem Cell Research, 8*, 388–402.

Comley, J. (2017). Spheroids: Rapidly becoming a preferred 3D culture format. *Drug Discovery World Spring, 2017*, 31–49.

Debbie King. (2019). *A new dimension of cell culture: The rise of spheroid culture systems.* Retrieved from https://cellculturedish.com/cell-culture-spheroid-culture-systems/.

Decker, M. L., et al. (1991). Cell shape and organization of the contractile apparatus in cultured adult cardiac myocytes. *Journal of Molecular and Cellular Cardiology, 23*, 817–832.

Desroches, B. R., et al. (2012). Functional scaffold-free 3-D cardiac microtissues: A novel model for the investigation of heart cells. *American Journal of Physiology-Heart and Circulatory Physiology, 302*, H2031–H2042.

Dubessy, C., Merlin, J. M., Marchal, C., & Guillemin, F. (2000). Spheroids in radiobiology and photodynamic therapy. *Critical Reviews in Oncology/Hematology, 36*, 179–192.

Duval, K., et al. (2017). Modeling physiological events in 2D vs. 3D cell culture. *Physiology (Bethesda, Md.), 32*, 266–277.

Eder, A., Vollert, I., Hansen, A., & Eschenhagen, T. (2016). Human engineered heart tissue as a model system for drug testing. *Advanced Drug Delivery Reviews, 96*, 214–224.

Edmondson, R., Broglie, J. J., Adcock, A. F., & Yang, L. (2014). Three-dimensional cell culture systems and their applications in drug discovery and cell-based biosensors. *Assay and Drug Development Technologies, 12*, 207–218.

Egashira, T., et al. (2012). Disease characterization using LQTS-specific induced pluripotent stem cells. *Cardiovascular Research, 95*, 419–429.

Eicke, D., et al. (2018). Large-scale production of megakaryocytes in microcarrier-supported stirred suspension bioreactors. *Scientific Reports, 8*, 10146.

Eschenhagen, T., et al. (1997). Three-dimensional reconstitution of embryonic cardiomyocytes in a collagen matrix: A new heart muscle model system. *The FASEB Journal, 11*, 683–694.

Falls, D. L. (2003). Neuregulins: Functions, forms, and signaling strategies. *Experimental Cell Research, 284*, 14–30.

Fan, D., Takawale, A., Lee, J., & Kassiri, Z. (2012). Cardiac fibroblasts, fibrosis and extracellular matrix remodeling in heart disease. *Fibrogenesis & Tissue Repair, 5*, 15.

Fennema, E., Rivron, N., Rouwkema, J., van Blitterswijk, C., & De Boer, J. (2013). Spheroid culture as a tool for creating 3D complex tissues. *Trends in Biotechnology, 31*, 108–115. https://doi.org/10.1016/j.tibtech.2012.12.003.

Ferri, N., et al. (2013). Drug attrition during pre-clinical and clinical development: Understanding and managing drug-induced cardiotoxicity. *Pharmacology & Therapeutics, 138*, 470–484.

Fey, S. J., & Wrzesinski, K. (2012). Determination of drug toxicity using 3D spheroids constructed from an immortal human hepatocyte cell line. *Toxicological Sciences, 127*, 403–411.

Feyen, D. A. M., Gaetani, R., Doevendans, P. A., & Sluijter, J. P. G. (2016). Stem cell-based therapy: Improving myocardial cell delivery. *Advanced Drug Delivery Reviews, 106*, 104–115.

Fink, C., et al. (2000). Chronic stretch of engineered heart tissue induces hypertrophy and functional improvement. *The FASEB Journal, 14*, 669–679.

Friedrich, J., et al. (2007). A reliable tool to determine cell viability in complex 3-D culture: The acid phosphatase assay. *Journal of Biomolecular Screening, 12*, 925–937.

Furtado, M. B., Nim, H. T., Boyd, S. E., & Rosenthal, N. A. (2016). View from the heart: Cardiac fibroblasts in development, scarring and regeneration. *Development, 143*, 387–397.

Gálvez-Montón, C., Prat-Vidal, C., Roura, S., Soler-Botija, C., & Bayes-Genis, A. (2013). Cardiac tissue engineering and the bioartificial heart. *Revista Española de Cardiología (English Edition), 66*, 391–399.

Garzoni, L. R., et al. (2009). Dissecting coronary angiogenesis: 3D co-culture of cardiomyocytes with endothelial or mesenchymal cells. *Experimental Cell Research, 315*, 3406–3418.

Gaspari, E., et al. (2018). Paracrine mechanisms in early differentiation of human pluripotent stem cells: Insights from a mathematical model. *Stem Cell Research, 32*, 1–7.

Giacomelli, E., Mummery, C. L., & Bellin, M. (2017a). Human heart disease: Lessons from human pluripotent stem cell-derived cardiomyocytes. *Cellular and Molecular Life Sciences, 74*, 3711–3739.

Giacomelli, E., et al. (2017b). Three-dimensional cardiac microtissues composed of cardiomyocytes and endothelial cells co-differentiated from human pluripotent stem cells. *Development, 144*, 1008–1017.

Gray, G., Toor, I., Castellan, R., Crisan, M., & Meloni, M. (2018). Resident cells of the myocardium: More than spectators in cardiac injury, repair and regeneration. *Current Opinion in Physiology, 1*, 46–51.

Griffith, L. G., & Swartz, M. A. (2006). Capturing complex 3D tissue physiology in vitro. *Nature Reviews. Molecular Cell Biology, 7*, 211–224.

Haraguchi, Y., et al. (2017). Three-dimensional human cardiac tissue engineered by centrifugation of stacked cell sheets and cross-sectional observation of its synchronous beatings by optical coherence tomography. *BioMed Research International, 5341702*, 2017.

Hashem, S. I., et al. (2015). Brief report: Oxidative stress mediates cardiomyocyte apoptosis in a human model of Danon disease and heart failure. *Stem Cells, 33*, 2343–2350.

Hirt, M. N., Hansen, A., & Eschenhagen, T. (2014). Cardiac tissue engineering. *Circulation Research, 114*, 354–367.

Hsieh, P. C. H., Davis, M. E., Lisowski, L. K., & Lee, R. T. (2006). Endothelial-cardiomyocyte interactions in cardiac development and repair. *Annual Review of Physiology, 68*, 51–66.

Huebsch, N., et al. (2016). Miniaturized iPS-cell-derived cardiac muscles for physiologically relevant drug response analyses. *Scientific Reports, 6*, 1–12.

Jiang, Y., Park, P., Hong, S.-M., & Ban, K. (2018). Maturation of cardiomyocytes derived from human pluripotent stem cells: Current strategies and limitations. *Molecules and Cells, 41*, 613–621.

Jones, S. P., et al. (2003). Endothelial nitric oxide synthase overexpression attenuates congestive heart failure in mice. *Proceedings of the National Academy of Sciences, 100*, 4891–4896.

Kabiri, M., et al. (2012). 3D mesenchymal stem/stromal cell osteogenesis and autocrine signalling. *Biochemical and Biophysical Research Communications, 419*, 142–147.

Kawagoe, S., et al. (2013). Morphological features of iPS cells generated from Fabry disease skin fibroblasts using Sendai virus vector (SeVdp). *Molecular Genetics and Metabolism, 109*, 386–389.

Kehat, I., et al. (2001). Human embryonic stem cells can differentiate into myocytes with structural and functional properties of cardiomyocytes. *The Journal of Clinical Investigation, 108*, 407–414.

Kelm, J. M., et al. (2004). Design of artificial myocardial microtissues. *Tissue Engineering, 10*, 201–214.

Kempf, H., & Zweigerdt, R. (2017). *Scalable Cardiac Differentiation of Pluripotent Stem Cells Using Specific Growth Factors and Small Molecules* (pp. 39–69). Cham: Springer. https://doi.org/10.1007/10_2017_30.

Kempf, H., et al. (2014). Controlling expansion and cardiomyogenic differentiation of human pluripotent stem cells in scalable suspension culture. *Stem Cell Reports, 3*, 1132–1146.

Kempf, H., Kropp, C., Olmer, R., Martin, U., & Zweigerdt, R. (2015). Cardiac differentiation of human pluripotent stem cells in scalable suspension culture. *Nature Protocols, 10*, 1345–1361.

Kempf, H., et al. (2016a). Bulk cell density and Wnt/TGFbeta signalling regulate mesendodermal patterning of human pluripotent stem cells. *Nature Communications, 7*, 13602.

Kempf, H., Andree, B., & Zweigerdt, R. (2016b). Large-scale production of human pluripotent stem cell derived cardiomyocytes. *Advanced Drug Delivery Reviews, 96*, 18–30.

Kensah, G., et al. (2013). Murine and human pluripotent stem cell-derived cardiac bodies form contractile myocardial tissue in vitro. *European Heart Journal, 34*, 1134–1146.

Kofron, C. M., & Mende, U. (2017). In vitro models of the cardiac microenvironment to study myocyte and non-myocyte crosstalk: Bioinspired approaches beyond the polystyrene dish. *Journal of Physiology, 595*, 3891–3905. https://doi.org/10.1113/JP273100.

Konze, S. A., et al. (2017). Proteomic analysis of human pluripotent stem cell cardiomyogenesis revealed altered expression of metabolic enzymes and PDLIM5 isoforms. *Journal of Proteome Research, 16*, 1133–1149.

Kropp, C., et al. (2016). Impact of feeding strategies on the scalable expansion of human pluripotent stem cells in single-use stirred tank bioreactors. *Stem Cells Translational Medicine, 5*, 1289–1301.

Kurokawa, Y. K., & George, S. C. (2016). Tissue engineering the cardiac microenvironment: Multicellular microphysiological systems for drug screening. *Advanced Drug Delivery Reviews, 96*, 225–233.

le Roux, L., et al. (2008). Optimizing imaging of three-dimensional multicellular tumor spheroids with fluorescent reporter proteins using confocal microscopy. *Molecular Imaging, 7*, 214–221.

Lee, M.-O., et al. (2019). Modelling cardiac fibrosis using three-dimensional cardiac microtissues derived from human embryonic stem cells. *Journal of Biological Engineering, 13*, 15.

Lei, J., McLane, L. T., Curtis, J. E., & Temenoff, J. S. (2014). Characterization of a multilayer heparin coating for biomolecule presentation to human mesenchymal stem cell spheroids. *Biomaterials Science, 2*, 666–673.

Leucker, T. M., & Jones, S. P. (2014). Endothelial dysfunction as a nexus for endothelial cell-cardiomyocyte miscommunication. *Frontiers in Physiology, 5*(328): 1–7.

Leucker, T. M., et al. (2013). Impairment of endothelial-myocardial interaction increases the susceptibility of cardiomyocytes to ischemia/reperfusion injury. PLoS One, 8, e70088.

Lian, X., et al. (2013). Directed cardiomyocyte differentiation from human pluripotent stem cells by modulating Wnt/β-catenin signaling under fully defined conditions. Nature Protocols, 8, 162–175.

Liang, P., et al. (2013). Drug screening using a library of human induced pluripotent stem cell–derived cardiomyocytes reveals disease-specific patterns of cardiotoxicity. Circulation, 127, 1677–1691.

Lim, J. J., et al. (2011). Development of nano- and microscale chondroitin sulfate particles for controlled growth factor delivery. Acta Biomaterialia, 7, 986–995.

Lin, B., et al. (2015). Modeling and study of the mechanism of dilated cardiomyopathy using induced pluripotent stem cells derived from individuals with Duchenne muscular dystrophy. Development, 142, e0905–e0905.

Ly, H. Q., Frangioni, J. V., & Hajjar, R. J. (2008). Imaging in cardiac cell-based therapy: In vivo tracking of the biological fate of therapeutic cells. Nature Clinical Practice. Cardiovascular Medicine, 5, S96–S102.

Ma, D., et al. (2013). Modeling type 3 long QT syndrome with cardiomyocytes derived from patient-specific induced pluripotent stem cells. International Journal of Cardiology, 168, 5277–5286.

Ma, D., et al. (2015). Characterization of a novel KCNQ1 mutation for type 1 long QT syndrome and assessment of the therapeutic potential of a novel IKs activator using patient-specific induced pluripotent stem cell-derived cardiomyocytes. Stem Cell Research & Therapy, 6, 39.

Masumoto, H., et al. (2016). The myocardial regenerative potential of three-dimensional engineered cardiac tissues composed of multiple human iPS cell-derived cardiovascular cell lineages. Scientific Reports, 6, 1–10.

Matsa, E., & Denning, C. (2012). In vitro uses of human pluripotent stem cell-derived cardiomyocytes. Journal of Cardiovascular Translational Research, 5, 581–592.

Matsa, E., et al. (2011). Drug evaluation in cardiomyocytes derived from human induced pluripotent stem cells carrying a long QT syndrome type 2 mutation. European Heart Journal, 32, 952–962.

Moran, A. E., et al. (2014). Temporal trends in ischemic heart disease mortality in 21 world regions, 1980 to 2010. Circulation, 129, 1483–1492.

Moretti, A., et al. (2010). Patient-specific induced pluripotent stem-cell models for long-QT syndrome. The New England Journal of Medicine, 363, 1397–1409.

Moscona, A. A. (1959). Tissues from dissociated cells. Scientific American, 200, 132–144.

Mummery, C. L. (2018). Perspectives on the use of human induced pluripotent stem cell-derived cardiomyocytes in biomedical research. Stem Cell Reports, 11, 1306–1311.

Naito, H. (2006). Optimizing engineered heart tissue for therapeutic applications as surrogate heart muscle. Circulation, 114, I-72-I-78.

Narmoneva, D. A., Vukmirovic, R., Davis, M. E., Kamm, R. D., & Lee, R. T. (2004). Endothelial cells promote cardiac myocyte survival and spatial reorganization. Circulation, 110, 962–968.

Navarrete, E. G., et al. (2013). Screening drug-induced arrhythmia [corrected] using human induced pluripotent stem cell-derived cardiomyocytes and low-impedance microelectrode arrays. Circulation, 128, S3–S13.

Nguyen, D. C., et al. (2014). Microscale generation of cardiospheres promotes robust enrichment of cardiomyocytes derived from human pluripotent stem cells. Stem Cell Reports, 3, 260–268.

Noguchi, R., et al. (2016). Development of a three-dimensional pre-vascularized scaffold-free contractile cardiac patch for treating heart disease. The Journal of Heart and Lung Transplantation, 35, 137–145.

Nunes, S. S., et al. (2013). Biowire: A platform for maturation of human pluripotent stem cell–derived cardiomyocytes. Nature Methods, 10, 781–787.

Olmer, R., et al. (2018). Differentiation of human pluripotent stem cells into functional endothelial cells in scalable suspension culture. Stem Cell Reports, 10, 1657–1672.

Pampaloni, F., Reynaud, E. G., & Stelzer, E. H. K. (2007). The third dimension bridges the gap between cell culture and live tissue. Nature Reviews. Molecular Cell Biology, 8, 839–845.

Parodi, E. M., & Kuhn, B. (2014). Editor's choice: Signalling between microvascular endothelium and cardiomyocytes through neuregulin. Cardiovascular Research, 102, 194.

Paulus, W. J., Vantrimpont, P. J., & Shah, A. M. (1994). Acute effects of nitric oxide on left ventricular relaxation and diastolic distensibility in humans. Assessment by bicoronary sodium nitroprusside infusion. Circulation, 89, 2070–2078.

Pinto, A. R., et al. (2016). Revisiting cardiac cellular composition. Circulation Research, 118, 400–409.

Pointon, A., et al. (2017). High-throughput imaging of cardiac microtissues for the assessment of cardiac contraction during drug discovery. Toxicological Sciences, 155, 444–457.

Polonchuk, L., et al. (2017). Cardiac spheroids as promising in vitro models to study the human heart microenvironment. *Scientific Reports, 7,* 1–12.

Pomeroy, J. E., Helfer, A., & Bursac, N. (2019). Biomaterializing the promise of cardiac tissue engineering. *Biotechnology Advances.* https://doi.org/10.1016/j.biotechadv.2019.02.009.

Rajala, K., Pekkanen-Mattila, M., & Aalto-Setälä, K. (2011). Cardiac differentiation of pluripotent stem cells. *Stem Cells International, 2011,* 1–12.

Raval, K. K., et al. (2015). Pompe disease results in a Golgi-based glycosylation deficit in human induced pluripotent stem cell-derived cardiomyocytes. *The Journal of Biological Chemistry, 290,* 3121–3136.

Ravenscroft, S. M., Pointon, A., Williams, A. W., Cross, M. J., & Sidaway, J. E. (2016). Cardiac non-myocyte cells show enhanced pharmacological function suggestive of contractile maturity in stem cell derived cardiomyocyte microtissues. *Toxicological Sciences, 152,* 99–112.

Robertson, C., Tran, D. D., & George, S. C. (2013). Concise review: Maturation phases of human pluripotent stem cell-derived cardiomyocytes. *Stem Cells, 31,* 829–837.

Rojas, S. V., et al. (2017). Transplantation of purified iPSC-derived cardiomyocytes in myocardial infarction. *PLoS One, 12,* e0173222.

Ronaldson-Bouchard, K., et al. (2018). Advanced maturation of human cardiac tissue grown from pluripotent stem cells. *Nature, 556,* 239–243.

Rother, J., et al. (2015). Crosstalk of cardiomyocytes and fibroblasts in co-cultures. *Open Biology, 5,* 150038.

Sala, L., et al. (2018). Musclemotion: A versatile open software tool to quantify cardiomyocyte and cardiac muscle contraction in vitro and in vivo. *Circulation Research, 122,* e5–e16.

Sanguinetti, M. C., & Tristani-Firouzi, M. (2006). hERG potassium channels and cardiac arrhythmia. *Nature, 440,* 463–469.

Slamon, D. J., et al. (2001). Use of chemotherapy plus a monoclonal antibody against HER2 for metastatic breast cancer that overexpresses HER2. *The New England Journal of Medicine, 344,* 783–792.

Souders, C. A., Bowers, S. L. K., & Baudino, T. A. (2009). Cardiac fibroblast: The renaissance cell. *Circulation Research, 105,* 1164–1176.

Stevens, K. R., et al. (2009a). Physiological function and transplantation of scaffold-free and vascularized human cardiac muscle tissue. *Proceedings of the National Academy of Sciences of the United States of America, 106,* 16568–16573.

Stevens, K. R., Pabon, L., Muskheli, V., & Murry, C. E. (2009b). Scaffold-free human cardiac tissue patch created from embryonic stem cells. *Tissue Engineering. Part A, 15,* 1211–1222.

Sutherland, R. M. (1988). Cell and environment interactions in tumor microregions: The multicell spheroid model. *Science, 240,* 177–184.

Technologies for Breakthrough in Heart Therapies: TECHNOBEAT. https://www.mh-hannover.de/technobeat.html

Templin, C., et al. (2012). Transplantation and tracking of human-induced pluripotent stem cells in a pig model of myocardial infarction. *Circulation, 126,* 430–439.

Thavandiran, N., et al. (2013). Design and formulation of functional pluripotent stem cell-derived cardiac microtissues. *Proceedings of the National Academy of Sciences, 110,* E4698–E4707.

Tiburcy, M., et al. (2011). Terminal differentiation, advanced organotypic maturation, and modeling of hypertrophic growth in engineered heart tissue. *Circulation Research, 109,* 1105–1114.

Tulloch, N. L., et al. (2011). Growth of engineered human myocardium with mechanical loading and vascular coculture. *Circulation Research, 109,* 47–59.

Ungrin, M. D., Joshi, C., Nica, A., Bauwens, C., & Zandstra, P. W. (2008). Reproducible, ultra high-throughput formation of multicellular organization from single cell suspension-derived human embryonic stem cell aggregates. *PLoS One, 3,* e1565.

Vrij, E. J., et al. (2016). 3D high throughput screening and profiling of embryoid bodies in thermoformed microwell plates. *Lab on a Chip, 16,* 734–742.

Vunjak-Novakovic, G., et al. (2010). Challenges in cardiac tissue engineering. *Tissue Engineering Part B: Reviews, 16,* 169–187.

Wallace, L., & Reichelt, J. (2013). Using 3D culture to investigate the role of mechanical signaling in keratinocyte stem cells. *Methods in Molecular Biology (Clifton, N.J.), 989,* 153–164.

Wang, G., et al. (2014). Modeling the mitochondrial cardiomyopathy of Barth syndrome with induced pluripotent stem cell and heart-on-chip technologies. *Nature Medicine, 20,* 616–623.

Wanjare, M., & Huang, N. F. (2017). Regulation of the microenvironment for cardiac tissue engineering. *Regenerative Medicine, 12,* 187–201.

Weber, N., et al. (2016). Stiff matrix induces switch to pure β-cardiac myosin heavy chain expression in human ESC-derived cardiomyocytes. *Basic Research in Cardiology, 111*, 68.

Xiu, Q. X., Set, Y. S., Sun, W., & Zweigerdt, R. (2009). Global expression profile of highly enriched cardiomyocytes derived from human embryonic stem cells. *Stem Cells, 27*, 2163–2174.

Yue, L., Xie, J., & Nattel, S. (2011). Molecular determinants of cardiac fibroblast electrical function and therapeutic implications for atrial fibrillation. *Cardiovascular Research, 89*, 744–753.

Zhang, J., et al. (2012). Extracellular matrix promotes highly efficient cardiac differentiation of human pluripotent stem cells: The matrix Sandwich method. *Circulation Research, 111*, 1125–1136.

Zhang, M., et al. (2015). Universal cardiac induction of human pluripotent stem cells in two and three-dimensional formats: Implications for in vitro maturation. *Stem Cells, 33*, 1456–1469.

Zhou, P., & Pu, W. T. (2016). Recounting cardiac cellular composition. *Circulation Research, 118*, 368–370.

Zimmermann, W.-H., & Eschenhagen, T. (2003). Cardiac tissue engineering for replacement therapy. *Heart Failure Reviews, 8*, 259–269.

Zimmermann, W.-H., et al. (2002). Cardiac grafting of engineered heart tissue in syngenic rats. *Circulation, 106*, 1151–1157.

Zimmermann, W.-H., et al. (2006). Engineered heart tissue grafts improve systolic and diastolic function in infarcted rat hearts. *Nature Medicine, 12*, 452–458.

Zuppinger, C. (2016). 3D culture for cardiac cells. *Biochimica et Biophysica Acta – Molecular Cell Research, 1863*, 1873–1881.

Zweigerdt, R. (2007). The art of cobbling a running pump-will human embryonic stem cells mend broken hearts? *Seminars in Cell & Developmental Biology, 18*, 794–804.

Induced Pluripotent Stem Cells

Holm Zaehres

© Springer Nature Switzerland AG 2020
B. Brand-Saberi (ed.), *Essential Current Concepts in Stem Cell Biology*,
Learning Materials in Biosciences, https://doi.org/10.1007/978-3-030-33923-4_7

What You Will Learn in This Chapter

To understand the biology and applications of embryonic or induced pluripotent stem cells, you first have to learn about molecular mechanism which mediate pluripotency. You can then learn how pluripotent stem cells can be induced by gene transfer of transcription factors and which assays are mandatory for their characterization. You will then learn about strategies to allow genome editing of iPS cells and somatic stem cells with CRISPR/Cas9 nuclease systems and finally about applications of these combined stem cell and genome editing techniques.

7.1 Pluripotency and Reprogramming

All cells of the human body develop from a fertilized egg. The developmental potential of a zygote is considered totipotent. After the first cell divisions via the morula stage the blastocyst develops. The inner cell mass of the blastocyst can be extracted by enzymatic digestion and explanted in a tissue culture dish in vitro to grow. Evans and Kaufmann as wells as Martin could establish embryonic stem (ES) cells from mouse blastocysts in 1981 (Evans and Kaufman 1981; Martin 1981). These cells are considered pluripotent, since they cannot develop by themselves into a complete organism but have the potential to differentiate into cells of the three germ layers ectoderm, mesoderm and endoderm as well as germ cells. Thomson et al. could explant the inner cell mass of in-vitro fertilized blastocysts developed from human zygotes and thereby establish human embryonic stem cell lines (Thomson et al. 1998). These human ES cell lines evoked high expectations among biomedical experts and in the general public, since they present a novel model system of natural as well as pathological development and can serve as a source of all cell types of the human body in regenerative medicine applications. The ethical controversy surrounding the use of these cells results from the fact that a form of human life, the early blastocyst, is destroyed, when establishing these cell lines. Thereby the generation of human ES cells is legally forbidden in some countries.

The pluripotency of murine embryonic stem cells is maintained by a comparatively simple network of transcription factors. The central acting factor is Oct4 (POU5F1), which belongs to the group of POU-domain-transcription factors (Schöler et al. 1990; Nichols et al. 1998). Oct4 is regulating its own expression by positive feed-back loops and binds to several promotors as a heterodimer with the transcription factor Sox2 (Yuan et al. 1995). Extensive in vivo studies in mice have confirmed the central role of Oct4 in maintaining pluripotency (Nichols et al. 1998), in which Oct4 acts dose-dependent (Niwa et al. 2000). As another factor to maintain pluripotency Nanog was identified (Chambers et al. 2003; Mitsui et al. 2003). Nanog overexpression stabilizes the pluripotent state (Darr et al. 2006). Oct4, Sox2 and Nanog maintain an autoregulatory network, which stabilizes the pluripotent state and is conserved between embryonic stem cells of mouse and human (Boyer et al. 2005).

Based on these findings it could be assumed, that a somatic cell, in which these pluripotency factors are experimentally induced, could gain pluripotent properties in vitro and in vivo. In such a way 'reprogramming' of somatic cells to a pluripotent state is feasible under certain culture conditions. Gurdon et al. have described almost 60 years ago, that somatic nuclei can be reprogrammed by transfer into enucleated oocytes of the claw frog *Xenopus laevis* (Gurdon 1962; Gurdon et al. 1958). Wilmut et al. achieved the reconstitution of a complete sheep after nuclear transfer of gland epithelial cells ('Dolly') (Wilmut et al. 1997). Afterwards, different groups could describe reprogramming of somatic nuclei after fusion with either murine or human embryonic stem cells (Cowan et al. 2005; Do and Scholer 2004; Tada et al. 2001).

7.2 iPS Cell Generation

In continuation of these lines of research Takahashi and Yamanaka have cloned 24 transcription factors, which are associated with pluripotency of embryonic stem cells into retroviral expression vectors and infected murine fibroblasts (Takahashi and Yamanaka 2006). The cells were further cultivated under culture conditions for embryonic stem cells. Within 3 weeks cell colonies appeared in the culture dish, which resembled mouse embryonic stem cells in morphology; almost indistinguishable in appearance. Thereupon different combinations of the 24 factors were tested in a combinatorial approach and the combination of the four factors Oct4, Sox2, Klf4 and cMyc resulted after infection of fibroblasts in the generation of ES-like colonies. Subsequent characterization demonstrated the high similarity of these *induced, pluripotent stem* (iPS) named cells with ES cells in their molecular signature (gene expression profile, surface marker expression), in their in vitro differentiation potential in cells of all three germ layers as well as in their teratoma formation potential in mice (◻ Fig. 7.1).

In the murine system, embryonic stem cells can be injected after genetic labelling in blastocysts as a further proof of pluripotency: These blastocysts can develop into chimeric mice after uterus transfer, in which the somatic cells as well as germ cells can be partially traced back to the newly injected cells, partially to the cells originating in the blastocyst. Transgenes of the injected cells can be found in the F1 generation of these chimera, when the cells have contributed to the germ line. Yamanaka's as well as Jaenisch's groups could prove germ line transmission of iPS derived cells (Okita et al. 2007; Wernig et al. 2007). Critical for reprogramming somatic cells to a pluripotent status is the activation of the endogenous pluripotency transcription network, which is thereafter maintained by auto-regulatory loops between Oct4, Sox2 and Nanog. At the reprogrammed stage the cells are not depending on the transgenes used for induction. In case of generating iPS cells with retroviral vectors the transgenes are getting 'silenced', which is a result of methylation of the promotor/enhancer regions in the LTR (long terminal repeat) of the retroviral provirus (Cherry et al. 2000; Laker et al. 1998). Thereby demonstration of silencing has to be part of the characterization of a retroviral-induced iPS cell line.

Klf4 and cMyc belong to the group of oncogenes, which function during reprogramming is mainly the immortalization of the somatic cells of origin and the acquisition of a higher proliferation potential (Yamanaka 2007). Oct4 and Sox2 then act to initiate the pluripotency status in the faster proliferating cells and stabilize it later. The function of these transcription is highly conserved between mouse and human (Boyer et al. 2005). Subsequently Takahashi and Yamanaka et al. could reprogram human fibroblasts to cells resembling human ES cells to a high degree with the same factor combination (Takahashi et al. 2007a). In parallel the group of James Thomson achieved the same by lentiviral expression of the pluripotency transcription factors Oct4, Sox2, Nanog and Lin28 in human fibroblasts and further cultivation under human ES culture conditions (Yu et al. 2007). The 'primed' pluripotent state of human iPS cell lines cultured in bFGF conditions (Thomson et al. 1998) corresponds to that of 'post-implantation' derived epiblast stem cells in mice; change of culture conditions (e.g. LIF, GSK3

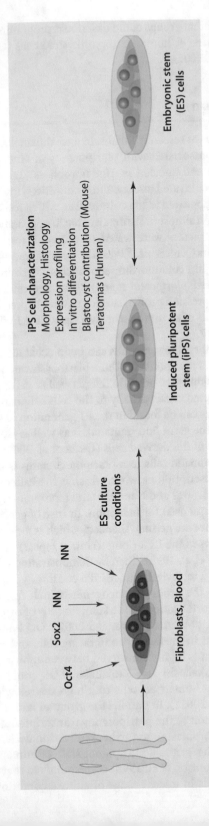

◧ Fig. 7.1   Generation and characterization of induced, pluripotent stem (iPS) cells. iPS cells are generated by gene transfer of pluripotency-associated transcription factors and further cultivation under embryonic stem (ES) cell culture conditions

inhibition) allows the derivation of 'naïve' human iPS cells with similar properties as mouse ES cells (Hanna et al. 2010).

A basic problem of iPS cells generated with retrovirally delivered Oct4, Sox2, Klf4 and cMyc for their use as a disease model system and potentially in cell therapy is, that Klf4 and cMyc are potent oncogenes and that the use of retroviral vector systems per se inherits the risks of insertional mutagenesis (Baum 2007; Hacein-Bey-Abina et al. 2003). Thereby there has been a great interest to renounce from the use of oncogenes during iPS cell generation and to introduce the factors by alternative gene transfer methods. Concerning the reduction of oncogenes, we and others have directed attention to cells to be reprogrammed, which already endogenously express amounts of the reprogramming factors. We were able to demonstrate that ectopic expression of Oct4 alone or of Oct4 together with either Klf4 or cMyc is sufficient to generate iPS cells from mouse and human neural stem cells (NSCs), which endogenously express Sox2, Klf4, and cMyc (Kim et al. 2008, 2009b, c). These one or two-factor iPS cells are similar to embryonic stem cells at the molecular level, can be efficiently differentiated *in vitro*, and are capable of teratoma formation *in vivo*. Mouse NSC iPS cells are also capable of germline contribution and transmission. We have proposed that the number of reprogramming factors required to generate pluripotent stem cells can be reduced when starting with somatic cells that endogenously express appropriate levels of complementing factors.

On a second side the diversity of expression vectors has been brought to use to generate iPS cells. Doxycyline-inducible lentiviral vectors can actively shut off the transferred reprogramming factors as demonstrated in mouse and human (Brambrink et al. 2008; Hockemeyer et al. 2008). Integrating vectors can be configurated for transient expression. Lentiviral expression cassettes with loxP recognition sites in their LTRs can be cut out from the genome by Cre recombinase expression (Soldner et al. 2009). PiggyBac based transposon systems allow the excision of reprogramming factors after successful reprogramming and transient transposase enzyme expression (Woltjen et al. 2009).

Among the viral gene transfer systems which do not integrate actively into the genome, adenoviral vectors were initially introduced to generate iPS cells (Stadtfeld et al. 2008). Expression vectors can be transfected as plasmids, which can further incorporate functions for replication e.g. the EBV (Epstein Barr Virus) origin of replication. These systems are commonly used to generate iPS cells (Okita et al. 2008; Yu et al. 2009). RNA transferring vectors based on Sendai virus are also of high interest for iPS cell generation (Ban et al. 2011; Fusaki et al. 2009). Treatment of murine and human fibroblasts for four to six cycles with Oct, Sox2, Klf4 and cMyc with protein transduction domains (e.g. like the HIV-1 TAT protein) resulted in outgrowth of ES-cell like colonies, which could be further characterized as iPS cells (Kim et al. 2009a; Zhou et al. 2009). The process seems so inefficient or laborious that almost no further studies where based on these protein transductions. Warren et al. have transfected modified RNAs for the generation of human iPS cells (Warren et al. 2010). In principle, small molecule compounds could activate signal transduction cascades, which contribute to reprogramming of target cells. Shi et al. could distinctly increase iPS reprogramming efficiencies after use of the histone-methyl-transferase-inhibitor BIX01294 (Shi et al. 2008). A similar effect was attributed to the histone-deacetylase-inhibitor VPA (valproic acid) (Huangfu et al. 2008). Targeted substitution of singular factors like Sox2 was described by the use of Tgf-ß and GSK3 signal inhibitors (Ichida et al. 2009; Li et al. 2009) as well as of Klf4 by 'kenpaullone' (Lyssiotis et al. 2009). However, there is no convincing report to induce human iPS cells just by activation of signal transduction pathways with small molecules.

Originally starting with fibroblasts and retroviral gene transfer, the iPS cell generation process has been diversified by the use of different starting somatic cell populations in mouse (fibroblasts, blood, liver, hepatocytes, pancreas, neural stem cells) and human (fibroblasts, blood, keratinocytes, neural stem cells), different transcription factor combinations, different gene transfer systems for the reprogramming factors, addition of small chemical compounds, and different iPS cell expansion and characterization conditions.

Step-by Step Protocols

- Induction of pluripotent stem cells with retroviral vectors from fibroblasts is written up as step-by step protocols in Takahashi et al. (2007b) and Park et al. (2008b).
- Induction of pluripotent stem cells with retroviral vectors from neural stem cells is written up as a step-by step protocol in Kim et al. (2009).
- Reprogramming of fibroblasts and blood cells to iPS cells with episomal vectors is written up as step-by step protocols in Okita et al. (2010, 2011b).

7.3 iPS Cell Characterization

The characterization of iPS cells comprises a molecular as well as developmental characterization. Concerning the morphology mouse iPS cells are indistinguishable from mouse ES cells, human iPS cells cannot be set apart from human ES cells. The culture conditions of murine ES cells correspond to the ones of murine ES cells, analogously the ones of human iPS cells correspond to human ES cells. Immunohistochemistry can detect the reactivated pluripotency marker Oct4, Sox2 and Nanog in the nucleus and the cell surface marker SSEA1 (for murine iPS lines) and SSEA3/4 and TRA-1-60/-81 (for human iPS lines). Quantitative RT-PCR analysis can be then further used for validation of these markers. Microarrays or next generation RNA sequencing are instrumental to monitor similarity of pluripotency marker expression on a genome-wide scale. If the iPS cells have been generated with integrating retro/lentiviral vectors, LTR promoter inactivation ('Silencing') should be demonstrated by PCR analysis with primer pairs for virus-specific expression of the reprogramming factors. This correlates with the pluripotent, fully reprogrammed status, whereby the expression of the pluripotency factors completely originates from the endogenous pluripotency loci.

For developmental characterization, the 'embryoid body' (EB) differentiation system is utilized. Thereby pluripotent ES or iPS cells are induced into the three germ layers by ablation of the signal transduction pathways for maintaining of pluripotency (e.g. LIF ('leukemia inhibitory factor') for murine cells and bFGF ('basic fibroblast growth factor') for human cells). Defined media conditions for the three germ layers allow further differentiation. Besides the three germ layers mouse and human pluripotent stem cells can be also differentiated in primordial germ cells (PGCs) and derivatives therefrom (Hübner et al. 2003; Kee et al. 2009; Sugawa et al. 2015; Yamashiro et al. 2018).

Genetically marked iPS cells can contribute to the development of chimaeric embryos and mice after blastocyst injection and also further in the F1 generation, when the iPS cells contributed to the germ line. The most advanced developmental assay is the tetraploid (4N) embryo aggregation, whereby the iPS cells are introduced in a blastocyst, which cells cannot develop further because of their tetraploidy. All diploid cells, which contribute to the development of mice, in this case all cells, must descend from the injected iPS cells. Different laboratories could reconstitute fertile mice completely from iPS cells (Boland et al. 2009; Wu et al. 2011; Zhao et al. 2009).

Another differentiation model is teratoma formation in immunocompromised mice. Teratomas are mixed tumors consisting of cells of all three germ layers originating from pluripotent cells. The teratoma assay is common to test pluripotency of murine and human cells in vivo (Lensch et al. 2007). Hereby histological expertise is essential to distinguish teratomas from e.g. tumors predominantly reconstituted from neuroectodermal cells (Daley et al. 2009).

Cell lines, which are similar to ES cells according to the criteria above are classified as iPS cell lines. Even if all pluripotency criteria are fulfilled, there is variation among iPS cell lines in their molecular signature as well as their differentiation capacity. This phenomenon has been termed 'epigenetic memory' of iPS cells (Kim et al. 2010). A high number of human ES and iPS cell lines were scored in their differentiation potential in the three germ layers in a quantitative EB differentiation assay (Bock et al. 2011). In these studies, certain iPS and ES cell lines demonstrated a preferred differentiation in certain germ layers or cell types. Since all lines were qualified as pluripotent according to their global gene expression profiles, the differences can be attributed to altered reactivation of cell- specific genes throughout the in vitro differentiation. Obviously cell-type specific genes could be reactivated in different pluripotent lines to variable degrees. In this context, global methylation profiles of different iPS lines were created, whereby iPS cell lines from somatic cells of the same origin appeared to be more similar to each other. The epigenetic memory of iPS cells could be critical for the use of iPS cells for disease modelling in vitro as well as concerning their use in potential cell therapies. Several studies have addressed the influence of the donor cell or origin on the differentiation capacity (Dorn et al. 2015; Kim et al. 2011). In this context, we have identified neural signature genes, which were differently expressed between neural progenitor cells differentiated either from neuroectoderm-derived iPSC or mesoderm-derived iPSC in vitro and after transplantation into the cortex of mice (Hargus et al. 2014).

7.4 iPS Cells and Genome Editing

Genome editing has been widely used to introduce sequence-specific alterations in the genome of cells to derive reporter or gene-knockout cell lines or create specific mutations for various purposes. Sequence-specific nuclease systems have been built on fusing artificial zinc finger and TAL effector DNA-binding domains with the nonspecific endonuclease domain of the restriction enzyme FokI (Boch et al. 2009; Hockemeyer et al. 2011; Urnov et al. 2005). Most current approaches entail application of the CRISPR/Cas9 system, which was originally discovered as a RNA-guided endonuclease bacterial immune response to foreign DNA and then further remodeled to allow double strand breaks in specific regions of mammalian genomes (Cong et al. 2013; Jinek et al. 2012; Mali et al. 2013). Previously homologous recombination in human pluripotent stem cells (PSC) has been feasible, albeit with low frequencies of homology directed repair (7 HDR/350 human ES clones (0.02%) in (Zwaka and Thomson 2003). The applications of CRISPR systems have increased the frequency of double strand breaks at specific genomic sites in pluripotent stem cells significantly (Cong et al. 2013; Mali et al. 2013) above 1% and higher. When combined with dual fluorescence selection strategies bi-allelic genome modifications can be achieved in almost all clones (Arias-Fuenzalida et al. 2017; Eggenschwiler et al. 2016).

Mandal et al. extended the application of CRISPR/Cas9 to directly ablate genes in human CD34+ blood stem / progenitor cells and CD4+ T cells (Mandal et al. 2014). We further expanded the utility of these genome editing strategy to neural progenitor cells (NPCs) by demonstrating efficient homology-directed repair (HDR) at the TAU locus from patients with frontotemporal dementia (Hallmann et al. 2017). Homologous recombination in neural progenitor cells (NPCs) after CRISPR/Cas9 vector application was achieved at a frequency of 12% (3 edited/25 clones) in line with the reported high CRISPR/Cas9-directed knockout frequencies ranging from 22% to 40% in human erythroleukemic K562 cells and 27% in human CD34+ stem/progenitor cells (Mandal et al. 2014). Genome editing in NPCs may provide certain advantages over genetic correction in PSCs. Human pluripotent stem cell derived NPCs offer easily, robustly expandable somatic, progenitor populations for neural disease modeling (Ehrlich et al. 2015; Hargus et al. 2014; Koch et al. 2009; Reinhardt et al. 2013). Genome-edited, robustly expandable NPCs offer better homogeneity towards the maturation of final neural cell types then genome-edited iPSCs and the clonal outgrowth after the puromycin selection step can be achieved in higher rates. NPC-directed genome editing expands the applicability of the CRISPR/Cas9 technology in neurodegenerative disease modeling and drug screening.

In principle genome editing of iPS cell derived cells can be efficiently undertaken on the level of pluripotent stem cells, but also on multipotent progenitor cells derived therefrom (◘ Fig. 7.2). For modelling correction of sickle cell anemia (SCA) we could either transfer the CRISPR/Cas9 hemoglobin beta gene targeting vectors and homology-directed repair constructs into SCA iPS cells or first pattern hematopoietic progenitor cells from these iPS cells and apply the genome editing constructs there. In both cases erythrocytes with wildtype hemoglobin beta chains could be differentiated thereafter (Dorn et al. 2015). Similarly, for modelling neural diseases, the genome editing could either take place on the level of iPS cells or neural progenitor cells derived therefrom (Bressan et al. 2017; Hallmann et al. 2017).

Combining iPS cells and genome editing allows us to study the contribution of single-point mutations to the pathogenesis of monogenetic as well as genetically more complex diseases: A disease candidate mutation can be rescued to the wildtype allele in patient iPS cell lines generating a so-called *isogenic control*. If two mutations are suspected to play a role in a certain pathogenesis each of them can be rescued individually to discriminate their contribution. Even multiple mutations can be introduced or rescued in different combinations at a certain gene locus to study the correlation of certain allele types with disease-associated phenotypes (Hockemeyer and Jaenisch 2016; Sterneckert et al. 2014). In all cases chemical compound treatments can be envisaged as part of the genome edited stem cell differentiations to identify new 'druggable' targets.

While human embryonic stem cell differentiations were initially considered as a novel tool to study early developmental processes and as a potential source of cells for allogenic transplantations, human induced pluripotent stem cells have paved the way for comprehensive disease modeling and drug discovery with somatic cells of individual patients (Park et al. 2008a, b; Takahashi et al. 2007a, b; Yu et al. 2007; Zaehres and Schöler 2007). Thereby the largest impact of iPS cell technology over short- to mid-term is on disease modelling and drug screening. Transgenic mice with constitutive, inducible, conditional, and cell type-specific gain-of-function or loss-of-function phenotypes have proven as advanced disease models for decades. Human pluripotent stem cells combined with genome editing will have an even greater impact as model systems for human diseases, as they are genuine human cells. In the years to come directed differentiation protocols for human pluripotent stem cells will be increasingly further developed from 2D cell culture

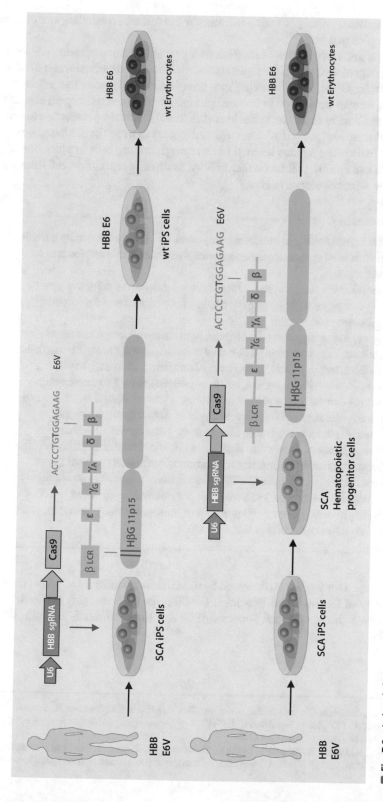

□ **Fig. 7.2** *Induced pluripotent stem cells and genome editing* to model genetic correction of sickle cell anemia (SCA) **a** CRISPR/Cas9 mediated homology-directed repair (HDR) at the hemoglobin beta (HBB) locus on the level of human SCA iPS cells and subsequent differentiation to wildtype (wt) erythrocytes. **b** CRISPR/Cas9 mediated HDR at the HBB locus on the level of human iPS-derived SCA hematopoietic stem/progenitor cells and subsequent differentiation to wildtype (wt) erythrocytes

systems to 3D organoid protocols, which will allow more real true modelling anatomy and physiology of organs in a dish.

In the long-term induced pluripotent stem cells provide a source of stem cell–derived autologous or allogenic cell grafts for regenerative medicine approaches, when produced and qualified under Good Manufacturing Practice (GMP) conditions. The suitability of reprogramming human cord blood to induced pluripotent stem cells (Giorgetti et al. 2009; Haase et al. 2009; Okita et al. 2011a, b; Zaehres et al. 2010) opens the perspective to generate *Human Leukocyte Antigen* (HLA)-matched pluripotent stem cell banks based on existing cord blood banks. The lessons learned from reprogramming cells to pluripotency or multipotency in cell culture will be further applied to directly reprogram cell lineages *in vivo* or to renew stem cell niches *in vivo*.

Take Home Message

1. Pluripotent embryonic and induced, pluripotent stem (iPS) cells can be differentiated in vitro in all ectodermal, mesodermal and endodermal cells as well as germ cells.
2. Patient-derived iPS cells and their differentiation derivatives carry the genetic mutations and potentially epigenetic marks of the patient and can thereby serve as disease models.
3. Induction of pluripotent stem cells can start from different somatic cell populations (e.g. fibroblasts, blood, keratinocytes, neural stem cells) using different transcription factor combinations (e.g. Oct4, Sox2, Nanog, oncogenes), different gene transfer systems (e.g. retro/lentiviral vectors, episomal plasmids, Sendai virus vectors, modified RNA transfection) and further culturing under conditions for mouse and human embryonic stem cells (e.g. naïve or primed state culture conditions).
4. iPS cell characterization embraces morphology, immunohistochemistry for pluripotency markers, expression profiling, in vitro differentiation, blastocyst contribution/transmission (mouse) and teratoma formation (human) assays.
5. Genome editing e.g. using the CRISPR/Cas9 nuclease system can be efficiently used to create or destroy (point) mutations after homology directed repair (HDR) on the level of pluripotent stem cells or multipotent (e.g. hematopoietic, neural) progenitor cells derived therefrom.

Acknowledgments   Our work was supported by financial contributions and grants from the German Ministry of Research and Education (BMBF), the German State of North Rhine Westphalia (NRW), the Max Planck Society (MPG) and Ruhr University Bochum (RUB), Medical Faculty.

References

Arias-Fuenzalida, J., Jarazo, J., Qing, X., Walter, J., Gomez-Giro, G., Nickels, S. L., Zaehres, H., Scholer, H. R., & Schwamborn, J. C. (2017). FACS-assisted CRISPR-Cas9 genome editing facilitates Parkinson's disease modeling. *Stem Cell Reports, 9*, 1423–1431.

Ban, H., Nishishita, N., Fusaki, N., Tabata, T., Saeki, K., Shikamura, M., Takada, N., Inoue, M., Hasegawa, M., Kawamata, S., et al. (2011). Efficient generation of transgene-free human induced pluripotent stem cells (iPSCs) by temperature-sensitive Sendai virus vectors. *Proceedings of the National Academy of Sciences of the United States of America, 108*, 14234–14239.

Baum, C. (2007). Insertional mutagenesis in gene therapy and stem cell biology. *Current Opinion in Hematology, 14*, 337–342.

Boch, J., Scholze, H., Schornack, S., Landgraf, A., Hahn, S., Kay, S., Lahaye, T., Nickstadt, A., & Bonas, U. (2009). Breaking the code of DNA binding specificity of TAL-type III effectors. *Science, 326*, 1509–1512.

Bock, C., Kiskinis, E., Verstappen, G., Gu, H., Boulting, G., Smith, Z. D., Ziller, M., Croft, G. F., Amoroso, M. W., Oakley, D. H., et al. (2011). Reference maps of human ES and iPS cell variation enable high-throughput characterization of pluripotent cell lines. *Cell, 144*, 439–452.

Boland, M. J., Hazen, J. L., Nazor, K. L., Rodriguez, A. R., Gifford, W., Martin, G., Kupriyanov, S., & Baldwin, K. K. (2009). Adult mice generated from induced pluripotent stem cells. *Nature, 461*, 91–94.

Boyer, L. A., Lee, T. I., Cole, M. F., Johnstone, S. E., Levine, S. S., Zucker, J. P., Guenther, M. G., Kumar, R. M., Murray, H. L., Jenner, R. G., et al. (2005). Core transcriptional regulatory circuitry in human embryonic stem cells. *Cell, 122*, 947–956.

Brambrink, T., Foreman, R., Welstead, G. G., Lengner, C. J., Wernig, M., Suh, H., & Jaenisch, R. (2008). Sequential expression of pluripotency markers during direct reprogramming of mouse somatic cells. *Cell Stem Cell, 2*, 151–159.

Bressan, R. B., Dewari, P. S., Kalantzaki, M., Gangoso, E., Matjusaitis, M., Garcia-Diaz, C., Blin, C., Grant, V., Bulstrode, H., Gogolok, S., et al. (2017). Efficient CRISPR/Cas9-assisted gene targeting enables rapid and precise genetic manipulation of mammalian neural stem cells. *Development, 144*, 635–648.

Chambers, I., Colby, D., Robertson, M., Nichols, J., Lee, S., Tweedie, S., & Smith, A. (2003). Functional expression cloning of Nanog, a pluripotency sustaining factor in embryonic stem cells. *Cell, 113*, 643–655.

Cherry, S. R., Biniszkiewicz, D., van Parijs, L., Baltimore, D., & Jaenisch, R. (2000). Retroviral expression in embryonic stem cells and hematopoietic stem cells. *Molecular and Cellular Biology, 20*, 7419–7426.

Cong, L., Ran, F. A., Cox, D., Lin, S., Barretto, R., Habib, N., Hsu, P. D., Wu, X., Jiang, W., Marraffini, L. A., et al. (2013). Multiplex genome engineering using CRISPR/Cas systems. *Science, 339*, 819–823.

Cowan, C. A., Atienza, J., Melton, D. A., & Eggan, K. (2005). Nuclear reprogramming of somatic cells after fusion with human embryonic stem cells. *Science, 309*, 1369–1373.

Daley, G. Q., Lensch, M. W., Jaenisch, R., Meissner, A., Plath, K., & Yamanaka, S. (2009). Broader implications of defining standards for the pluripotency of iPSCs. *Cell Stem Cell, 4*, 200–201; author reply 202.

Darr, H., Mayshar, Y., & Benvenisty, N. (2006). Overexpression of NANOG in human ES cells enables feeder-free growth while inducing primitive ectoderm features. *Development, 133*, 1193–1201.

Do, J. T., & Scholer, H. R. (2004). Nuclei of embryonic stem cells reprogram somatic cells. *Stem Cells, 22*, 941–949.

Dorn, I., Klich, K., Arauzo-Bravo, M. J., Radstaak, M., Santourlidis, S., Ghanjati, F., Radke, T. F., Psathaki, O. E., Hargus, G., Kramer, J., et al. (2015). Erythroid differentiation of human induced pluripotent stem cells is independent of donor cell type of origin. *Haematologica, 100*, 32–41.

Eggenschwiler, R., Moslem, M., Fraguas, M. S., Galla, M., Papp, O., Naujock, M., Fonfara, I., Gensch, I., Wahner, A., Beh-Pajooh, A., et al. (2016). Improved bi-allelic modification of a transcriptionally silent locus in patient-derived iPSC by Cas9 nickase. *Scientific Reports, 6*, 38198.

Ehrlich, M., Hallmann, A. L., Reinhardt, P., Arauzo-Bravo, M. J., Korr, S., Ropke, A., Psathaki, O. E., Ehling, P., Meuth, S. G., Oblak, A. L., et al. (2015). Distinct neurodegenerative changes in an induced pluripotent stem cell model of frontotemporal dementia linked to mutant TAU protein. *Stem Cell Reports, 5*, 83–96.

Evans, M. J., & Kaufman, M. H. (1981). Establishment in culture of pluripotential cells from mouse embryos. *Nature, 292*, 154–156.

Fusaki, N., Ban, H., Nishiyama, A., Saeki, K., & Hasegawa, M. (2009). Efficient induction of transgene-free human pluripotent stem cells using a vector based on Sendai virus, an RNA virus that does not integrate into the host genome. *Proceedings of the Japan Academy. Series B, Physical and Biological Sciences, 85*, 348–362.

Giorgetti, A., Montserrat, N., Aasen, T., Gonzalez, F., Rodriguez-Piza, I., Vassena, R., Raya, A., Boue, S., Barrero, M. J., Corbella, B. A., et al. (2009). Generation of induced pluripotent stem cells from human cord blood using OCT4 and SOX2. *Cell Stem Cell, 5*, 353–357.

Gurdon, J. B. (1962). The developmental capacity of nuclei taken from intestinal epithelium cells of feeding tadpoles. *Journal of Embryology and Experimental Morphology, 10*, 622–640.

Gurdon, J. B., Elsdale, T. R., & Fischberg, M. (1958). Sexually mature individuals of Xenopus laevis from the transplantation of single somatic nuclei. *Nature, 182*, 64–65.

Haase, A., Olmer, R., Schwanke, K., Wunderlich, S., Merkert, S., Hess, C., Zweigerdt, R., Gruh, I., Meyer, J., Wagner, S., et al. (2009). Generation of induced pluripotent stem cells from human cord blood. *Cell Stem Cell, 5,* 434–441.

Hacein-Bey-Abina, S., Von Kalle, C., Schmidt, M., McCormack, M. P., Wulffraat, N., Lebouch, P., Lim, A., Osborne, C. S., Pawliuk, R., Morillon, E., et al. (2003). LMO2-associated clonal T cell proliferation in two patients after gene therapy for SCID-X1. *Science, 302,* 415–419.

Hallmann, A. L., Arauzo-Bravo, M. J., Mavrommatis, L., Ehrlich, M., Ropke, A., Brockhaus, J., Missler, M., Sterneckert, J., Scholer, H. R., Kuhlmann, T., et al. (2017). Astrocyte pathology in a human neural stem cell model of frontotemporal dementia caused by mutant TAU protein. *Scientific Reports, 7,* 42991.

Hanna, J., Cheng, A. W., Saha, K., Kim, J., Lengner, C. J., Soldner, F., Cassady, J. P., Muffat, J., Carey, B. W., & Jaenisch, R. (2010). Human embryonic stem cells with biological and epigenetic characteristics similar to those of mouse ESCs. *Proceedings of the National Academy of Sciences of the United States of America, 107,* 9222–9227.

Hargus, G., Ehrlich, M., Arauzo-Bravo, M. J., Hemmer, K., Hallmann, A. L., Reinhardt, P., Kim, K. P., Adachi, K., Santourlidis, S., Ghanjati, F., et al. (2014). Origin-dependent neural cell identities in differentiated human iPSCs in vitro and after transplantation into the mouse brain. *Cell Reports, 8,* 1697–1703.

Hockemeyer, D., & Jaenisch, R. (2016). Induced pluripotent stem cells meet genome editing. *Cell Stem Cell, 18,* 573–586.

Hockemeyer, D., Soldner, F., Cook, E. G., Gao, Q., Mitalipova, M., & Jaenisch, R. (2008). A drug-inducible system for direct reprogramming of human somatic cells to pluripotency. *Cell Stem Cell, 3,* 346–353.

Hockemeyer, D., Wang, H., Kiani, S., Lai, C. S., Gao, Q., Cassady, J. P., Cost, G. J., Zhang, L., Santiago, Y., Miller, J. C., et al. (2011). Genetic engineering of human pluripotent cells using TALE nucleases. *Nature Biotechnology, 29,* 731–734.

Huangfu, D., Maehr, R., Guo, W., Eijkelenboom, A., Snitow, M., Chen, A. E., & Melton, D. A. (2008). Induction of pluripotent stem cells by defined factors is greatly improved by small-molecule compounds. *Nature Biotechnology, 26,* 795–797.

Hübner, K., Fuhrmann, G., Christenson, L. K., Kehler, J., Reinbold, R., De La Fuente, R., Wood, J., Strauss, J. F., 3rd, Boiani, M., & Schöler, H. R. (2003). Derivation of oocytes from mouse embryonic stem cells. *Science, 300,* 1251–1256.

Ichida, J. K., Blanchard, J., Lam, K., Son, E. Y., Chung, J. E., Egli, D., Loh, K. M., Carter, A. C., Di Giorgio, F. P., Koszka, K., et al. (2009). A small-molecule inhibitor of tgf-Beta signaling replaces sox2 in reprogramming by inducing nanog. *Cell Stem Cell, 5,* 491–503.

Jinek, M., Chylinski, K., Fonfara, I., Hauer, M., Doudna, J. A., & Charpentier, E. (2012). A programmable dual-RNA-guided DNA endonuclease in adaptive bacterial immunity. *Science, 337,* 816–821.

Kee, K., Angeles, V. T., Flores, M., Nguyen, H. N., & Reijo Pera, R. A. (2009). Human DAZL, DAZ and BOULE genes modulate primordial germ-cell and haploid gamete formation. *Nature, 462,* 222–225.

Kim, J.B.*, Zaehres, H.*, Wu, G., Gentile, L., Ko, K., Sebastiano, V., Arauzo-Bravo, M.J., Ruau, D., Han, D.W., Zenke, M., *et al.* (2008). Pluripotent stem cells induced from adult neural stem cells by reprogramming with two factors. Nature 454, 646–650.

Kim, D., Kim, C. H., Moon, J. I., Chung, Y. G., Chang, M. Y., Han, B. S., Ko, S., Yang, E., Cha, K. Y., Lanza, R., et al. (2009a). Generation of human induced pluripotent stem cells by direct delivery of reprogramming proteins. *Cell Stem Cell, 4,* 472–476.

Kim, J. B., Greber, B., Arauzo-Bravo, M. J., Meyer, J., Park, K. I., Zaehres, H., & Scholer, H. R. (2009b). Direct reprogramming of human neural stem cells by OCT4. *Nature, 461,* 649–643.

Kim, J. B., Sebastiano, V., Wu, G., Arauzo-Bravo, M. J., Sasse, P., Gentile, L., Ko, K., Ruau, D., Ehrich, M., van den Boom, D., et al. (2009c). Oct4-induced pluripotency in adult neural stem cells. *Cell, 136,* 411–419.

Kim, K., Doi, A., Wen, B., Ng, K., Zhao, R., Cahan, P., Kim, J., Aryee, M. J., Ji, H., Ehrlich, L. I., et al. (2010). Epigenetic memory in induced pluripotent stem cells. *Nature, 467,* 285–290.

Kim, K., Zhao, R., Doi, A., Ng, K., Unternaehrer, J., Cahan, P., Huo, H., Loh, Y. H., Aryee, M. J., Lensch, M. W., et al. (2011). Donor cell type can influence the epigenome and differentiation potential of human induced pluripotent stem cells. *Nature Biotechnology, 29,* 1117–1119.

Koch, P., Opitz, T., Steinbeck, J. A., Ladewig, J., & Brustle, O. (2009). A rosette-type, self-renewing human ES cell-derived neural stem cell with potential for in vitro instruction and synaptic integration. *Proceedings of the National Academy of Sciences of the United States of America, 106,* 3225–3230.

Laker, C., Meyer, J., Schopen, A., Friel, J., Heberlein, C., Ostertag, W., & Stocking, C. (1998). Host cis-mediated extinction of a retrovirus permissive for expression in embryonal stem cells during differentiation. *Journal of Virology, 72,* 339–348.

Lensch, M. W., Schlaeger, T. M., Zon, L. I., & Daley, G. Q. (2007). Teratoma formation assays with human embryonic stem cells: A rationale for one type of human-animal chimera. *Cell Stem Cell, 1*, 253–258.

Li, W., Zhou, H., Abujarour, R., Zhu, S., Young Joo, J., Lin, T., Hao, E., Scholer, H. R., Hayek, A., & Ding, S. (2009). Generation of human-induced pluripotent stem cells in the absence of exogenous Sox2. *Stem Cells, 27*, 2992–3000.

Lyssiotis, C. A., Foreman, R. K., Staerk, J., Garcia, M., Mathur, D., Markoulaki, S., Hanna, J., Lairson, L. L., Charette, B. D., Bouchez, L. C., et al. (2009). Reprogramming of murine fibroblasts to induced pluripotent stem cells with chemical complementation of Klf4. *Proceedings of the National Academy of Sciences of the United States of America, 106*, 8912–8917.

Mali, P., Yang, L., Esvelt, K. M., Aach, J., Guell, M., DiCarlo, J. E., Norville, J. E., & Church, G. M. (2013). RNA-guided human genome engineering via Cas9. *Science, 339*, 823–826.

Mandal, P. K., Ferreira, L. M., Collins, R., Meissner, T. B., Boutwell, C. L., Friesen, M., Vrbanac, V., Garrison, B. S., Stortchevoi, A., Bryder, D., et al. (2014). Efficient ablation of genes in human hematopoietic stem and effector cells using CRISPR/Cas9. *Cell Stem Cell, 15*, 643–652.

Martin, G. R. (1981). Isolation of a pluripotent cell line from early mouse embryos cultured in medium conditioned by teratocarcinoma stem cells. *Proceedings of the National Academy of Sciences of the United States of America, 78*, 7634–7638.

Mitsui, K., Tokuzawa, Y., Itoh, H., Segawa, K., Murakami, M., Takahashi, K., Maruyama, M., Maeda, M., & Yamanaka, S. (2003). The homeoprotein Nanog is required for maintenance of pluripotency in mouse epiblast and ES cells. *Cell, 113*, 631–642.

Nichols, J., Zevnik, B., Anastassiadis, K., Niwa, H., Klewe-Nebenius, D., Chambers, I., Scholer, H., & Smith, A. (1998). Formation of pluripotent stem cells in the mammalian embryo depends on the POU transcription factor Oct4. *Cell, 95*, 379–391.

Niwa, H., Miyazaki, J., & Smith, A. G. (2000). Quantitative expression of Oct-3/4 defines differentiation, dedifferentiation or self-renewal of ES cells. *Nature Genetics, 24*, 372–376.

Okita, K., Ichisaka, T., & Yamanaka, S. (2007). Generation of germline-competent induced pluripotent stem cells. *Nature, 448*, 313–317.

Okita, K., Nakagawa, M., Hyenjong, H., Ichisaka, T., & Yamanaka, S. (2008). Generation of mouse induced pluripotent stem cells without viral vectors. *Science, 322*, 949–953.

Okita, K., Matsumura, Y., Sato, Y., Okada, A., Morizane, A., Okamoto, S., Hong, H., Nakagawa, M., Tanabe, K., Tezuka, K., et al. (2011a). A more efficient method to generate integration-free human iPS cells. *Nature Methods, 8*, 409–412.

Park, I. H., Arora, N., Huo, H., Maherali, N., Ahfeldt, T., Shimamura, A., Lensch, M. W., Cowan, C., Hochedlinger, K., & Daley, G. Q. (2008a). Disease-specific induced pluripotent stem cells. *Cell, 134*, 877–886.

Reinhardt, P., Glatza, M., Hemmer, K., Tsytsyura, Y., Thiel, C. S., Hoing, S., Moritz, S., Parga, J. A., Wagner, L., Bruder, J. M., et al. (2013). Derivation and expansion using only small molecules of human neural progenitors for neurodegenerative disease modeling. *PLoS One, 8*, e59252.

Schöler, H. R., Ruppert, S., Suzuki, N., Chowdhury, K., & Gruss, P. (1990). New type of POU domain in germ line-specific protein Oct-4. *Nature, 344*, 435–439.

Shi, Y., Do, J. T., Desponts, C., Hahm, H. S., Scholer, H. R., & Ding, S. (2008). A combined chemical and genetic approach for the generation of induced pluripotent stem cells. *Cell Stem Cell, 2*, 525–528.

Soldner, F., Hockemeyer, D., Beard, C., Gao, Q., Bell, G. W., Cook, E. G., Hargus, G., Blak, A., Cooper, O., Mitalipova, M., et al. (2009). Parkinson's disease patient-derived induced pluripotent stem cells free of viral reprogramming factors. *Cell, 136*, 964–977.

Stadtfeld, M., Nagaya, M., Utikal, J., Weir, G., & Hochedlinger, K. (2008). Induced pluripotent stem cells generated without viral integration. *Science, 322*, 945–949.

Sterneckert, J. L., Reinhardt, P., & Scholer, H. R. (2014). Investigating human disease using stem cell models. *Nature Reviews. Genetics, 15*, 625–639.

Sugawa, F., Arauzo-Bravo, M. J., Yoon, J., Kim, K. P., Aramaki, S., Wu, G., Stehling, M., Psathaki, O. E., Hubner, K., & Scholer, H. R. (2015). Human primordial germ cell commitment in vitro associates with a unique PRDM14 expression profile. *The EMBO Journal, 34*, 1009–1024.

Tada, M., Takahama, Y., Abe, K., Nakatsuji, N., & Tada, T. (2001). Nuclear reprogramming of somatic cells by in vitro hybridization with ES cells. *Current Biology, 11*, 1553–1558.

Takahashi, K., & Yamanaka, S. (2006). Induction of pluripotent stem cells from mouse embryonic and adult fibroblast cultures by defined factors. *Cell, 126*, 663–676.

Takahashi, K., Tanabe, K., Ohnuki, M., Narita, M., Ichisaka, T., Tomoda, K., & Yamanaka, S. (2007a). Induction of pluripotent stem cells from adult human fibroblasts by defined factors. *Cell, 131*, 861–872.

Thomson, J. A., Itskovitz-Eldor, J., Shapiro, S. S., Waknitz, M. A., Swiergiel, J. J., Marshall, V. S., & Jones, J. M. (1998). Embryonic stem cell lines derived from human blastocysts. *Science, 282*, 1145–1147.

Urnov, F. D., Miller, J. C., Lee, Y. L., Beausejour, C. M., Rock, J. M., Augustus, S., Jamieson, A. C., Porteus, M. H., Gregory, P. D., & Holmes, M. C. (2005). Highly efficient endogenous human gene correction using designed zinc-finger nucleases. *Nature, 435*, 646–651.

Warren, L., Manos, P. D., Ahfeldt, T., Loh, Y. H., Li, H., Lau, F., Ebina, W., Mandal, P. K., Smith, Z. D., Meissner, A., et al. (2010). Highly efficient reprogramming to pluripotency and directed differentiation of human cells with synthetic modified mRNA. *Cell Stem Cell, 7*, 618–630.

Wernig, M., Meissner, A., Foreman, R., Brambrink, T., Ku, M., Hochedlinger, K., Bernstein, B. E., & Jaenisch, R. (2007). In vitro reprogramming of fibroblasts into a pluripotent ES-cell-like state. *Nature, 448*, 318–324.

Wilmut, I., Schnieke, A. E., McWhir, J., Kind, A. J., & Campbell, K. H. (1997). Viable offspring derived from fetal and adult mammalian cells. *Nature, 385*, 810–813.

Woltjen, K., Michael, I. P., Mohseni, P., Desai, R., Mileikovsky, M., Hamalainen, R., Cowling, R., Wang, W., Liu, P., Gertsenstein, M., et al. (2009). piggyBac transposition reprograms fibroblasts to induced pluripotent stem cells. *Nature, 458*, 766–770.

Wu, G., Liu, N., Rittelmeyer, I., Sharma, A. D., Sgodda, M., Zaehres, H., Bleidissel, M., Greber, B., Gentile, L., Han, D. W., et al. (2011). Generation of healthy mice from gene-corrected disease-specific induced pluripotent stem cells. *PLoS Biology, 9*, e1001099.

Yamanaka, S. (2007). Strategies and new developments in the generation of patient-specific pluripotent stem cells. *Cell Stem Cell, 1*, 39–49.

Yamashiro, C., Sasaki, K., Yabuta, Y., Kojima, Y., Nakamura, T., Okamoto, I., Yokobayashi, S., Murase, Y., Ishikura, Y., Shirane, K., Sasaki, H., Yamamoto, T., & Saitou, M. (2018). Generation of human oogonia from induced pluripotent stem cells in vitro. *Science, 362*, 356–360.

Yu, J., Vodyanik, M. A., Smuga-Otto, K., Antosiewicz-Bourget, J., Frane, J. L., Tian, S., Nie, J., Jonsdottir, G. A., Ruotti, V., Stewart, R., et al. (2007). Induced pluripotent stem cell lines derived from human somatic cells. *Science, 318*, 1917–1920.

Yu, J., Hu, K., Smuga-Otto, K., Tian, S., Stewart, R., Slukvin, I. I., & Thomson, J. A. (2009). Human induced pluripotent stem cells free of vector and transgene sequences. *Science, 324*, 797–801.

Yuan, H., Corbi, N., Basilico, C., & Dailey, L. (1995). Developmental-specific activity of the FGF-4 enhancer requires the synergistic action of Sox2 and Oct-3. *Genes & Development, 9*, 2635–2645.

Zaehres, H., & Schöler, H. R. (2007). Induction of pluripotency: From mouse to human. *Cell, 131*, 834–835.

Zaehres, H., Kogler, G., Arauzo-Bravo, M. J., Bleidissel, M., Santourlidis, S., Weinhold, S., Greber, B., Kim, J. B., Buchheiser, A., Liedtke, S., et al. (2010). Induction of pluripotency in human cord blood unrestricted somatic stem cells. *Experimental Hematology, 38*, 809–818, e801-802.

Zhao, X. Y., Li, W., Lv, Z., Liu, L., Tong, M., Hai, T., Hao, J., Guo, C. L., Ma, Q. W., Wang, L., et al. (2009). iPS cells produce viable mice through tetraploid complementation. *Nature, 461*, 86–90.

Zhou, H., Wu, S., Joo, J. Y., Zhu, S., Han, D. W., Lin, T., Trauger, S., Bien, G., Yao, S., Zhu, Y., et al. (2009). Generation of induced pluripotent stem cells using recombinant proteins. *Cell Stem Cell, 4*, 381–384.

Zwaka, T. P., & Thomson, J. A. (2003). Homologous recombination in human embryonic stem cells. *Nature Biotechnology, 21*, 319–321.

Step-by Step Protocols

Kim, J. B., Zaehres, H., Araúzo-Bravo, M. J., & Schöler, H. R. (2009). Generation of induced pluripotent stem cells from neural stem cells. *Nature Protocols, 4*, 1464–1470.

Okita, K., Hong, H., Takahashi, K., & Yamanaka, S. (2010). Generation of mouse-induced pluripotent stem cells with plasmid vectors. *Nature Protocols, 5*, 418–428.

Okita, K., Matsumura, Y., Sato, Y., Okada, A., Morizane, A., Okamoto, S., Hong, H., Nakagawa, M., Tanabe, K., Tezuka, K., Shibata, T., Kunisada, T., Takahashi, M., Takahashi, J., Saji, H., & Yamanaka, S. (2011b). A more efficient method to generate integration-free human iPS cells. *Nature Methods, 8*, 409–412.

Park, I. H., Lerou, P. H., Zhao, R., Huo, H., & Daley, G. Q. (2008b). Generation of human-induced pluripotent stem cells. *Nature Protocols, 3*, 1180–1186.

Takahashi, K., Okita, K., Nakagawa, M., & Yamanaka, S. (2007b). Induction of pluripotent stem cells from fibroblast cultures. *Nature Protocols, 2*, 3081–3089.

Epigenetics of Somatic Cell Reprogramming

Yixuan Wang, Jianfeng Zhou, and Shaorong Gao

© Springer Nature Switzerland AG 2020
B. Brand-Saberi (ed.), *Essential Current Concepts in Stem Cell Biology*,
Learning Materials in Biosciences, https://doi.org/10.1007/978-3-030-33923-4_8

What You Will Learn in This Chapter
This chapter gives a brief introduction of common epigenetic modifications during somatic reprogramming process (we only focus iPSCs reprogramming here) including histone methylation/acetylation, histone variants substitution, and DNA/RNA modification. In addition, we will discuss how these epigenetic modifications affect the production and quality of iPSCs, and thus obtain more efficient reprogramming methods.

8.1 Introduction

A key point in stem cell research is how to obtain cells with multi-directional differentiation potential for clinical study, which is also the study focus in somatic cell reprogramming research field. So far, there are several approaches to achieve reprogramming, including somatic cell nuclear transfer (SCNT), cell fusion and transcriptional factor-induced pluripotency. Nuclear transfer is to transplant the nucleus of an adult somatic cell to an enucleated oocyte. This *tour de force*, developed around 1960s, is the first to prove that there are some substances in the oocyte cytoplasm that can successfully induce somatic reprogramming (Gurdon 1962). However, due to the complexity of cytoplasmic composition, it is still not clear what kinds of the key factors or the involved mechanisms that lead to reprogramming. Fusion of a somatic cell with an embryonic stem cell (ESC) by electroporation or chemical treatment could also induce somatic cell reprogramming, which highlights the importance of ESC-specific transcriptional factors for genome reprogramming and cell fate determination (Tada et al. 1997). This concept inspired Takahashi and Yamanaka to screen the key pluripotency-related transcriptional factors for somatic cell reprogramming, and finally led to the generation of induced pluripotent stem cells (iPSCs) by enforced over-expression of the four key genes including OCT4, SOX2, KLF4 and c-MYC (Takahashi and Yamanaka 2006). Moreover, the full pluripotency of the iPSCs was proved by generating full-term embryos via tetraploid complementation experiments (Kang et al. 2009; Zhao et al. 2009), suggesting that the iPSCs were similar to ESCs. This exciting discovery not only greatly advances our knowledge of somatic cell reprogramming field, but also provides new ideas and materials for regenerative medicine and personalized clinical treatment.

Both differentiated somatic cells and pluripotent stem cells share essentially the same genome but have distinct morphologies and characteristics, which are shaped by the developmental program. It is well known that epigenetic regulation is a key process in determining the different cell functional output from the genetic information. Comparison of the somatic state with the pluripotent state reveals that the somatic cells show a dense chromatin state (heterochromatin) while most stem cells exhibit an open and loose chromatin state (euchromatin) which is more feasible to accommodate quick changes on transcriptome. These epigenetic modifications include histone methylation/acetylation, histone variants substitution, and DNA/RNA modification. While such epigenetic barriers are gradually built up to sustain specific characteristics of cells during development or differentiation, reprogramming to pluripotency needs to overcome a series of epigenetic barriers. Thus, understanding the dynamics of epigenetic modifications and deciphering the roles of these modifications will help to further optimize the iPSC technique and to generate more qualified iPSCs during reprogramming.

8.2 Histone Modifications and Responsible Modifiers

During the differentiation of stem cells into terminally differentiated cells, some inhibitory histone modifications are gradually established, and chromatin becomes more condensed. In the process of somatic cell reprogramming, the somatic related epigenetic modifications are erased, and new epigenetic features are established, and chromatin becomes decondensed. As we all know, the basic unit of chromatin is nucleosome, which is composed mainly of octamer formed by four histones H2A, H2B, H3, H4 and DNA entangled thereon. Many post-translational modifications (PTMs) occur at the ends of the histones, such as methylation, acetylation, phosphorylation, ubiquitination, SUMOylation, *etc.* These modifications and their interactions are widely involved in the regulation of many important biological events such as gene expression, cell growth and lineage differentiation. Histone modifications serve to recruit other proteins or protein complexes by specific recognition of the modified histones via specific domains, rather than through simply stabilizing or destabilizing the interaction between histones and the underlying DNA. Different kinds of precise recognitions between these modifications and protein complexes constitute the biochemical molecular basis of histone modification-mediated epigenetic regulation, and a variety of enzymes play key roles in the reprogramming process.

Some histone lysine residues can be acetylated at the α-amino group, which neutralizes the charge of lysine and alters the overall electrostatic properties of histones. It weakens the interaction of histones with negatively charged DNA, making the chromatin structure more decondensed. Therefore, Histone acetylation is considered as a marker of active transcription in the euchromatin region, which mainly includes promoters, enhancers and gene bodies. Lysine acetylation has been found to occur on H3 (K4, K9, K14, K18, K23, K27, K36 and K56), H4 (K5, K8, K12, K16, K20 and K91), H2A (K5 and K9) and H2B (K5, K12, K15, K16, K20 and K120). It is catalyzed by lysine acetyltransferases (KATs) and counteracted by histone deacetylases (HDACs), and the interaction between the two histone modifiers determines the level of acetylation in any given genomic regions (Zhou et al. 2011). KATs are broadly classified into three categories: the GNAT family represented by Gcn5, MYST family represented by MOF and p300/CBP family. Higher organisms have more complex HDAC sets that are mainly divided into four categories: Class I, II, III and IV HDACs. In iPSCs and ESCs, chromatin is relatively loose, and histone acetylation level is high; however, in terminally differentiated cells, chromatin is often condensed and the overall acetylation level is low.

Histone lysine acetyltransferase Gcn5 is a critical regulator for early reprogramming initiations. Upon initiation of somatic reprogramming, Gcn5 strongly associates with Myc and form a positive feed-forward loop that activates a distinct alternative splicing network and coregulates a group of RNA splicing and RNA processing genes that are needed for somatic cell reprogramming (Hirsch et al. 2015).

MOF, a specific H4K16 acetyltransferase, is an integral part of the ESC core transcriptional network and plays an important role in maintaining ESC self-renewal and pluripotency. Deletion of MOF results in abnormal expression of NANOG, OCT4 and SOX2 in ESCs, thus leading to loss of characteristic clonal morphology, alkaline phosphatase (AP) staining, and differentiation potential. In addition, MOF is a key factor in effective reprogramming. iPSCs expresses high levels of MOF and this expression is significantly up-regulated with reprogramming (Li et al. 2012). p300 is shown to promote acetylation of

OCT4, SOX2, and KLF4 at multiple sites to change their transcription activity, thus regulating stem cell reprogramming (Dai et al. 2014). p300 and CBP can be recruited to the NANOG locus to help maintain the undifferentiated state of ESCs (Fang et al. 2014).

A large number of natural and synthetic HDAC inhibitors (HDACi) have shown outstanding utility in the ESCs differentiation pathway, which may be caused by the remodeling of ESCs chromatin. Thus, HDACi can reverse the epigenetic features that characterize genes involved in self-renewal or differentiation regulation. HDACi can be used in somatic cell reprogramming process via different approaches. Treatment of donor cells before nuclear transfer or treatment of cloned embryos following nuclear transfer resulted in facilitation of embryo cloning and improvement of embryo developmental potential. In mouse SCNT experiments, HDACi significantly increased the efficiency and quality of cloned embryos by promoting the activation of embryo-specific genes (Bui et al. 2011). In fusion experiments, HDACi treatment increased the capacity of ESCs with a relatively low acetylation level to achieve reprogramming of MEFs (Hezroni et al. 2011). In the transcriptional factor-induced reprogramming experiment of human fibroblasts, HDACi treatment can significantly increase the efficiency of iPSC production and replace two of the Yamanaka four factors - c-MYC and KLF4 (Huangfu et al. 2008a, b; Mali et al. 2010). These effects are mainly due to enhanced histone acetylation, chromatin decondensation, increased RNA synthesis, and inhibition of apoptosis. However, the specificity or selectivity of HDACi and the involved molecular mechanism of the corresponding complex are still unclear.

In addition to histone acetylation, methylation of histone lysine, arginine and other sites is also common for histone modifications. Histone methylation modifiers mainly include histone lysine methyltransferase (HKMT), protein arginine methyltransferase (PRMT), and histone lysine demethylase (HKDM). There are six families in HKMT: KMT1 family catalyzes H3K9me3; KMT2 family catalyzes H3K4me3; KMT3 family catalyzes H3K36me3; KMT4 family catalyzes H3K79me3; KMT5 family catalyzes H4K20me3; KMT6 family catalyzes H3K27me3. All of above contain SET domain to perform catalytic function; except for methylation of H3K79 by DOT1L. Not only methylation at different sites exercise different biological functions, different degrees of methylation at the same site also produce different biological effects. For example, a mono-methylation modification of H3K4 (H3K4me1) can be used to label an enhancer region, whereas a tri-methylation modification at the same position (H3K4me3) is present at the transcription start site.

Some H3K9me3 regions, in addition to hindering the rate or efficiency of reprogramming, continue to affect the final reprogrammed iPSC state, resulting in incomplete ESC-like state transitions. The H3K9me3 heterochromatin region is considered to be a major obstacle in the early stage of somatic cell reprogramming, the removal of which may be an effective strategy to improve the reprogramming efficiency. Indeed, knockdown of SUV39H1/H2 methyltransferase results in a decrease in H3K9me3 level, which in turn accelerates the reprogramming process and increases the number of human iPSCs colonies (Soufi et al. 2012). Also, depletion of other main H3K9 methyltransferases including EHMT1, EHMT2, and SETDB1 in fibroblast cells increases the efficiency of iPSC formation from both fibroblasts and pre-iPSCs (Onder et al. 2012; Sridharan et al. 2013). However, it is still unclear which methyltransferase is most responsible for stabilizing the differentiated state. Vitamin C can accelerate reprogramming and knockdown of lysine-specific demethylase Kdm3b in pre-iPSCs blocks vitamin C-induced further reprogramming, indicating the cooperative network between vitamin C and H3K9me3 demethylase to reduce the H3K9me3 level in pre-iPSCs. While BMPs contributes to build H3K9me3 modification, arresting the reprogramming at pre-iPSC stage, knockdown of the H3K9 methyltransferase SETDB1 rescues the inhibitory effect of BMP. Therefore, it can be concluded that H3K9 methyltransferases are the downstream targets of BMPs, the critical signaling molecules. They serve as the on/off switch for the pre-iPSCs to fully reprogrammed iPSCs transition by regulating H3K9 methylation status at the core pluripotency loci (Esteban et al. 2010; Chen et al. 2013a; Koche et al. 2011).

Methylation of histone H3K4 is generally considered as a marker for transcriptional activation. H3K4me/me2 normally deposits in the enhancer region of the transcriptional activated gene, while H3K4me3 modification always occurs in the promoter region of the pluripotency-related transcriptional activated gene. H3K4 methylation is regulated by TrxG complex, and its core component WDR5 can promote tri-methylation of H3K4. WDR5 is highly expressed in ESCs/iPSCs, binding to the promoters of pluripotency-related genes, regulating pluripotency and self-renewal of ESCs/iPSCs. As cells start the differentiation program, the expression level of WDR5 gradually decreases. Moreover, knockdown of WDR5 during the reprogramming process significantly impairs the reprogramming efficiency (Ang et al. 2011). On the other hand, H3K4 demethylases also play important roles in somatic cell reprogramming. Knockdown of H3K4 demethylase LSD1, accompanied by elevated levels of H3K4me2/3, can induce differentiation of human ESCs, which is associated with de-repression of developmental genes (Whyte et al. 2012). In addition, Lsd1$^{-/-}$ES cells have potent potential to generate embryonic tissues from the embryoid bodies, further suggesting its role in differentiation inhibition and self-renewal maintenance (Macfarlan et al. 2011). LSD1 can maintain the methylation status balance between H3K4 and H3K27 in the regulatory region of some important developmental genes. In mouse ESCs, LSD1 stabilizes DNA methyltransferase 1 (DNMT1), and deletion of Lsd1 results in loss of DNA methylation (Wang et al. 2009). Moreover, LSD1 and its associated nucleosome remodeling and histone deacetylase (NuRD) complex are recruited to OCT4-occupied enhancers at active pluripotent genes in ESCs. LSD1–NuRD complex is required for silencing of ESC enhancers during differentiation, which is essential for complete shutdown of the ESC gene expression program and the transition to new cell states by removing H3K4me1 (Whyte et al. 2012). These studies provide a new approach for finding more secure and more efficient reprogramming methods. The combination of LSD1 inhibitors with other small molecule compounds can reduce the number of reprogramming factors (Wang et al. 2014a), even without any exogenous transcriptional factors (Hou et al. 2013).

Methylation of H3K27 also plays an important role not only in pluripotent stem cells induction, but also in pluripotency maintenance of stem cells. The Polycomb Repressive Complex PRC2 that is responsible for catalyzing the methylation of H3K27, can downregulate the expression of somatic cell-related genes in the early stage of reprogramming. In ESCs, the loss of any components of the PRC2 complex will cause the loss of H3K27 tri-methylation, and further significantly impair the self-renewal capacity of the cells. Also, silencing of PRC2 complex can significantly reduce the efficiency of reprogramming. On the other hand, H3K27 demethylase Utx (also known as Kdm6a) regulates the efficient induction, rather than maintenance, of pluripotency (Mansour et al. 2012). Somatic cells lacking Utx fail to robustly reprogram to the ground state of pluripotency (Mansour et al. 2012). Utx directly interacts with the reprogramming factors of OSK, and uses its histone demethylase catalytic activity to facilitate iPSC formation (Mansour et al. 2012). Utx depletion results in aberrant dynamics of H3K27me3 repressive chromatin demethylation patterns in somatic cells undergoing reprogramming, which directly hampers the de-repression of potent pluripotency-related gene modules (including Sall1, Sall4 and Utf1), which can cooperatively substitute for exogenous OSK supplementation in iPSC formation (Mansour et al. 2012). Taken together, Utx ensures efficient, robust and timely demethylation of H3K27 during the reprogramming induction stages, and plays an important regulatory role in somatic cell reprogramming to pluripotent state (Mansour et al. 2012).

Moreover, H3K79 methylation could be required for differentiation and thus serving as a barrier in somatic cell reprogramming. H3K79me3 is enriched in pericentromeric heterochromatic regions, whereas H3K79me2 is more diffusely located (Ooga et al. 2008). Both marks are removed after fertilization.

To address how chromatin-modifier proteins affect reprogramming and identify key proteins that have a positive or negative effect on the induced reprogramming process in detail, the researchers used short hairpin RNA (shRNA) pool to target the genes in both DNA and histone methylation pathways systematically. Studies have shown that inhibiting the core components of the PRC1 and PRC2, including the histone H3K27 methyltransferase EZH2, reduces reprogramming efficiency; while inhibiting SUV39H1, YY1 and DOT1L significantly improves reprogramming efficiency. Among them, the histone H3K79 methyltransferase DOT1L was inhibited by shRNA or small molecule, which significantly accelerated reprogramming and increased the yield of iPSC colonies, and also replaced Klf4 and c-Myc. Genome-wide analysis of the H3K79me2 distribution revealed that fibroblast-specific genes associated with mesenchymal transition to epithelial cells (MET) lost their H3K79me2 mark during the initial stages of reprogramming, and these genes were also silenced in pluripotent cells. Inhibition of DOT1L contributes to the loss of this modification in genes that are inhibited in the pluripotent state. These findings suggest that specific chromatin-modifying enzymes may act as barriers or facilitators of reprogramming, and by regulating these chromatin modifiers, iPSCs can be more efficiently generated with less dependence on exogenous transcription factors (Onder et al. 2012).

H3K36 methylation status is also important for somatic cell reprogramming efficiency. H3K36me2-specific demethylase, Kdm2b, has the capacity to promote iPSC generation, which depends on its demethylase and DNA-binding activities, however, independent of its role in antagonizing senescence. Kdm2b functions at the beginning of the reprogramming process and enhances in part by promoting cell-cycle progression and overcoming senescence through repression of the Ink4/Arf locus and/or facilitating the early transcriptional response to the reprogramming factors. Kdm2b contributes to gene activation by targeting and demethylating the gene promoters (Liang et al. 2012; Wang et al. 2011). Moreover, Kdm2a is important for Oct4 reactivation in cell-fusion-mediated reprogramming (Ma et al. 2008). The overexpression of Kdm2a/2b potently enhances reprogramming with three (OSK) or fewer factors in the presence of vitamin C (Wang et al. 2011).

8.3 Histone Chaperones

Chromatin remodeling changes chromatin structure dynamically, helping functional proteins or protein complexes bind to specific sites in the genome to regulate gene expression. During somatic cell reprogramming, the gene expression program and chromatin structure change dramatically. There must be different types of chromatin remodelers such as histone chaperones playing various roles in this process.

Previously considered as a "dumb" histone carrier, it is now known that histone chaperone plays a key role in the various stages of histone existence. Chaperones bind to histones during synthesis, escorting them into the nucleus and helping histones to specifically bind to DNA during different processes such as DNA replication, repair or transcription. Histone chaperones directly or indirectly regulate histone PTMs, which are functionally critical to the above process.

However, we still do not have many clues about the role of histone chaperones in reprogramming. To clarify the regulatory pathways that protect the state of somatic cells, RNAi screens targeting chromatin factors during transcription factor-mediated reprogramming of mouse fibroblasts to iPSCs were performed. The subunits of chromatin assembly factor-1 (CAF-1) complex, including Chaf1a and Chaf1b, appear as the most prominent hits in screening (Cheloufi et al. 2015). Suppression of CAF-1 makes the chromatin structure of enhancer elements more accessible early in reprogramming, which is achieved by lowering the level of H3K9 methylation. These changes are accompanied by a decrease in somatic heterochromatin domains, increased binding of Sox2 to pluripotency-specific targets, and activation of related genes (Cheloufi et al. 2015). Optimal regulation of CAF-1 and transcription factor levels increased reprogramming efficiency by several orders of magnitude and promoted iPSC formation in as little as 4 days. In addition, it was found that CAF-1 suppression not only improved the efficiency of reprogramming, but also promoted direct lineage conversion, suggesting that histone chaperone can modulate cellular plasticity in a regenerative setting (Cheloufi et al. 2015).

The histone chaperone Asf1a is also involved in cell reprogramming and maintenance of human ESCs (Gonzalez-Munoz et al. 2014). Asf1a acts upstream of CAF-1 by transferring newly synthesized and acetylated histones to CAF-1 (Ransom et al. 2010). In contrast to CAF-1, overexpression of Asf1a, but not downregulation, enhances human iPSCs formation. This phenotype is presumed to be mediated by increased deposition of acetylated histones, thereby inducing a more accessible chromatin state (Ransom et al. 2010).

The downregulation of Aprataxin PNK-like factor (APLF) promotes reprogramming by augmenting the expression of E-cadherin (Cdh1), which is implicated in the mesenchymal-to-epithelial transition (MET) involved in the generation of iPSCs. Downregulation of APLF in MEFs expedites the loss of the repressive MacroH2A.1 (encoded by H2afy) histone variant from the Cdh1 promoter and enhances the incorporation of active histone H3K4me2 marks at the promoters of the pluripotency genes *Nanog* and *Klf1*, thereby accelerating the process of cellular reprogramming and increasing the efficiency of iPSC generation (Syed et al. 2016).

Histone cell cycle regulator (HIRA) is a histone variant H3.3-specific chaperone, which preferentially introduces the histone variant H3.3 into the nucleosome. The HIRA complex is conserved across species, such as yeast, Drosophila, plants, *etc.*, mediating the deposition of H3.3 into euchromatic regions. In ESCs, HIRA directly interacts with the PRC2 complex, facilitating recruitment of PRC2 complex and establishing the correct H3K27me3 in the promoter regions of the developmental regulatory genes (Banaszynski et al. 2013). HIRA-dependent histone deposition facilitates transcriptional recovery after genotoxic stress. For example, HIRA deposits newly synthesized H3.3 to damaged DNA sites in response to UV irradiation (Adam et al. 2013). Zygotes from Hira mutant mouse exhibit loss of H3.3 incorporation in both paternal and maternal genomes. Moreover, DNA replication and rRNA transcription are also impaired (Lin et al. 2014; Inoue and Zhang 2014). Specific deletion of Hira in developing mouse oocytes results in altered chromatin homeostasis, including decreased DNA methylation, increased DNase I sensitivity, and accumulated DNA damage, along with a severe fertility phenotype (Nashun et al. 2015). Hira is critical for mouse embryonic development; insufficient production of the gene may disrupt normal embryonic development (Roberts et al. 2002).

Recent studies show that death domain-associated protein (DAXX) is an H3.3-specific histone chaperone unique to metazoans. Distinct from HIRA that deposits H3.3 in euchromatin, DAXX deposits H3.3 into telomeres and pericentric heterochromatin (Goldberg et al. 2010; Drane et al. 2010; Lewis et al. 2010). DAXX forms an H3.3 preassembly complex with the α-thalassemia/mental retardation syndrome protein (ATRX), a SNF2-like and ATP-dependent chromatin remodeling factor. In mouse ESCs, H3K9Ac is enriched in purified HIRA-associated H3.3 complexes but not in DAXX-associated H3.3 complexes. While the DAXX–ATRX complex is enriched for H3K9me3 (Elsaesser and Allis 2010). The DAXX-ATRX complex and the HIRA complex control the deposition of the corresponding H3.3 at different locations in the genome, and each deposit may have a different PTM to play a different role.

8.4   Histone Variants

In addition to several common histones, some histone variants can also participate in the formation of nucleosomes, greatly increasing the diversity and complexity of nucleosomes and even chromosome structures.

A small group of non-canonical variants of histones emerged from canonical histones with one or a few amino acid differences. These histone variants are expressed at relatively low levels but have distinct biological functions by altering the conformation of nucleosomes. Massive replacements of canonical histones by non-canonical histone variants (or vice versa) can be observed during fertilization or germ cell generation, which are correlated with cell reprogramming and cell fate alteration.

Two histone variants TH2A and TH2B, highly expressed in the mouse oocyte, play important roles in activation of paternal genome during fertilization and OSKM-induced somatic reprogramming (Shinagawa et al. 2014). Combinational transduction of Oct4, Klf4 and Th2a/b can reprogram somatic cells into the pluripotent state efficiently. Notably, the roles of TH2A/B in reprogramming are very likely due to their deposition of chaperone NPM. Moreover, NPM can lead to global de-condensation of sperm chromatin during fertilization and of somatic cell nuclei during SCNT, thus significantly increase the success rates of reprogramming, suggesting that the TH2A/B replacement is a general phenomenon for both *in vitro* and *in vivo* reprogramming and genome reactivation (Shinagawa et al. 2014). Recent studies have also shown that TH2A/B as well as NPM can reprogram human somatic cells and improve the quality of human iPSCs. However, different from TH2A/B playing positive roles in pluripotency induction, some histone variants have some negative effects on iPSC generation, such as marcoH2A. MarcoH2A is highly expressed in human somatic cells and is down-regulated during reprogramming. Knocking down marcoH2A in human keratinocytes can increase reprogramming efficiency, while over-expressing it will inhibit reprogramming. Further studies have found that in human keratinocytes, marcoH2A is mainly distributed on genes associated with pluripotency or development. During reprogramming, marcoH2A inhibits H3K4 methylation and makes the genes at relevant sites at very low expression levels (Barrero et al. 2013). Thus, while some histone variants are important for the activation of quiescent genomes, some play important roles in maintaining the somatic cell epigenome at the repressive state.

The distribution of genomic histone variants can serve as an epigenetic marker for evaluating the quality of iPSCs, as well as providing new ideas for discovering more efficient reprogramming methods. H2A.X is a variant of histone H2A that serves as a functional marker to distinguish the developmental potential of different mouse iPSCs. In mouse ESCs, H2A.X specifically binds to genes involved in the regulation of extra-embryonic lineage development, inhibiting its expression, thereby preventing the differentiation of ESCs into trophoblast cells. For somatic reprogramming, if iPSCs faithfully recapitulated the specific H2A.X distribution patterns similar to ESCs, such iPSCs can obtain the capacity to support the development of 'all-iPSC' mice via tetraploid complementation, which are completely developed from iPSCs. However, iPSC clones with aberrant deposition of H2A.X exhibit upregulation of genes in extra-embryonic lineage and a pre-disposition to extra-embryonic differentiation bias, resulting in failure to support the 'all-iPSC' mice development (Wu et al. 2014).

Recent study suggests that the histone variant H3.3 is a mark of transcriptionally active chromatin needs to be reconsidered. In mammals, histone variant H3.3 is encoded by two different genes (h3f3a and h3f3b) that are translated to produce the same protein product (Frank et al. 2003; Wellman et al. 1987). Unlike classical histone H3, which expresses and incorporates chromatin in S phase, H3.3 expression is not cell cycle-regulated. It is expressed throughout the cell cycle of quiescent cells, mitotic cells and proliferating cells. H3.3 may deposit in a manner independent of DNA synthesis during and outside the S phase (Skene and Henikoff 2013). During somatic cell reprogramming, the transient binding of the reprogramming factors to their corresponding binding sites results in the incorporation of histone variant H3.3, which makes the chromatin looser and enhances the accessibility of the binding sites. These chromatin changes increase the possibility of activation of Oct4, Sox2, Klf4 and other pluripotency genes. The activating gene is labeled with a high level of histone variant H3.3. The replication-dependent histone variant H3.1 is incorporated during DNA replication, diluting H3.3. Due to some unknown mechanism, redistributed H3.3 causes the gene to be reactivated or silenced (Gonzalez-Munoz et al. 2014). H3.3 is an essential parent factor for somatic cell reprogramming in oocytes. The knockdown of the parental H3.3 will make reprogramming difficult and will not reactivate many key pluripotency genes; this impaired reprogramming can be rescued by injecting exogenous H3.3 mRNA into oocytes of SCNT embryos. The parent H3.3 participates in reprogramming by substituting the donor core-derived H3 (H3.3 and canonical H3) with de novo synthesis of H3.3, and the substitution is not a global effect, but depends on the identity of the donor core (Wen et al. 2014a).

Mammalian oocytes contain the maternal-specific linker histone H1foo, which is a homolog of the *Xenopus* linker histone B4. H1foo is specifically expressed during the germinal vesicle (GV) stage until the late two-cell stage or early four-cell stage, and is required for oocyte maturation (Furuya et al. 2007; Gao et al. 2004; Tanaka et al. 2003, 2005). H1foo can replace c-Myc in the Yamanaka four-factor induced reprogramming process and is more likely to generate qualified iPSCs (Kunitomi et al. 2016). Expression of H1foo in ESCs allows continuous activation of pluripotency-related genes and prevents differentiation *in vitro* (Hayakawa et al. 2012). In the early stage of reprogramming, H3K27me3 modification is globally lost, affecting the state of heterochromatin. H1foo reduces heterochromatin regions, making chromatin more decondensed than somatic H1 and other linker histone states, resulting in a chromatin state that is more suitable for reprogramming (Hayakawa et al. 2012; Saeki et al. 2005). ChIP analysis in H1foo-ESCs revealed that H1foo targeted selectively to a set of hypomethylated genomic loci, making these target loci decondensed. Thus, H1foo also has an impact on the genome-wide, locus specific epigenetic status (Hayakawa et al. 2012).

8.5 DNA Methylation/Demethylation

DNA methylation/demethylation and histone modifications both serve as important epigenetic regulators, precisely controlling the genes activation and inactivation during development in mammals. DNA methylation typically occurs in cytosines in CpG dinucleotides, forming 5-methylcytosine cytosine residues (5mC). The distribution of this modification on the genome is asymmetric, with CpG-rich regions (also referred as CpG islands) and CpG-deficient regions (including gene bodies, intergenic regions). About 70% of the gene promoters are present in CpG islands, especially the promoters of many house-keeping genes (Saxonov et al. 2006; Gardiner-Garden and Frommer 1987). DNA methylation can regulate and stabilize chromatin structure. It can also regulate gene expression by recruiting related protein complexes, or by preventing the transcription machinery access to certain regions of DNA. Thus, a prevalent view holds that DNA methylation restrict transcription, whereas DNA demethylation allow the gene to be expressed (Saxonov et al. 2006; Gardiner-Garden and Frommer 1987).

The state of DNA methylation changes dynamically throughout development. The most prominent events are the successive occurrences of methylation and demethylation, including genome-wide DNA demethylation during mammalian pre-implantation development after fertilization and tissue-specific re-methylation in post-implantation development (Saxonov et al. 2006; Gardiner-Garden and Frommer 1987). Indeed, the pattern of DNA methylation varies between cell types and growth conditions. And it seems that different cell types use their unique and robust DNA methylation patterns to regulate expression patterns of tissue-specific genes (Gardiner-Garden and Frommer 1987). Similar to other epigenetic modifications mentioned above, DNA methylation also plays an important role in somatic cell reprogramming regulation (Gardiner-Garden and Frommer 1987). However, DNA methylation is an inefficient and slow process compared to histone PTMs such as acetylation, methylation, etc.

DNA methylation is catalyzed by DNA methyltransferases (DNMTs), which are divided into two categories: maintenance DNMT1 and de novo DNMT3A and DNMT3B. These three DNMTs are broadly involved in embryonic development, and the expression of DNMTs is greatly reduced in terminally differentiated cells.

The function of DNMT1 depends on the DNA replication process. During DNA replication, DNMT1 is in the position of the replication fork, and here newly formed hemi-methylated DNA is formed (Leonhardt et al. 1992). DNMT1 binds to newly synthesized hemi-methylated DNA to help methylation precisely in accordance with methylation patterns prior to DNA replication (Hermann et al. 2004). Because DNMT1 maintains the original methylation pattern in the cell lineage, it is called maintenance methyltransferase. Conversely, the de novo methyltransferases DNMT3A and DNMT3B can methylate DNA sites that are not methylated, which is independent of DNA replication. In addition, DNMT3L - a homologous protein of DNMT3A/DNMT3B, but without the conserved catalytic domain common to DNA methyltransferases – binds to DNMT3A/DNMT3B and activates their methyltransferase activity (Suetake et al. 2004).

DNA demethylation always adopts two approaches to remove the methyl group on cytosine residue. One is passive demethylation; the other is active demethylation. Passive DNA demethylation occurs during cell division. During DNA replication, the cytosine on the newly synthesized DNA strand is not methylated, which may be caused by inhibition or inactivation of DNMT1. Thus, the methylation level of the whole genome is diluted after each cycle of cell division, resulting in the decrease in the overall methylation state. Active DNA demethylation occurs in both mitotic and non-dividing cells, relying on the corresponding enzyme to catalyze the conversion of 5mC to unmethylated cytosine. This process is completed by the following DNA dioxygenase, the Ten-Eleven-Translocation family proteins including TET1, TET2, TET3, thymine DNA glycosylase (TDG) and DNA base excision repair (BER) pathway. The TET proteins oxidize 5-methylcytosine (5mC) to 5-hydroxymethylcytosine (5hmC), and then oxidize to 5-formylcytosine (5fC) and 5-carboxycytosine (5caC), which are further catalytically modified by TDG and ultimately converted to unmodified cytosine through BER pathway (Wu and Zhang 2011).

DNA methylation is a robust epigenetic modification that can lead to sustained silencing of genes. All of the three DNMTs mentioned above are enriched in ESCs, suggesting their important roles in supporting pluripotency of ESCs. During somatic cell reprogramming, DNA methylation occurs mainly after histone modification and chromatin structure changes, serving as an extremely important epigenetic barrier to cellular reprogramming. Treatment of reprogrammed cells with the DNA methylation inhibitor 5-Aza-CR, or reduction of Dnmt1 expression by siRNA or shRNA, induces rapid and stable conversion of these cells to fully reprogrammed iPSCs, greatly improves the efficiency more than 30 folds (Mikkelsen et al. 2008). Moreover, although somatic cells of DNMT3A/DNMT3B knock-out mouse can still normally induce the formation of iPSCs, they have a restricted developmental potential (Pawlak and Jaenisch 2011).

In addition, demethylation of pluripotency-related genes plays important roles in development. TET3 is the first identified enzyme in methyl cytosine conversion by active DNA demethylation pathway during zygotic activation. During zygotic activation, TET3 is predominantly enriched in the male pronucleus, and knockout of Tet3 in zygote can impede the active DNA demethylation of paternal genome, and thus delay the subsequent activation of paternal pluripotency-related genes during early embryonic development (Gu et al. 2011). However, both Tet1 and Tet3 knockout mice are partially lethal, suggesting that redundant functions may exist in Tet family proteins (Gu et al. 2011; Dawlaty et al. 2011). Tet family protein-dependent DNA modifications are also important for reprogramming. Both Tet2 and Parp1 are recruited to Nanog and Esrrb loci for establishment of early epigenetic modifications, which are essential for transcriptional activation at the pluripotency genes to complete the reprogramming process (Doege et al. 2012). The core pluripotent gene Nanog can interact with Tet1 and Tet2, binding to the target of pluripotent genes, increasing the reprogramming efficiency to generate the fully reprogrammed iPSCs (Costa et al. 2013). Oct4, Nanog, and Tet1 can form a positive feedback regulatory network, reducing methylation levels in the Oct4 and Tet1 promoter regions (Costa et al. 2013). Interestingly, Tet1 can work as a double-edged sword for somatic cell reprogramming regulation. It can either positively or negatively regulates reprogramming depending on the absence or presence of vitamin C (Chen et al. 2013b). In the context of vitamin C, knockout of Tet1 can enhance reprogramming, and its overexpression can impair the reprogramming by modulating the obligatory mesenchymal-to-epithelial transition (MET). On the other hand, Tet1 can boost somatic cell reprogramming independent of MET in the absence of vitamin C (Chen et al. 2013b). This study also found that Tet1 regulates 5hmC formation at loci critical for MET in a vitamin C-dependent fashion, strongly suggesting the role of vitamin C in determination of TET1 protein biological outcome during reprogramming. However, another study showed that depletion of Tet1 or its downstream TDG could impaire reprogramming by blocking MET, suggesting that

Tet1 may serve as a booster for mid-stage or late-stage of reprogramming (Hu et al. 2014). Furthermore, Tet1 (T) is capable of replacing essential reprogramming factors during reprogramming. It can replace Oct4 and fulfill the somatic cell reprogramming in combination with Sox2 (S), Klf4 (K) and c-Myc (M) (Gao et al. 2013). Analyzing the efficient TSKM secondary reprogramming system reveals that both 5mC and 5hmC modifications increase at an intermediate stage of the reprogramming process, correlating with a transition in the transcriptional profile. 5hmC enrichment is involved in the demethylation and reactivation of genes and regulatory regions that are important for pluripotency, indicating that changes in DNA methylation and hydroxymethylation play important roles in genome-wide epigenetic remodeling during reprogramming. Importantly, the combination of Tet1 with Oct4 is enough to reprogram the cells toward a high-quality pluripotent state with normal 5hmC levels. These OT (Oct4-Tet1)-iPSCs can also efficiently generate 'all-iPSC' mice with a normal life span and no obvious tumorigenicity compared to the OSKM-derived 'all-iPSC' mice (Chen et al. 2015a). In summary, Tet1 can replace multiple Yamanaka factors to achieve reprogramming, further elucidating the important roles of epigenetic modifiers during this process.

Epigenetic information can be passed from one generation to another through DNA methylation, histone modifications, and small RNA changes, a process known as epigenetic memory. Different types of somatic cells have different transcriptional and epigenetic characteristics. During the process of induced reprogramming, the epigenetic memory of donor cells often has a certain influence on the differentiation potential of the finally obtained iPSCs. One of the main causes of epigenetic memory residue is that DNA of the derived reprogramming cells is not completely demethylated or remethylated. Rather than being a strictly dynamic mechanism for regulating gene expression, DNA methylation state can serve as a long-term memory of previous gene expression decisions that were mediated by transcriptional factors which might no longer present in the cell (Dor and Cedar 2018). In somatic cells, many pluripotency-related genes and their regulatory regions are covered by H3K9me3 modification and DNA methylation to sustain the silencing. During reprogramming, some regions that cannot be completely demethylated are still enriched in H3K9me3 mark, therefore impeding the accessibility of OSKM transcription factors for gene activation. Such regions are named as reprogramming resistance regions (RRRs) (Matoba et al. 2014). The lower generation of iPSCs will leave some methylation characteristics of donor cells, making iPSCs easy to differentiate into donor cell-associated germ layer cells, limiting the differentiation potential.

Genomic imprinting is also closely related to DNA methylation, making the genes differentially expressed in a parental allele-specific manner and playing a key regulatory role in mammalian growth and development. During induced reprogramming, DNA methylation on some imprinting loci is erased, while some not. Comparison of genetically identical mouse ESCs and iPSCs reveals that the imprinted Dlk1-Dio3 gene cluster on chromosome 12qF1 is aberrantly silenced in some iPSC clones, which contributed poorly to chimaeras and failed to support the development of entirely iPSC-derived mice, consistent with its developmental roles. However, treatment of these iPSC clones with a histone deacetylase inhibitor reactivated this locus and restored its capacity to support the full-term development of the all-iPSC mice (Stadtfeld et al. 2010). Further study by the same lab showed ascorbic acid could prevent imprinting loss of Dlk1-Dio3 and thus facilitate generation of all-iPS cell mice from terminally differentiated B cells (Stadtfeld et al. 2012). In addition, high-throughput sequencing analysis between the iPSC lines with and without the ability to support all-iPSC mice identified another imprinted gene Zrsr1 important for reprogramming quality. Methylation of Zrsr1 was consistently disrupted in the iPSCs with reduced pluripotency, which could not support the generation of all-iPSC mice. Furthermore, the disrupted methylation on Zrsr1 could not be rescued by improving culture conditions or subcloning of iPSCs (Chang et al. 2014).

DNA methylation generally represses transcription. However, recent studies in *Arabidopsis* find that it can activate gene transcription in some instances. In this case, a protein complex is recruited to chromatin by DNA methylation, which specifically activates the transcription of genes that are already mildly transcribed but not those that are transcriptionally silent such as transposable elements (Harris et al. 2018). Thus, by counteracting the repression effect caused by transposon insertion in neighboring genes while leaving transposons silent, DNA methylation fine-tunes gene expression.

In some cases, DNA methylation can activate gene transcription (Harris et al. 2018). In addition, another study using deep whole-genome bisulfite sequencing revealed sequence-dependent CpG methylation imbalances at thousands of heterozygous regulatory loci, which are enriched for stochastic switching between fully methylated and unmethylated states of DNA (Onuchic et al. 2018). These above novel findings provide new perspectives for understanding the biological functions of DNA methylation and demethylation in reprogramming process.

8.6 RNA Methylation/Demethylation

There are increasing evidences showing that a wide variety of RNA modifications play crucial roles in the regulation of gene expression. Among them, RNA methylation, with N(6)-methyl adenosine (m^6A) and 5-methylcytosine (m5C) as the most representative, is the one of most important in RNA modifications. In this part, we will discuss the m^6A modification in detail since it is the most prevalent and reversible internal modification in mammalian mRNAs and non-coding RNAs with in-depth study so far.

The formation of m^6A modification is catalyzed by a specific methyltransferase, which uses SAM as a methyl group donor and transfers the methyl group to N^6-adnosine of RNA. Both METTL3 and METTL14 contain a SAM binding site and a DPPW (Asp-Pro-Pro-Trp) functional domain. These two proteins form a stable heterodimer core complex, mediating m^6A deposition on mammalian nuclear RNAs. WTAP, a mammalian splicing factor, interacts with the METTL3-METTL14 complex, serving as a regulatory subunit. Although WTAP does not possess methylation activity, it is required for the methyltransferase complex localization into nuclear speckles enriched with pre-mRNA processing factors and thus significantly affects the catalytic activity of the m^6A methyltransferase in vivo. The majority of RNAs bound by WTAP and METTL3-METTL14 complex contain the consensus m^6A motif RRACH (R = G or A; H = A, C or U), the highly conserved regions where the m^6A modification occurs (Liu et al. 2014; Ping et al. 2014). However, there are some m^6A modifications that do not contain the RRACH conserved motif, suggesting that there may be more methyltransferases directly involved in the m6A depositions.

For m^6A demethylases, only two have been identified in mammals: FTO and ALKBH5. FTO catalyzed m6A demethylation requires several intermediate steps. It oxidizes m^6A to generates N^6-hydroxymethyladenosine (hm^6A) as an intermediate modification and N^6-formyladenosine (f^6A) as a further oxidized product, which was finally catalyzed to generate A (Fu et al. 2013). Both hm^6A and f^6A have half-life times of ~3 h in aqueous solution under physiological relevant conditions, and are present in isolated mRNA from human cells as well as mouse tissues (Fu et al. 2013). Unlike FTO, ALKBH5 directly catalyzes the conversion of m6A to A, with no intermediates found yet. Due to the complexity of RNA processing and the diversity of RNA substrates, m6A demethylase other than FTO and ALKBH5 is likely to be present.

m6A modification exerts biological effects mainly through binding proteins comprising a YTH domain such as YTHDF proteins (Dominissini et al. 2012). Recent study showed that YTHDF2 protein preferentially binds to m^6A-RNA, while its binding affinities to hm^6A or f^6A were attenuated to a level similar to A (Fu et al. 2013).

Several studies demonstrated strong correlations between m^6A modification and pluripotency. Genetic inactivation or depletion of mouse and human *Mettl3*, one of the m^6A methyltransferases, led to the depletion of m^6A modification on several target genes and affected mRNA stability, which further prolonged Nanog expression upon differentiation, thus impairing ESC exit from self-renewal program toward differentiation into several lineages both *in vitro* and *in vivo* (Batista et al. 2014). Knockdown of Mettl3 and Mettl14 in mouse ESCs led to similar phenotypes, characterized by lack of m^6A modification and self-renewal capacity lost (Wang et al. 2014b). Thus, m^6A is a mark of transcriptome flexibility required for stem cell identity maintenance and differentiation to specific lineages.

Knockout of Mettl3 in pre-implantation epiblasts and naïve embryonic stem cells also exhibit depletion of m^6A modification in mRNAs. Yet the cells are viable, however, they fail to adequately terminate their naïve state and, subsequently, undergo aberrant and restricted lineage priming at the post-implantation stage, which leads to early embryonic lethality. Deep investigation found that m^6A predominantly and directly reduces mRNA stability, including that of key naïve pluripotency-promoting transcripts (Geula et al. 2015).

ZFP217 interacts with several epigenetic regulators, activates transcription of key pluripotency genes, and modulates m^6A deposition on their transcripts by sequestering the m^6A methyltransferase METTL3. Consistently, Zfp217 depletion compromises ESC self-renewal and somatic cell reprogramming, globally increases m^6A RNA levels, and enhances m^6A modification of Nanog, Sox2, Klf4, and c-Myc mRNAs, promoting their degradation. Moreover, ZFP217 binds its own target gene mRNAs, which are also METTL3-associated, and is enriched at promoters of m^6A-modified transcripts (Aguilo et al. 2015). Profiling m6A modification in the mRNA transcriptomes of four cell types with different degrees of pluripotency identified gene- and cell-type-specific m^6A modifications. Moreover, manipulation of microRNA (miRNA) expression or sequences, which regulate m^6A modification via a sequence pairing mechanism, can affect the binding of METTL3 to the miRNA target mRNAs and alter m^6A modification levels. They could also observe that increased m6A abundance promotes the reprogramming of MEFs to pluripotent stem cells; conversely, reduced m^6A levels impede reprogramming, providing some clues for future functional studies of m6A modification in somatic cell reprogramming (Chen et al. 2015b).

Unlike the dense study of m6A modification, little is known about the deposition and de-deposition of m^5C modification. Studies found that the NSUN family of proteins and DNMT2 are candidates for m^5C methyltransferase. Reprogramming of cells to pluripotency can be achieved using m^5C and pseudouridine-modified mRNAs encoding the four Yamanaka factors (Warren et al. 2010) (◘ Fig. 8.1, ◘ Table 8.1).

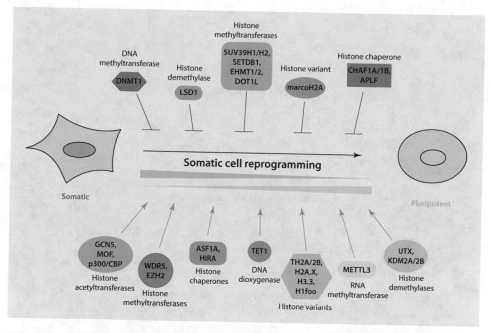

Fig. 8.1 Epigenetic modifiers in induced reprogramming of somatic cells to pluripotent cells. As shown, the upper modifiers such as DNMT1 and SETDB1 can inhibit the somatic cell reprogramming process, while the below ones including GCN5, WDR5, etc. can facilitate the reprogramming process

Table 8.1 Epigenetic modifications and modifiers in induced somatic reprogramming

Modifiers/ Regulators	Function	Roles	Reference
GCN5	Histone lysine acetyltransferase	Activates a distinct alternative splicing network and coregulates a group of RNA splicing and RNA processing genes that are needed for somatic cell reprogramming.	Hirsch et al. (2015)
MOF	H4K16 acetyltransferase	An integral part of the ESC core transcriptional network and plays an important role in maintaining ESC self-renewal and pluripotency.	Li et al. (2012)
SUV39H1/H2	H3K9 methyltransferase	Suppress reprogramming.	Onder et al. (2012)
EHMT1/2	H3K9 methyltransferase	Associated with transcriptional repression.	Onder et al. (2012)
SETDB1	H3K9 methyltransferase	Suppress reprogramming.	Sridharan et al. (2013)

(continued)

◘ **Table 8.1** (continued)

Modifiers/ Regulators	Function	Roles	Reference
WDR5	H3K4me3 reader	Interacts with the pluripotency transcription factor Oct4.	Ang et al. (2011)
LSD1	H3K4/K9 demethylase	Inhibition of it promotes the MET and pluripotency gene activation.	Cacchiarelli et al. (2015)
UTX	H3K27 demethylase	Interacts with OSK to activate potent pluripotency-promoting gene modules.	Mansour et al. (2012)
EZH2	H3K27 methyltrans-ferase	Inhibition of it reduces reprogramming efficiency.	Onder et al. (2012)
DOT1L	H3K79 methyltrans-ferase	Inhibition of it contributes to the loss of H3K79me2 in genes that are repressed in the pluripotent state.	Onder et al. (2012)
KDM2A/B	H3K36 demethylase	Contributes to gene activation by targeting and demethylating the gene promoters.	Liang et al. (2012)
CHAF1A and CHAF1B	Histone chaperone	Components of chromatin assembly factor-1 (CAF-1) complex; leads to a more accessible chromatin structure at pluripo-tency enhancers early during reprogram-ming.	Cheloufi et al. (2015)
ASF1A	Histone chaperone	Affects the expression of core pluripotency genes.	Gonzalez-Munoz et al. (2014)
APLF	Histone chaperone	Accelerates the process of cellular reprogramming.	Syed et al. (2016)
HIRA	Histone variant H3.3-specific chaperone	Directly interacts with the PRC2 complex, facilitating recruitment of PRC2 complex and establishing the correct H3K27me3 in the promoter regions of the developmental regulatory genes.	Banaszynski et al. (2013)
TH2A/2B	Histone variant	Plays important roles in activation of paternal genome during fertilization and OSKM-induced somatic reprogramming.	Shinagawa et al. (2014)
MacroH2A	Histone variant	Distributes on genes associated with pluripotency or development, inhibits H3K4 methylation and makes the genes at relevant sites at very low expression levels during reprogramming.	Barrero et al. (2013)
H2A.X	Histone variant	Serves as a functional marker to distinguish the developmental potential of different mouse iPSCs.	Wu et al. (2014)

◻ **Table 8.1** (continued)

Modifiers/ Regulators	Function	Roles	Reference
H3.3	Histone variant	Makes the chromatin looser and enhances the accessibility of the binding sites. Then, increasing the possibility of activation of Oct4, Sox2, Klf4 and other pluripotency genes.	Wen et al. (2014b)
H1foo	Histone variant	Makes chromatin more decondensed, more for reprogramming. Expression of it allows continuous activation of pluripotency-related genes and prevents differentiation in vitro.	Hayakawa et al. (2012)
DNMT1	DNA methyltrans-ferase	Inhibition of it promotes DNA demethyl-ation in the late phase of reprogramming.	Mikkelsen et al. (2008)
TET1	DNA dioxygenase	"Promotes Oct4 demethylation and reactivation; capable of replacing essential reprogramming factors during reprogram-ming; regulates 5hmC formation at loci critical for MET in a Vc-dependent fashion."	Costa et al. (2013), Hu et al. (2014), Gao et al. (2013), Chen et al. (2013b, 2015a)
TET2	DNA dioxygenase	Recruited to Nanog and Esrrb loci, essential for transcriptional activation at the pluripotency genes; interacts with NANOG to enhance reprogramming	Doege et al. (2012), Costa et al. (2013)
METTL3	RNA m6A methyltrans-ferase	Positive influence for reprogramming	Chen et al. (2015b)

8.7 Conclusion

The development of somatic cell reprogramming techniques not only provide a reliable platform for in vitro stem cell biology research, but also in vivo regenerative medicine studies. As the pioneer study of iPSC reprogramming technique, the forced expression of key transcriptional factors can not only reprogram cell into ESC-like pluripotent state, but also facilitate the study of cell fate determination during trans-differentiation. Although recent studies have demonstrated the important roles of some epigenetic modifications and remodeling involved in reprogramming, the molecular mechanisms of epigenetic events in the 'black box' of reprogramming are largely unknown. Plus, the crosstalk among the epigenetic events is also unclear. Deep investigation of the modifiers and remodeling factors will shed light on improvements of both reprogramming efficiency and the outputs qualities of iPSCs.

┌─ **Take Home Message** ───┐

In the past decade, the method of somatic cell induced reprogramming has been greatly developed, and there are clinical trials underway. Recent studies have continually elucidated the specific mechanisms of some epigenetic events during reprogramming, but the dynamics of many epigenetic modifications throughout the reprogramming process and the interactions between them remain unclear; even for some already known epigenetic events, we are still not sure whether they are the drivers or the consequences of reprogramming. In addition, whether the method of regulating the reprogramming process by interfering with one or several epigenetic modifications is safe for clinical applications remains to be studied.

└──┘

References

Adam, S., Polo, S. E., & Almouzni, G. (2013). Transcription recovery after DNA damage requires chromatin priming by the H3.3 histone chaperone HIRA. *Cell, 155*(1), 94–106.

Aguilo, F., et al. (2015). Coordination of m(6)A mRNA methylation and gene transcription by ZFP217 regulates pluripotency and reprogramming. *Cell Stem Cell, 17*(6), 689–704.

Ang, Y. S., et al. (2011). Wdr5 mediates self-renewal and reprogramming via the embryonic stem cell core transcriptional network. *Cell, 145*(2), 183–197.

Banaszynski, L. A., et al. (2013). Hira-dependent histone H3.3 deposition facilitates PRC2 recruitment at developmental loci in ES cells. *Cell, 155*(1), 107–120.

Barrero, M. J., et al. (2013). Macrohistone variants preserve cell identity by preventing the gain of H3K4me2 during reprogramming to pluripotency. *Cell Reports, 3*(4), 1005–1011.

Batista, P. J., et al. (2014). m(6)A RNA modification controls cell fate transition in mammalian embryonic stem cells. *Cell Stem Cell, 15*(6), 707–719.

Bui, H. T., et al. (2011). Histone deacetylase inhibition improves activation of ribosomal RNA genes and embryonic nucleolar reprogramming in cloned mouse embryos. *Biology of Reproduction, 85*(5), 1048–1056.

Cacchiarelli, D., et al. (2015). Integrative analyses of human reprogramming reveal dynamic nature of induced pluripotency. *Cell, 162*(2), 412–424.

Chang, G., et al. (2014). High-throughput sequencing reveals the disruption of methylation of imprinted gene in induced pluripotent stem cells. *Cell Research, 24*(3), 293–306.

Cheloufi, S., et al. (2015). The histone chaperone CAF-1 safeguards somatic cell identity. *Nature, 528*(7581), 218–224.

Chen, J., et al. (2013a). H3K9 methylation is a barrier during somatic cell reprogramming into iPSCs. *Nature Genetics, 45*(1), 34–42.

Chen, J., et al. (2013b). Vitamin C modulates TET1 function during somatic cell reprogramming. *Nature Genetics, 45*(12), 1504–1509.

Chen, J., et al. (2015a). The combination of Tet1 with Oct4 generates high-quality mouse-induced pluripotent stem cells. *Stem Cells, 33*(3), 686–698.

Chen, T., et al. (2015b). m(6)A RNA methylation is regulated by microRNAs and promotes reprogramming to pluripotency. *Cell Stem Cell, 16*(3), 289–301.

Costa, Y., et al. (2013). NANOG-dependent function of TET1 and TET2 in establishment of pluripotency. *Nature, 495*(7441), 370–374.

Dai, X., et al. (2014). Acetylation-dependent regulation of essential iPS-inducing factors: A regulatory crossroad for pluripotency and tumorigenesis. *Cancer Medicine, 3*(5), 1211–1224.

Dawlaty, M. M., et al. (2011). Tet1 is dispensable for maintaining pluripotency and its loss is compatible with embryonic and postnatal development. *Cell Stem Cell, 9*(2), 166–175.

Doege, C. A., et al. (2012). Early-stage epigenetic modification during somatic cell reprogramming by Parp1 and Tet2. *Nature, 488*(7413), 652–655.

Dominissini, D., et al. (2012). Topology of the human and mouse m6A RNA methylomes revealed by m6A-seq. *Nature, 485*(7397), 201–206.

Dor, Y., & Cedar, H. (2018). Principles of DNA methylation and their implications for biology and medicine. *Lancet, 392*(10149), 777–786.

Drane, P., et al. (2010). The death-associated protein DAXX is a novel histone chaperone involved in the replication-independent deposition of H3.3. *Genes & Development, 24*(12), 1253–1265.

Elsaesser, S. J., & Allis, C. D. (2010). HIRA and Daxx constitute two independent histone H3.3-containing predeposition complexes. *Cold Spring Harbor Symposia on Quantitative Biology, 75*, 27–34.

Esteban, M. A., et al. (2010). Vitamin C enhances the generation of mouse and human induced pluripotent stem cells. *Cell Stem Cell, 6*(1), 71–79.

Fang, F., et al. (2014). Coactivators p300 and CBP maintain the identity of mouse embryonic stem cells by mediating long-range chromatin structure. *Stem Cells, 32*(7), 1805–1816.

Frank, D., Doenecke, D., & Albig, W. (2003). Differential expression of human replacement and cell cycle dependent H3 histone genes. *Gene, 312*, 135–143.

Fu, Y., et al. (2013). FTO-mediated formation of N6-hydroxymethyladenosine and N6-formyladenosine in mammalian RNA. *Nature Communications, 4*, 1798.

Furuya, M., et al. (2007). H1foo is indispensable for meiotic maturation of the mouse oocyte. *The Journal of Reproduction and Development, 53*(4), 895–902.

Gao, S., et al. (2004). Rapid H1 linker histone transitions following fertilization or somatic cell nuclear transfer: Evidence for a uniform developmental program in mice. *Developmental Biology, 266*(1), 62–75.

Gao, Y., et al. (2013). Replacement of Oct4 by Tet1 during iPSC induction reveals an important role of DNA methylation and hydroxymethylation in reprogramming. *Cell Stem Cell, 12*(4), 453–469.

Gardiner-Garden, M., & Frommer, M. (1987). CpG islands in vertebrate genomes. *Journal of Molecular Biology, 196*(2), 261–282.

Geula, S., et al. (2015). Stem cells. m6A mRNA methylation facilitates resolution of naive pluripotency toward differentiation. *Science, 347*(6225), 1002–1006.

Goldberg, A. D., et al. (2010). Distinct factors control histone variant H3.3 localization at specific genomic regions. *Cell, 140*(5), 678–691.

Gonzalez-Munoz, E., et al. (2014). Cell reprogramming. Histone chaperone ASF1A is required for maintenance of pluripotency and cellular reprogramming. *Science, 345*(6198), 822–825.

Gu, T. P., et al. (2011). The role of Tet3 DNA dioxygenase in epigenetic reprogramming by oocytes. *Nature, 477*(7366), 606–610.

Gurdon, J. B. (1962). Adult frogs derived from the nuclei of single somatic cells. *Developmental Biology, 4*, 256–273.

Harris, C. J., et al. (2018). A DNA methylation reader complex that enhances gene transcription. *Science, 362*(6419), 1182–1186.

Hayakawa, K., et al. (2012). Oocyte-specific linker histone H1foo is an epigenomic modulator that decondenses chromatin and impairs pluripotency. *Epigenetics, 7*(9), 1029–1036.

Hermann, A., Goyal, R., & Jeltsch, A. (2004). The Dnmt1 DNA-(cytosine-C5)-methyltransferase methylates DNA processively with high preference for hemimethylated target sites. *The Journal of Biological Chemistry, 279*(46), 48350–48359.

Hezroni, H., et al. (2011). H3K9 histone acetylation predicts pluripotency and reprogramming capacity of ES cells. *Nucleus, 2*(4), 300–309.

Hirsch, C. L., et al. (2015). Myc and SAGA rewire an alternative splicing network during early somatic cell reprogramming. *Genes & Development, 29*(8), 803–816.

Hou, P., et al. (2013). Pluripotent stem cells induced from mouse somatic cells by small-molecule compounds. *Science, 341*(6146), 651–654.

Hu, X., et al. (2014). Tet and TDG mediate DNA demethylation essential for mesenchymal-to-epithelial transition in somatic cell reprogramming. *Cell Stem Cell, 14*(4), 512–522.

Huangfu, D., et al. (2008a). Induction of pluripotent stem cells from primary human fibroblasts with only Oct4 and Sox2. *Nature Biotechnology, 26*(11), 1269–1275.

Huangfu, D., et al. (2008b). Induction of pluripotent stem cells by defined factors is greatly improved by small-molecule compounds. *Nature Biotechnology, 26*(7), 795–797.

Inoue, A., & Zhang, Y. (2014). Nucleosome assembly is required for nuclear pore complex assembly in mouse zygotes. *Nature Structural & Molecular Biology, 21*(7), 609–616.

Kang, L., et al. (2009). iPS cells can support full-term development of tetraploid blastocyst-complemented embryos. *Cell Stem Cell, 5*(2), 135–138.

Koche, R. P., et al. (2011). Reprogramming factor expression initiates widespread targeted chromatin remodeling. *Cell Stem Cell, 8*(1), 96–105.

Kunitomi, A., et al. (2016). H1foo has a pivotal role in qualifying induced pluripotent stem cells. *Stem Cell Reports, 6*(6), 825–833.

Leonhardt, H., et al. (1992). A targeting sequence directs DNA methyltransferase to sites of DNA replication in mammalian nuclei. *Cell, 71*(5), 865–873.

Lewis, P. W., et al. (2010). Daxx is an H3.3-specific histone chaperone and cooperates with ATRX in replication-independent chromatin assembly at telomeres. *Proceedings of the National Academy of Sciences of the United States of America, 107*(32), 14075–14080.

Li, X., et al. (2012). The histone acetyltransferase MOF is a key regulator of the embryonic stem cell core transcriptional network. *Cell Stem Cell, 11*(2), 163–178.

Liang, G., He, J., & Zhang, Y. (2012). Kdm2b promotes induced pluripotent stem cell generation by facilitating gene activation early in reprogramming. *Nature Cell Biology, 14*(5), 457–466.

Lin, C. J., et al. (2014). Hira-mediated H3.3 incorporation is required for DNA replication and ribosomal RNA transcription in the mouse zygote. *Developmental Cell, 30*(3), 268–279.

Liu, J., et al. (2014). A METTL3-METTL14 complex mediates mammalian nuclear RNA N6-adenosine methylation. *Nature Chemical Biology, 10*(2), 93–95.

Ma, D. K., et al. (2008). G9a and Jhdm2a regulate embryonic stem cell fusion-induced reprogramming of adult neural stem cells. *Stem Cells, 26*(8), 2131–2141.

Macfarlan, T. S., et al. (2011). Endogenous retroviruses and neighboring genes are coordinately repressed by LSD1/KDM1A. *Genes & Development, 25*(6), 594–607.

Mali, P., et al. (2010). Butyrate greatly enhances derivation of human induced pluripotent stem cells by promoting epigenetic remodeling and the expression of pluripotency-associated genes. *Stem Cells, 28*(4), 713–720.

Mansour, A. A., et al. (2012). The H3K27 demethylase Utx regulates somatic and germ cell epigenetic reprogramming. *Nature, 488*(7411), 409–413.

Matoba, S., et al. (2014). Embryonic development following somatic cell nuclear transfer impeded by persisting histone methylation. *Cell, 159*(4), 884–895.

Mikkelsen, T. S., et al. (2008). Dissecting direct reprogramming through integrative genomic analysis. *Nature, 454*(7200), 49–55.

Nashun, B., et al. (2015). Continuous histone replacement by Hira is essential for normal transcriptional regulation and De novo DNA methylation during mouse oogenesis. *Molecular Cell, 60*(4), 611–625.

Onder, T. T., et al. (2012). Chromatin-modifying enzymes as modulators of reprogramming. *Nature, 483*(7391), 598–602.

Onuchic, V., et al. (2018). Allele-specific epigenome maps reveal sequence-dependent stochastic switching at regulatory loci. *Science, 361*(6409), eaar3146.

Ooga, M., et al. (2008). Changes in H3K79 methylation during preimplantation development in mice. *Biology of Reproduction, 78*(3), 413–424.

Pawlak, M., & Jaenisch, R. (2011). De novo DNA methylation by Dnmt3a and Dnmt3b is dispensable for nuclear reprogramming of somatic cells to a pluripotent state. *Genes & Development, 25*(10), 1035–1040.

Ping, X. L., et al. (2014). Mammalian WTAP is a regulatory subunit of the RNA N6-methyladenosine methyltransferase. *Cell Research, 24*(2), 177–189.

Ransom, M., Dennehey, B. K., & Tyler, J. K. (2010). Chaperoning histones during DNA replication and repair. *Cell, 140*(2), 183–195.

Roberts, C., et al. (2002). Targeted mutagenesis of the Hira gene results in gastrulation defects and patterning abnormalities of mesoendodermal derivatives prior to early embryonic lethality. *Molecular and Cellular Biology, 22*(7), 2318–2328.

Saeki, H., et al. (2005). Linker histone variants control chromatin dynamics during early embryogenesis. *Proceedings of the National Academy of Sciences of the United States of America, 102*(16), 5697–5702.

Saxonov, S., Berg, P., & Brutlag, D. L. (2006). A genome-wide analysis of CpG dinucleotides in the human genome distinguishes two distinct classes of promoters. *Proceedings of the National Academy of Sciences of the United States of America, 103*(5), 1412–1417.

Shinagawa, T., et al. (2014). Histone variants enriched in oocytes enhance reprogramming to induced pluripotent stem cells. *Cell Stem Cell, 14*(2), 217–227.

Skene, P. J., & Henikoff, S. (2013). Histone variants in pluripotency and disease. *Development, 140*(12), 2513–2524.

Soufi, A., Donahue, G., & Zaret, K. S. (2012). Facilitators and impediments of the pluripotency reprogramming factors' initial engagement with the genome. *Cell, 151*(5), 994–1004.

Sridharan, R., et al. (2013). Proteomic and genomic approaches reveal critical functions of H3K9 methylation and heterochromatin protein-1gamma in reprogramming to pluripotency. *Nature Cell Biology, 15*(7), 872–882.

Stadtfeld, M., et al. (2010). Aberrant silencing of imprinted genes on chromosome 12qF1 in mouse induced pluripotent stem cells. *Nature, 465*(7295), 175–181.

Stadtfeld, M., et al. (2012). Ascorbic acid prevents loss of Dlk1-Dio3 imprinting and facilitates generation of all-iPS cell mice from terminally differentiated B cells. *Nature Genetics, 44*(4), 398–405, S1-2.

Suetake, I., et al. (2004). DNMT3L stimulates the DNA methylation activity of Dnmt3a and Dnmt3b through a direct interaction. *The Journal of Biological Chemistry, 279*(26), 27816–27823.

Syed, K. M., et al. (2016). Histone chaperone APLF regulates induction of pluripotency in murine fibroblasts. *Journal of Cell Science, 129*(24), 4576–4591.

Tada, M., et al. (1997). Embryonic germ cells induce epigenetic reprogramming of somatic nucleus in hybrid cells. *The EMBO Journal, 16*(21), 6510–6520.

Takahashi, K., & Yamanaka, S. (2006). Induction of pluripotent stem cells from mouse embryonic and adult fibroblast cultures by defined factors. *Cell, 126*(4), 663–676.

Tanaka, M., et al. (2003). H1oo: A pre-embryonic H1 linker histone in search of a function. *Molecular and Cellular Endocrinology, 202*(1–2), 5–9.

Tanaka, M., et al. (2005). H1FOO is coupled to the initiation of oocytic growth. *Biology of Reproduction, 72*(1), 135–142.

Wang, J., et al. (2009). The lysine demethylase LSD1 (KDM1) is required for maintenance of global DNA methylation. *Nature Genetics, 41*(1), 125–129.

Wang, T., et al. (2011). The histone demethylases Jhdm1a/1b enhance somatic cell reprogramming in a vitamin-C-dependent manner. *Cell Stem Cell, 9*(6), 575–587.

Wang, H., et al. (2014a). Small molecules enable cardiac reprogramming of mouse fibroblasts with a single factor, Oct4. *Cell Reports, 6*(5), 951–960.

Wang, Y., et al. (2014b). N6-methyladenosine modification destabilizes developmental regulators in embryonic stem cells. *Nature Cell Biology, 16*(2), 191–198.

Warren, L., et al. (2010). Highly efficient reprogramming to pluripotency and directed differentiation of human cells with synthetic modified mRNA. *Cell Stem Cell, 7*(5), 618–630.

Wellman, S. E., et al. (1987). Characterization of mouse H3.3-like histone genes. *Gene, 59*(1), 29–39.

Wen, D., et al. (2014a). H3.3 replacement facilitates epigenetic reprogramming of donor nuclei in somatic cell nuclear transfer embryos. *Nucleus, 5*(5), 369–375.

Wen, D. C., et al. (2014b). Histone variant H3.3 is an essential maternal factor for oocyte reprogramming. *Proceedings of the National Academy of Sciences of the United States of America, 111*(20), 7325–7330.

Whyte, W. A., et al. (2012). Enhancer decommissioning by LSD1 during embryonic stem cell differentiation. *Nature, 482*(7384), 221–225.

Wu, H., & Zhang, Y. (2011). Mechanisms and functions of Tet protein-mediated 5-methylcytosine oxidation. *Genes & Development, 25*(23), 2436–2452.

Wu, T., et al. (2014). Histone variant H2A.X deposition pattern serves as a functional epigenetic mark for distinguishing the developmental potentials of iPSCs. *Cell Stem Cell, 15*(3), 281–294.

Zhao, X. Y., et al. (2009). iPS cells produce viable mice through tetraploid complementation. *Nature, 461*(7260), 86–90.

Zhou, V. W., Goren, A., & Bernstein, B. E. (2011). Charting histone modifications and the functional organization of mammalian genomes. *Nature Reviews. Genetics, 12*(1), 7–18.

Human Pluripotent Stem Cells and Neural Regeneration

Xiaoqing Zhang

© Springer Nature Switzerland AG 2020
B. Brand-Saberi (ed.), *Essential Current Concepts in Stem Cell Biology*,
Learning Materials in Biosciences, https://doi.org/10.1007/978-3-030-33923-4_9

What You Can Learn in This Chapter
What are the cellular and molecular processes underlying neural developmental events during embryogenesis? How can we manipulate a human pluripotent stem cell to a neural lineage of interest? Can the *in vitro* generated neural cells hold the potentiality for replacement therapy in neurological disorders?

9.1 Summary

Neurological disorders always end up with neuronal loss and neural circuit dysfunction, which could not self-repair since local neural progenitors only generate restricted neuronal subtypes, and these progenitor cells also decline remarkably during aging. Human pluripotent stem cells (hPSCs) are a valuable cell source to produce almost the entire spectrum of regional neural progenitors and then different neuronal subtypes, which showed promising potentials to replenish defined neuronal loss and restore functional neural circuits in animal models (Thomson et al. 1998; Zhang et al. 2001). Stem cell-based replacement therapy for several neurological disorders is now undergoing intensive clinical observations. This chapter mainly focuses on cellular and molecular aspects of neural development, strategies of converting human pluripotent stem cells to desired neuronal subtypes, and exemplifications of applying human pluripotent stem cells to treat neurological disorders, such as Alzheimer's disease, Huntington's disease and Parkinson's disease.

9.2 Human Pluripotent Stem Cells

Human pluripotent stem cells (hPSCs) include human embryonic stem cells (hESCs) and human induced pluripotent stem cells (hiPSCs), which can self-renew or differentiate into all three germ layers under proper culture conditions. In 1998, Thomson et al., successfully isolated the cells within the inner cell mass from the early human blastocyst and long-term maintained them as hESCs on mouse embryonic fibroblast feeder layer in the presence of serum replacements supplied with basic fibroblast growth factors (FGF2) (Thomson et al. 1998). Further studies identified the essential roles of transcription factors, mostly Nanog, Oct4 and Sox2, which activate the network of pluripotency genes and repress lineage differentiation genes, and therefore keep the hESCs in an undifferentiated state. Later on, through expressing of Oct4, Sox2, Klf4 and c-Myc in differentiated somatic cells, hiPSCs were established after being given the powerful reprogramming capabilities of these pluripotency factors. Genetic, epigenetic and functional assays confirmed that hiPSCs largely resemble the nature of hESCs.

As *in vivo*, a fundamental aspect of hPSCs is their potency to generate the whole spectrum of fully specified functional cell lineages of ectodermal, mesodermal or endodermal origin. Dynamically monitoring the *in vitro* differentiation process in combination with gene targeting techniques is now an alternative way to study key developmental events during cell fate conversion, which adds enormously to the traditional paradigm in studying development via animal models. *In vitro* specified human cells also provide a previously unachieved cell model for pathologic study and phenotypic drug screening for genetic disorders. Eventually, yielding functional cell types and treating patients bearing currently incurable diseases will be one of the most important applications for hPSCs.

9.3 Neural Development

In order to fully apply hPSCs for studying development and disease, it is crucial to build differentiation protocols for various lineages of all three germ layers. The ideal protocols for lineage specification should be chemically-defined, highly efficient and faithfully mirror general principles of *in vivo* development.

Mammalian neural development could be roughly discriminated into three stages, neural induction, regional patterning as well as neurogenesis or gliogenesis. Neural induction is a process happening in a gastrulating embryo, where the upper tissue layer (epiblast) becomes thickened and flattened, and develops into the neural plate comprising columnar neural epithelial cells (Sasai and De Robertis 1997). The neural plate is the anlage to form the future entire central nervous system (CNS), which later on will become the brain and the spinal cord. On the bilateral edges of the neural plate, there are neural plate borders. During neurulation, the neural plate borders lift upwards and converge at the dorsal midline to form the neural tube. Regional patterning happens concomitantly when the neural tube forms. Regional neural progenitors (NPs) will be specified from the primitive neural epithelial cells in the neural plate along both the anteroposterior (A-P) and dorsoventral (D-V) axes. In the dorsal midline region of a closed neural tube, converged neural plate borders become the roof plate. Meanwhile, following an epithelial to mesenchymal transition, neural crest cells delaminate from the roof plate, which are primary origins of the peripheral nervous system (PNS), including the cranial, spinal and autonomic nerves as well as Schwann cells and pigment cells (◻ Fig. 9.1).

The notochord, a transient rod-like structure of mesoderm origin, locates ventrally to the midline of the posterior half of the neural tube. The notochord plays an important role in maintaining the left-right asymmetry and development of adjacent tissues. Given signals secreted from the notochord, the adjacent ventral midline neural epithelial cells of the neural plate or neural tube will be specified into the floor plate (FP). The FP of the midbrain harbors neurogenic activity and is the major region to generate dopaminergic neurons.

Converting pluripotent epiblast cells to neural epithelia, and specifying neural epithelia into various regional neural progenitors during early neural development could be summarized into an activation-transformation paradigm (Chi et al. 2017). The stem cell located at a higher hierarchy will adopt or activate a prominent cell fate which does not require additional inductive signals, while additional developmentally related signals must be in place in order to guide or transform the stem cell to other different cell fates. During neural induction, the epiblast will adopt a neural fate by default, and Wnts and

◻ **Fig. 9.1** Neural induction and regional patterning are two major cellular events accompanying neural plate and neural tube formation during embryogenesis

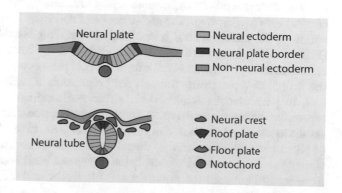

Neural plate

☐ Neural ectoderm
■ Neural plate border
☐ Non-neural ectoderm

Neural tube

◗ Neural crest
▼ Roof plate
◖ Floor plate
● Notochord

transforming growth factor β (TGFβ) superfamily members drive the epiblast to meso-dermal, endodermal, and non-neural ectodermal tissues (Chambers et al. 2009). As to A-P patterning, neural ectodermal cells will take a prospective forebrain regional identity, while caudalization signals, including fibroblast growth factor 8 (FGF8), Wnts and reti-noic acid (RA) generate midbrain, hindbrain and cervical spinal cord progenitors accord-ingly (Metzis et al. 2018). It has also been well demonstrated that in human cells, D-V patterning of the forebrain and spinal cord also follows the activation/transformation model (Chi et al. 2017). Both the human forebrain and spinal cord adopt dorsal telen-cephalon or dorsal spinal cord fate in the absence of additional inductive cues, and sonic hedgehog (Shh) is a robust and required signal morphogen to ventralize the forebrain and spinal primordium to ventral telencephalon and the ventral region of the spinal cord (Li et al. 2005; Li et al. 2009).

The activation-transformation paradigm can be partly explained by preset expression of key intrinsic transcription factors in uncommitted stem cells. Inner cell mass and epi-blast cells express Sox2. During the neural induction stage, Sox2 is maintained in neural ectodermal cells within the neural plate, but shuts off when the pluripotent cells are sig-naled to a mesodermal or endodermal fate (Ying et al. 2003). Otx2 is a hallmark gene expressed restrictedly in the anterior neural tube, including the forebrain and midbrain, but not the hindbrain. Otx2 is also early expressed in the epiblast and neural ectodermal cells, and caudalization signals downregulate Otx2 expression and are indispensable for midbrain and hindbrain regional specification. Another striking example is related to the D-V patterning of human forebrain. Human neural ectodermal cells uniformly express Pax6, a powerful transcription factor for specification of the dorsal fate of the mammalian telencephalon (Zhang et al. 2010). Without Shh, Pax6 represses developmentally related ventral genes and therefore specifies and maintains the neuroectodermal cells to a dorsal telencephalic fate (Gaspard et al. 2008; Chi et al. 2017). The ventralization morphogen Shh represses Pax6 expression and in turn activates ventral gene expression to specify the ven-tral telencephalon via this repression-release model.

Within the activation-transformation paradigm, inductive signals to transform a pre-set cell fate to another are mostly secreted from the patterning centers, such as the roof plate, the FP, and the notochord. The notochord is a major patterning center for producing Shh. Shh emitted from notochord induces the FP in a neural tube, which also produces Shh to strengthen the ventralization magnitude (Ericson et al. 1995). In Shh null mouse, all ventral telencephalic NPs are missing as assessed by a loss of expression of all ventral markers including Nkx2.1, Dlx2 and Gsx2. Moreover, Shh represses the expression of various transcription factor genes related to the dorsal development. After neural tube closure, Wnts and bone morphogenetic proteins (BMPs), derived from the roof plate and cortical hem participate in the maintenance of the dorsal identity. Activation of Wnt sig-naling in the mouse ventral telencephalon also represses the subpallium development (Wilson and Rubenstein 2000). It is thought that the combined effects of the gradient concentrations of Wnts and BMPs from the roof plate as well as Shh from the FP regulate the overall D-V patterning through tight regulation of region-specific transcription fac-tors, in particular, Pax6 and Nkx2.1. Transgenic animal studies also reveal a mutual repression between these region-specific transcription factors. Pax6 loss of function results in abnormal expression of ventral marker genes in the dorsal territory, whereas loss of Nkx2.1 results in a ventral to dorsal respecification as evidenced by extended expres-sion of Pax6 in the ventral telencephalon. FGF8 and Wnt1 secreted from the isthmic orga-nizer located between the midbrain and hindbrain, and the RA synthesized within the

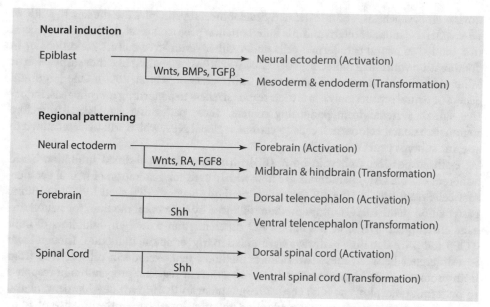

Neural induction

Epiblast ——————————→ Neural ectoderm (Activation)

Wnts, BMPs, TGFβ ——→ Mesoderm & endoderm (Transformation)

Regional patterning

Neural ectoderm ——————————→ Forebrain (Activation)

Wnts, RA, FGF8 ——→ Midbrain & hindbrain (Transformation)

Forebrain ——————————→ Dorsal telencephalon (Activation)

Shh ——→ Ventral telencephalon (Transformation)

Spinal Cord ——————————→ Dorsal spinal cord (Activation)

Shh ——→ Ventral spinal cord (Transformation)

Fig. 9.2 Cell fate conversion during neural induction and regional patterning follows an activation/transformation paradigm

hindbrain are two potent caudalization signals, which are crucial for the normal patterning and development of the midbrain and hindbrain (◻ Fig. 9.2).

Neurogenesis continues throughout embryonic development and postnatal life. Usually, a neuronal subtype is determined by the regional identity of its parental NP. NPs can self-renew for several rounds and then exit the cell cycle and differentiate into neurons and glia, including astrocytes and oligodendrocytes, sequentially. Differentiated neurons will then migrate to their destinations and integrate to form functional neural circuits, which is the basic unit to conduct a specific neurological activity. As to the cortical development, the rapidly dividing NPs locate in the ventricular zone/subventricular zone of the brain during the neural development, and subsequently differentiate into various cells in the cortical plate of the cortex. The neurogenesis in the cortex follows an "inside-out" pattern of morphogenesis: The neurons born from cortical progenitors with early birthdays tend to migrate shorter distances and those with late birthdays migrate further. During the migration, neurons are guided by the radial glial cells, which extend processes from the inner to the outer surface of the cortex. In the ventral part of the telencephalon, NPs in the medial ganglionic eminence (MGE) are the major source to produce GABAergic interneurons. GABAergic interneurons migrate tangentially in a long distance to the dorsal telencephalon, where they rearrange and mature, and form inhibitory synapses with local excitory cortical neurons.

9.4 Targeted Neural Differentiation

Both hESCs and hiPSCs are able to sequentially differentiate into neural ectoderm, regional NPs and various neurons and glia *in vitro* (Zhang et al. 2001). More importantly, the differentiation processes *in vitro* mirror exactly the developmental events happened during embryogenesis, and the defined developmental principles are the key to guide *in vitro* differentiation. Though multiple differentiation protocols exist, hPSCs are in general first guided to neural ectodermal cells under either serum-free culture conditions or in combination with small molecules to inhibit TGFβ/BMP signaling in either suspension or adherent culture conditions (Zhang et al. 2001; Chambers et al. 2009). One important aspect of neural ectodermal cells is their responsiveness to patterning morphogens, inductive signals secreted from patterning centers. These patterning signals will therefore regionalize neural ectodermal cells to various regional NPs, which will be determined to specific subtype neurons and glia.

Both suspension embryoid body (EB) formation and dual-Smad inhibition-based adherent culture (AD) paradigms are now widely used for generation of neural ectodermal cells from hPSCs. The EB formation method suspends detached hPSCs to mimic gastrulation in the hPSC culture medium followed by the neural medium for neural lineage enrichment. While in the hPSCs-AD differentiation paradigm, inhibitors of both TGFβ and BMP signaling pathways were added to trigger neural induction. Though there is a difference between these two distinct methods, neuroectoderm cells derived from both protocols show high potency to generate different regional progenitors in response to patterning morphogens (Chi et al. 2016). Similar to the *in vivo* development, neural ectodermal cells will automatically adopt an anterior dorsal fate in the absence of additional patterning morphogens. As a potent caudalization patterning morphogen, RA efficiently caudalized neural ectodermal cells to a hindbrain and cervical spinal cord identity in both EB and AD differentiation paradigms (Li et al. 2005).

Recent research indicates that the caudal part of the spinal cord develops from neuromesodermal progenitors (NMP), which adopt a posterior regional identity even before the neural ectodermal fate has been initiated (Metzis et al. 2018). It seems that the entire epiblast could be subdivided into an anterior and a posterior compartment. It is more likely that the EB differentiation protocol favors the generation of an anterior neural ectoderm. Shh ventralizes this anterior neural ectoderm to a ventral telencephalic fate, such as the lateral ganglionic eminence (LGE) and MGE (Ma et al. 2012; Liu et al. 2013). FGF8 and Wnts regionalize the neural ectoderm generated from the EB method to a midbrain fate. However, hPSCs are easily guided to the posterior developmental structures under the AD differentiation paradigm. Under AD differentiation, early exposure of Shh causes targeted differentiation of hPSCs to a FP fate (Nkx2.1+/Sox1−/FoxA2+), while under the EB conditions, Shh activation determines a MGE fate (Nkx2.1+/Sox1+/FoxA2−) (Fasano et al. 2010; Chi et al. 2016). Spinal motor neurons are generated from the AD conditions with a much higher efficiency as compared with the EB conditions. One can expect that under the caudalization signal, hPSCs will be guided to the posterior epiblast and NMPs, which will generate more caudal spinal cord progenitors as well. Since obtaining disease related NPs and neuronal subtypes is the key for modeling or treating specific neurological disorders, EB *vs* AD differentiation paradigms need to be carefully selected.

The gradient of the patterning morphogens are crucial for establishing a regional identity of the neural progenitors. In *in vitro* neural differentiation, concentrations and durations of the patterning signals applied are equally critical. For example, medium level of Shh (200 ng/ml) treatment in neural ectodermal cells yielded from the EB method will end up with a LGE fate, while a higher level of Shh (500–1000 ng/ml) treatment in the same cells will generate more ventral MGE fate (Ma et al. 2012; Liu et al. 2013; Maroof et al. 2013). As aforementioned, NPs will sequentially generate neurons and glia. Oligo2+ ventral spinal NPs generated from hPSCs will first differentiate into the spinal motor neu-

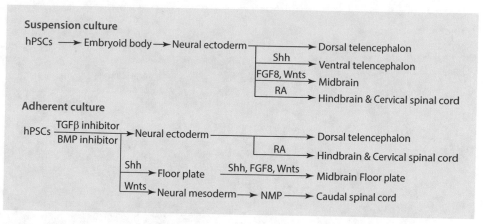

Fig. 9.3 Targeted neuronal differentiation through suspension culture or adherent culture in combination with patterning morphogens

rons in the first month of differentiation, but will generate oligodendrocytes when the progenitors are maintained for 3–6 months *in vitro* (□ Fig. 9.3).

9.5 LGE Progenitor Differentiation and Huntington's Disease

Huntington's disease (HD) is an autosomal dominant neurodegenerative disorder with distinct symptoms, including chorea and dystonia, sleep disorders, motor dysfunction, cognitive impairment, and psychiatric abnormalities. Genetically, HD is caused by the expansion of Cytosine, Adenine, and Guanine (CAG) repeats near the start of exon 1 in the gene encoding the protein Huntingtin (HTT). The mutant HTT (mHTT) proteins are detrimental for the GABAergic striatal medium spiny neurons (MSNs) in the basal ganglia, which constitute 95% of all neurons in the striatum.

Recently, cell replacement therapies represent a promising direction for the treatment of HD (Ma et al. 2012). As a proof-of-concept study, fetal neural tissue specimens isolated from donor fetal brain had been used as neuro-grafts into HD patients in clinical trials, which showed moderate improvement in motor function. Though promising, this cell replacement therapy based on fetal tissues is technically limited because of a lack of enough donor tissues, poorly defined cell types and ethics issues. Several pioneering studies have successfully specified hPSCs into LGE progenitors and functional striatal MSNs. This is mostly achieved by applying medium level of Shh in human neural ectodermal cells under EB culture conditions. Cocktailing the Activin A and Wnts signaling shows an improvement in generating LGE progenitors. The LGE progenitors yielded from hPSCs are positive for the forebrain marker FoxG1, and LGE marker Meis2 and Gsx2. MSNs generated from these LGE progenitors *in vitro* have typical spiny morphology and uniformly express DARPP32 and GABA. More importantly, MSNs differentiated from hPSCs show gradual maturation and have activity-related neurotransmitter release, spontaneous action potentials and synaptic connections *in vitro*. Grafting of hPSCs differentiated LGE progenitors in either surgically or genetically modeled HD mice re-populated GABAergic MSNs neuronal loss in the striatum, and significantly improved cognitive and motor deficits. In a recent study, Schaffer and colleagues developed a scalable biomaterial-based 3D

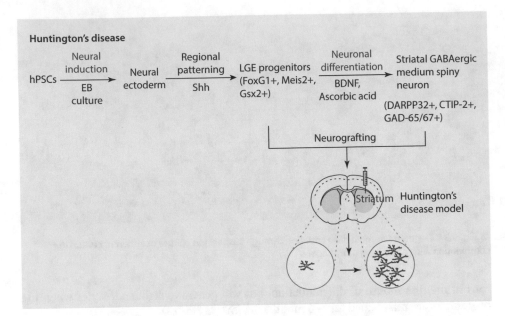

Fig. 9.4   Targeted LGE progenitor and MSN differentiation for cell replacement therapy for HD

platform to generate LGE progenitors and MSNs from hPSCs. LGE progenitors generated from this 3D system showed a better survival of transplanted cells and functional recovery in transgenic mouse HD model, suggesting the robustness of combined material science and stem cell techniques in cell-based replacement therapy (◘ Fig. 9.4).

9.6   MGE Progenitor Differentiation and Alzheimer's Disease

Alzheimer's disease (AD) is a neurodegenerative disease associated with serious loss of presynaptic cholinergic functions. AD patients usually suffer from a progressive decline in memory and cognitive function as well as behavioral symptoms, such as disorientation and hallucinations. Previous studies indicated that the decreased acetylcholine (ACh) release is a major feature of AD, which results from the declined number and functionality of basal forebrain cholinergic neurons (BFCNs) in a relatively early stage of AD. Several strategies exist in order to alleviate AD symptoms. For example, additional supply of neurotrophic factors (NTFs) directly or through a cell carrier is a way to improve the survival and function of cholinergic neurons. Acetylcholinesterase Inhibitor (AchEI) and N-methyl-D-aspartate receptor agonist treatment have also been applied in AD disease models or even patients, which show moderate benefits in cognition and memory recovery. However, all these therapeutic strategies only showed mild and temporary effectiveness.

Cell replacement therapy has also been proposed to be an ultimate way to cure AD. MGE is the sole origin to generate BFCNs during embryogenesis. Both hESCs and hiPSCs have been efficiently differentiated toward a MGE fate and then BFCNs (Liu et al. 2013; Yue et al. 2015). High concentrations of Shh under the EB differentiation conditions is the key to regionalize neural ectodermal cells to MGE progenitors, which belongs to the ventral most part of telencephalon. Blocking the Wnts pathway concomitantly with the presence of the ventralization morphogen Shh will facilitate ventral patterning. Though Shh treatment in AD differentiated hPSCs will guide the cells otherwise to the FP fate, small molecules inhibiting the Wnts/P38/JAK-STAT pathways generates MGE progenitors under the AD differentiation conditions. The MGE progenitors differentiated in vitro uniformly express Nkx2.1 and FoxG1. Matured BFCNs differentiated from these MGE progenitors express MAP2 and Synapsin, key functional neuronal markers, and ChAT, the rate limited enzyme for synthesize ACh. The in vitro produced BFCNs also showed active action potentials and synaptic activities over long periods of maturation in the culture. Both with surgical lesion and transgenic AD mouse models, transplantation of MGE progenitors into the bilateral hippocampus or basal nuclear shows clear and striking cognitive functional recovery. Histological and electrophysiological studies have also confirmed BFCN differentiation, maturation, long-term survival, and forming defined neural circuits with local neurons (◻ Fig. 9.5).

9.7 FP Progenitor Differentiation and Parkinson's Disease

As one of the most prominent neurodegenerative disorders, Parkinson's disease (PD) bearing patients usually suffer from tremor, hypokinesia, rigidity and abnormal gait and posture. Pathological studies show that the cellular events underlying PD is the progressive death of dopaminergic (DA) neurons reside in the Substantia nigra, which causes insufficient release of dopamine in the striatum where DA neurons project.

The most commonly used drug for treating PD in the clinic is L-DOPA, which is selectively transported into the DA neurons and where it is readily converted to dopamine to compensate the reduced dopamine release in the caudate nucleus and putamen. L-DOPA has been observed with significant clinical benefit, but long-term use of L-DOPA showed ineffectiveness and has side effects, such as the on–off fluctuations and the emergence of dyskinesias. Therefore, there is a need for a more complete and long lasting method for restoring dopamine neurotransmission.

In 1992, human fetal mesencephalic tissues were transplanted into the caudate nucleus and putamen of Parkinson's patients for clinical trials. Symptoms, such as periods of dyskinesia and off episodes, were improved in patients after transplantation. This proof-of-concept clinical study suggests that cell transplantation therapy holds great potential for treating PD. However, several clinical studies failed to reveal statistically significant outcomes of fetal mesencephalic tissues for replacement therapy of PD, probably because of the variable quality of the donor tissues.

Midbrain DA neurons have now efficiently been generated from hPSCs (Kriks et al. 2011; Xi et al. 2012; Steinbeck et al. 2015; Wu et al. 2015; Chen et al. 2016; Kikuchi et al. 2017). It is now well acknowledged that the AD differentiation procedure is the most appropriate way to generate FP progenitors and DA neurons. Through applying Shh and FGF8 to ventralize and caudalize the adherently cultured hPSCs to induce a midbrain FP progenitor fate (EN1+/OTX2+ /FOXA2+/ LMX1A+) and adjusting Wnts signaling to promote dopaminergic differentiation, midbrain DA neurons (TH+/EN1+/OTX2+/FOXA2+/LMX1A+/NURR1+) can be efficiently produced. Moreover, these differentiated DA neurons displayed spontaneous action potential spikes, and this spiking was accompanied by a slow, subthreshold oscillatory potential resembling midbrain DA neurons in vivo. In addition, after transplantation of these midbrain DA neuronal progenitors into the striatum of the PD mouse or rat model, a complete restoration of amphetamine-

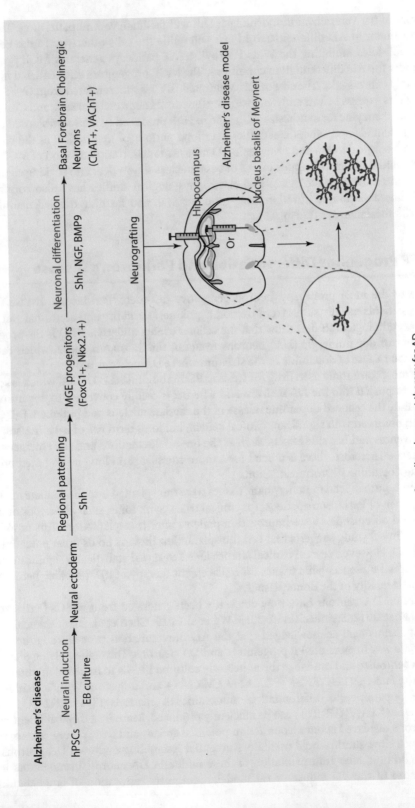

Fig. 9.5 Targeted MGE progenitor and BFCN differentiation for cell replacement therapy for AD

induced rotation behavior and improvements in tests of forelimb use and akinesia are observed. Notably, the efficacy of hPSCs-based dopaminergic progenitor cells transplantation in PD has now been also proved in monkeys and under clinical trials (Wang et al. 2018). A pioneering study conducted by Zhang and colleagues developed a system that enables precise regulation of hPSCs-derived neuronal activity for *in vivo* transplantation through chemogenetics (Chen et al. 2016). The midbrain DA neurons differentiated from hPSCs engineered to express DREADDs (designer receptors exclusively activated by designer drug) showed tight regulation of the activity of engrafted neurons, thus offering more accurate tools for future PD treatment (Fig. 9.6).

9.8 Conclusions and Perspectives

In this chapter, basic neural developmental concepts, especially related to neural induction and regional specification are introduced. Importantly, the basic neural developmental principles also apply to hPSCs when they are differentiated toward a neural fate. Both neural induction and regional patterning follows an activation/transformation model, that is, the cells will first take a cell fate by default and additive inductive signals are required to guide the cells to other fates instead. Both hESCs and hiPSCs hold great potentials in generating bona fide regional human NPs and according neuronal subtypes, which will offer unlimited cell sources for cell based replacement therapy for those publicly concerned neurodegenerative diseases, such as HD, AD and PD. Though exciting and promising evidence is accumulating by applying hPSCs in treating neurological diseases in animal models, careful and systematic clinical trials are urgently needed in order to fully prove the effectiveness and safety of applying hPSCs in treating patients. Integrating material science and genetic engineering techniques in cell replacement therapy in a plus to safeguard and move forward this promising approach to an ultimate success in the clinic.

> **Take Home Message**
>
> 1. The entire process of neural development can be categorized into three stages in sequential, neural induction, regional patterning as well as neurogenesis and gliogenesis.
> 2. The activation/transformation paradigm applies to most of the cell fate determination events during *in vivo* neural development and *in vitro* neural differentiation.
> 3. Extracellular cues, such as TGFβ, BMPs, Wnts, Shh, RA and FGFs signaling molecules, are crucial cell fate inducers or blockers for either neural induction or regional patterning of neural progenitors.
> 4. Human pluripotent stem cells could be efficiently targeted to neural ectoderm cells, various regional neural progenitors, and different neuronal subtypes, which mirrors *in vivo* neural development.
> 5. A series of proof-of-concept laboratory studies have validated the efficacy of human pluripotent stem cell-based replacement therapy for currently incurable neurological diseases.

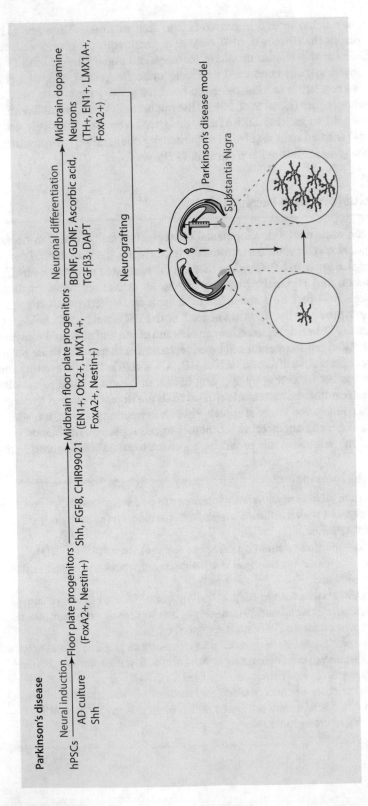

Fig. 9.6   Targeted midbrain FP progenitor and DA neuron differentiation for cell replacement therapy for PD

References

Chambers, S. M., Fasano, C. A., Papapetrou, E. P., Tomishima, M., Sadelain, M., & Studer, L. (2009). Highly efficient neural conversion of human ES and iPS cells by dual inhibition of SMAD signaling. *Nature Biotechnology, 27,* 275–280.

Chen, Y., Xiong, M., Dong, Y., Haberman, A., Cao, J., Liu, H., Zhou, W., & Zhang, S.-C. (2016). Chemical control of grafted human PSC-derived neurons in a mouse model of Parkinson's disease. *Cell Stem Cell, 18,* 817–826.

Chi, L., Fan, B., Zhang, K., Du, Y., Liu, Z., Fang, Y., Chen, Z., Ren, X., Xu, X., Jiang, C., et al. (2016). Targeted differentiation of regional ventral neuroprogenitors and related neuronal subtypes from human pluripotent stem cells. *Stem Cell Reports, 7,* 941–954.

Chi, L., Fan, B., Feng, D., Chen, Z., Liu, Z., Hui, Y., Xu, X., Ma, L., Fang, Y., Zhang, Q., et al. (2017). The dorsoventral patterning of human forebrain follows an activation/transformation model. *Cerebral Cortex, 27,* 2941–2954.

Ericson, J., Muhr, J., Placzek, M., Lints, T., Jessel, T. M., & Edlund, T. (1995). Sonic hedgehog induces the differentiation of ventral forebrain neurons: A common signal for ventral patterning within the neural tube. *Cell, 81,* 747–756.

Fasano, C. A., Chambers, S. M., Lee, G., Tomishima, M. J., & Studer, L. (2010). Efficient derivation of functional floor plate tissue from human embryonic stem cells. *Cell Stem Cell, 6,* 336–347.

Gaspard, N., Bouschet, T., Hourez, R., Dimidschstein, J., Naeije, G., van den Ameele, J., Espuny-Camacho, I., Herpoel, A., Passante, L., Schiffmann, S. N., et al. (2008). An intrinsic mechanism of corticogenesis from embryonic stem cells. *Nature, 455,* 351.

Kikuchi, T., Morizane, A., Doi, D., Magotani, H., Onoe, H., Hayashi, T., Mizuma, H., Takara, S., Takahashi, R., Inoue, H., et al. (2017). Human iPS cell-derived dopaminergic neurons function in a primate Parkinson's disease model. *Nature, 548,* 592.

Kriks, S., Shim, J.-W., Piao, J., Ganat, Y. M., Wakeman, D. R., Xie, Z., Carrillo-Reid, L., Auyeung, G., Antonacci, C., Buch, A., et al. (2011). Dopamine neurons derived from human ES cells efficiently engraft in animal models of Parkinson's disease. *Nature, 480,* 547.

Li, X.-J., Du, Z.-W., Zarnowska, E. D., Pankratz, M., Hansen, L. O., Pearce, R. A., & Zhang, S.-C. (2005). Specification of motoneurons from human embryonic stem cells. *Nature Biotechnology, 23,* 215.

Li, X.-J., Zhang, X., Johnson, M. A., Wang, Z.-B., LaVaute, T., & Zhang, S.-C. (2009). Coordination of sonic hedgehog and Wnt signaling determines ventral and dorsal telencephalic neuron types from human embryonic stem cells. *Development, 136,* 4055–4063.

Liu, Y., Weick, J. P., Liu, H., Krencik, R., Zhang, X., Ma, L., Zhou, G.-m., Ayala, M., & Zhang, S.-C. (2013). Medial ganglionic eminence–like cells derived from human embryonic stem cells correct learning and memory deficits. *Nature Biotechnology, 31,* 440.

Ma, L., Hu, B., Liu, Y., Vermilyea, S. C., Liu, H., Gao, L., Sun, Y., Zhang, X., & Zhang, S.-C. (2012). Human embryonic stem cell-derived GABA neurons correct locomotion deficits in Quinolinic acid-lesioned mice. *Cell Stem Cell, 10,* 455–464.

Maroof, A. M., Keros, S., Tyson, J. A., Ying, S.-W., Ganat, Y. M., Merkle, F. T., Liu, B., Goulburn, A., Stanley, E. G., Elefanty, A. G., et al. (2013). Directed differentiation and functional maturation of cortical interneurons from human embryonic stem cells. *Cell Stem Cell, 12,* 559–572.

Metzis, V., Steinhauser, S., Pakanavicius, E., Gouti, M., Stamataki, D., Ivanovitch, K., Watson, T., Rayon, T., Mousavy Gharavy, S. N., Lovell-Badge, R., et al. (2018). Nervous system regionalization entails axial allocation before neural differentiation. *Cell, 175,* 1105–1118.e1117.

Sasai, Y., & De Robertis, E. M. (1997). Ectodermal patterning in vertebrate embryos. *Developmental Biology, 182,* 5–20.

Steinbeck, J. A., Choi, S. J., Mrejeru, A., Ganat, Y., Deisseroth, K., Sulzer, D., Mosharov, E. V., & Studer, L. (2015). Optogenetics enables functional analysis of human embryonic stem cell–derived grafts in a Parkinson's disease model. *Nature Biotechnology, 33,* 204–209.

Thomson, J. A., Itskovitz-Eldor, J., Shapiro, S. S., Waknitz, M. A., Swiergiel, J. J., Marshall, V. S., & Jones, J. M. (1998). Embryonic stem cell lines derived from human blastocysts. *Science, 282,* 1145–1147.

Wang, Y.-K., Zhu, W.-W., Wu, M.-H., Wu, Y.-H., Liu, Z.-X., Liang, L.-M., Sheng, C., Hao, J., Wang, L., Li, W., et al. (2018). Human clinical-grade parthenogenetic ESC-derived dopaminergic neurons recover locomotive defects of nonhuman primate models of Parkinson's disease. *Stem Cell Reports, 11,* 171–182.

Wilson, S. W., & Rubenstein, J. L. R. (2000). Induction and dorsoventral patterning of the telencephalon. *Neuron, 28,* 641–651.

Wu, J., Sheng, C., Liu, Z., Jia, W., Wang, B., Li, M., Fu, L., Ren, Z., An, J., Sang, L., et al. (2015). Lmx1a enhances the effect of iNSCs in a PD model. *Stem Cell Research, 14*, 1–9.

Xi, J., Liu, Y., Liu, H., Chen, H., Emborg, M. E., & Zhang, S.-C. (2012). Specification of midbrain dopamine neurons from primate pluripotent stem cells. *Stem Cells, 30*, 1655–1663.

Ying, Q.-L., Stavridis, M., Griffiths, D., Li, M., & Smith, A. (2003). Conversion of embryonic stem cells into neuroectodermal precursors in adherent monoculture. *Nature Biotechnology, 21*, 183–186.

Yue, W., Li, Y., Zhang, T., Jiang, M., Qian, Y., Zhang, M., Sheng, N., Feng, S., Tang, K., Yu, X., et al. (2015). ESC-derived basal forebrain cholinergic neurons ameliorate the cognitive symptoms associated with Alzheimer's disease in mouse models. *Stem Cell Reports, 5*, 776–790.

Zhang, S.-C., Wernig, M., Duncan, I. D., Brüstle, O., & Thomson, J. A. (2001). In vitro differentiation of transplantable neural precursors from human embryonic stem cells. *Nature Biotechnology, 19*, 1129–1133.

Zhang, X., Huang, C. T., Chen, J., Pankratz, M. T., Xi, J., Li, J., Yang, Y., LaVaute, T. M., Li, X.-J., Ayala, M., et al. (2010). Pax6 is a human neuroectoderm cell fate determinant. *Cell Stem Cell, 7*, 90–100.

Dopaminergic Neuron-Related Stem Cells

Chengzhong Wang and Yu-Qiang Ding

© Springer Nature Switzerland AG 2020
B. Brand-Saberi (ed.), *Essential Current Concepts in Stem Cell Biology*,
Learning Materials in Biosciences, https://doi.org/10.1007/978-3-030-33923-4_10

What You Will Learn in This Chapter

Progressive degeneration of midbrain dopaminergic (mDA) neurons causes Parkinson's disease (PD), the second most common neurodegenerative disease. No cures are available for PD at this time, and healthy mDA neuron transplantation provides a promising disease-modifying therapy. In this chapter, we focus on dopaminergic neuron-related stem cells and their potential application. Specifically, this includes the molecular mechanisms underlying mDA neuron morphogenesis, such as early patterning of mesencephalon (midbrain), and the genes involved in mDA neuron induction and maturation. A variety of strategies to generate mDA neurons for cell-based therapies are summarized, including generation of mDA neurons from neural stem cells (NSCs) and pluripotent stem cells (PSCs), and direct reprogramming of somatic cells into mDA neurons. Lastly, several first-in-human clinical trials using hPSCs-derived mDA neurons are starting to recruit patients or planned to initiate shortly, thus challenges and promises for cell-based therapies for PD are discussed.

10.1 Introduction

Dopaminergic (DA) neurons are distributed in several regions in adult mammalian brain. Among them, DA neurons resided in the substantia nigra pars compacta (SNc) of the ventral midbrain (referred to as mDA neurons) draw much more attention, because the mDA neurons are required for the control of voluntary movement via the nigrostriatal pathway. It is well known that the gradual loss of mDA neurons in SNc results in an aged-related neurodegenerative Parkinson's disease (PD) (Lees et al. 2009), which is the second most common neurodegenerative disorder and affects about 1% of the general population over the age of 65 (de Lau and Breteler 2006). The disease is debilitating, and characterized as a movement disorder consisting of bradykinesia, rigidity, rest tremor and postural instability. Although drug treatment such as using L-Dopa can alleviate PD symptoms, it cannot stop or slow the pathological progress of PD.

Given the huge progresses in identifying the molecular mechanisms underlying induced pluripotent stem cells (iPS), intensive efforts have been made to generate mDA neurons from stem cells for application of cell replacement therapy for PD. The knowledge about the mechanism underlying mDA neuron development is important for engineering functional mDA neurons. In this chapter, we provide an overall description of mDA neuron morphogenesis and put more efforts on progress of mDA neuron-related stem cells.

10.2 Early Patterning of Mesencephalon (Midbrain)

The floor plate (FP), a small group of cells located at the ventral midline of neural tube during early development, has long been considered to be non-neurogenic and functions as induction center controlling neural tube patterning and axonal guidance (Jessell 2000; Placzek and Briscoe 2005). Recent studies have demonstrated midbrain FP has distinct characteristics with the neurogenic capability. mDA neurons are generated in the FP region of the midbrain during embryonic development, mostly after embryonic day (E) 11 in mice (Ono et al. 2007).

However, the early patterning midbrain in the anterio-posterior (AP) and dorso-ventral (DV) axes also plays important roles in the generation of mDA neurons, and this pattern-

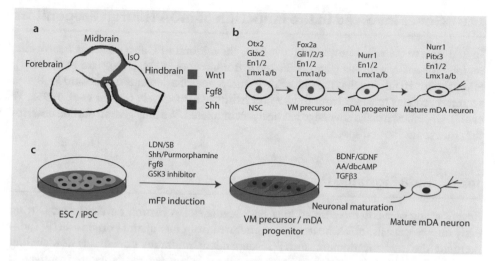

Fig. 10.1 *In vivo* and *in vitro* generation of mDA neurons. **a** Sagittal section of mouse E11.5 embryonic brain, showing the expression of some key morphogens important for mDA neuron development *in vivo*. White stars indicate the midbrain floor plate (mFP), where mDA neurons originate. IsO isthmic organizer. **b** A series of stages of mDA neuron development. A network of transcription regulatory factors essential for mDA neuron specification and maturation are indicated. **c** An example of the protocols for the differentiation of PSCs into mDA neurons. The PSCs including ESC and iPSC in culture are treated with small molecules or patterning morphogens for mFP induction and ventral mesencephalic (VM) precursors and mDA progenitors. The neuronal maturation is further promoted by the addition of several factors including BDNF, GDNF and TGFβ3. AA ascorbic acid, dbcAMP dibutyryl cyclic adenosine monophosphate

ing is controlled by induction signals arising from the ▶ midbrain FP and the isthmic organizer during E7-9. In the DV patterning of midbrain neural tube, sonic hedgehog (Shh), fibroblast growth factor (Fgf) 8 and Wnt1 are essential and sufficient for the induction of mDA neurons (Hynes et al. 1995a, b; Ye et al. 1998; Prakash et al. 2006) (▶ Fig. 10.1). The AP patterning is achieved by the isthmic organizer, the boundary region between the midbrain and hindbrain, and correct positioning of the organizer in the AP axis is dependent on the mutual repression of two opposing transcription factors (Otx2 and Gbx2), and disruption of this mutual repression affects the size of mDA neuron population as well as that of serotonergic neurons located caudal to mDA neurons (Liu and Joyner 2001; Brodski et al. 2003; Shi et al. 2012).

Fgf8 is the major player for the induction activity of isthmic organizer (Crossley and Martin 1995). The defective inductive activity not only affects the morphogenesis of tectum (dorsal midbrain) and cerebellum (rostral hindbrain) but also the generation of mDA neurons. The initiation and maintenance of Fgf8 expression depends on the genes expressed by the isthmic organizer. For example, although the Lim-homeodomain factor Lmx1b is expressed in mDA neurons, this expression alone is dispensable for mDA neuron development. The loss of mDA neurons in Lmx1b mutants (Smidt et al. 2000) is due to the failure of initiation of Fgf8 expression in the absence of Lmx1b in the isthmic organizer (Guo et al. 2007). A special caution should be paid in studying the role of genes in mDA neuron development when it probably participates or has been known to be involved in the regulation of isthmic inductive activity.

10.3   Genes Involved in the Induction of mDA Neuron Progenitors

In addition to the well-known inductive role, the midbrain FP also serves as neurogenic regions for mDA neurons (Ono et al. 2007; Bonilla et al. 2008; Hebsgaard et al. 2009). Since the initial finding that the induction of mDA neurons depends on Shh and Fgf8, two secreted proteins released from the FP and isthmus, respectively (Hynes et al. 1995a; Ye et al. 1998), substantial knowledge has been accumulated. We only give an outline description due to the limited space.

10.3.1   Fgf8 Signaling

Recent studies have demonstrated how Fgf8 regulates DA neuron development. During early embryogenesis, mDA neurons are generated from two distinct origins: FGF- independent diencephalic domain and FGF-dependent midbrain domain. FGF signaling operates in proliferative midbrain DA progenitors. When Fgf receptors are inactivated, the midbrain DA domain adopts the diencephalic characteristics, with altered patterning of ventral midbrain region and failure of mDA neuron maturation (Saarimäki-Vire et al. 2007; Lahti et al. 2012). The intracellular mechanism of Fgf8 signaling in regulating mDA neuron progenitors and precursors remains elusive.

10.3.2   Shh Signaling

Lmx1a has been proved to be a key downstream gene of Shh signals in the determination of mDA progenitors in the midbrain FP region, as evidenced by the findings that Shh induces the expression of Lmx1a, which subsequently induces the expression of its downstream effector Msx1 (Andersson et al. 2006). These two transcription factors are determinant genes in the establishment of mDA progenitor fate in the ventral midbrain. Other genes (e.g. Ngn2, Hes, Otx2) may not specifically function in the determination of mDA neuron progenitors in response to Shh signals, instead they functions as either general proneural genes, maintaining neural progenitor pool, or early A-P patterning factors for regionalization of mesencephalon.

FoxA2 is a well-known FP marker and has been considered to be regulated by Shh in regulating mDA neurons morphogenesis (Chung et al. 2009). The downstream effectors of Shh signaling are Gli1-3 genes, whose mutations results in defective mDA neurogenesis (Park et al. 2000). Shh and Gli1 induces the expression of FoxA2 (Hynes et al. 1997; Park et al. 2000). However, FoxA2 seems to be more effective in patterning midbrain neural tube via both Shh-dependent and independent mechanisms (Bayly et al. 2012). A mutual regulatory loop between Shh and FoxA2 expression in the patterning as well as consequent mDA neurogenesis has been proposed (Lin et al. 2009; Nissim-Eliraz et al. 2013). In addition, a bHLH transcription factor Nato3 has been shown to contribute to mDA neurogenesis, in which Shh–FoxA2 regulatory loop is involved (Ono et al. 2010; Nissim-Eliraz et al. 2013). Data from conditional deletion of FoxA1 and FoxA2 mice show that they positively regulate Lmx1a and Lmx1b expression and inhibit Nkx2.2 expression in the ventricular zone, and regulate mDA neuron differentiation cooperatively with Lmx1a and Lmx1b (Lin et al. 2009).

10.3.3 Wnt Signaling

It is well accepted that Wnt family are key regulators during DA neuron induction. Several Wnts (e.g. Wnt1, Wnt2, Wnt3a and Wnt5a) and their downstream gene β-catenin are implicated in multiple steps in mDA morphogenesis, including the acquisition of DA identity, proliferation of mDA progenitors, progression of progenitors into post-mitotic DA neurons and terminal differentiation of post-mitotic mDA neurons, likely via Wnt1–Lmx1a and Shh–FoxA2 regulatory loops (Hegarty et al. 2013).

10.4 Genes Involved in the Development of Post-mitotic mDA Neurons

10.4.1 Lmx1b

Lmx1b expression is initiated in the isthmus around E8.5 and present in the midbrain FP, which contains mDA progenitors at E10.5 (Guo et al. 2007). Its expression is turned off in mDA progenitors, but turned on in post-mitotic mDA neurons at E11.5 and co-expressed with tyrosine hydroxylase (TH) into adulthood like Lmx1a (Dai et al. 2008; Zou et al. 2009). Although Lmx1b mutant mice show a loss of mDA neurons (Smidt et al. 2000), our unpublished data with loss and gain of function evidence demonstrate that the loss of mDA neurons is caused by impaired inductive activity of isthmic organizer in Lmx1b mutants (Guo et al. 2007). Lmx1b itself is dispensable for the development of mDA neurons, but it plays a compensatory role with Lmx1a in controlling mDA neuron development and the process of mDA neuron degeneration in aged mice (Ono et al. 2007; Yan et al. 2011; Laguna et al. 2015).

10.4.2 Nurr1

Nurr1, a steroid-thyroid hormone-activated transcription factors, is expressed in the post-mitotic mDA neurons prior to the initiation of TH and its expression in mDA neurons is maintained into adulthood (Zetterström et al. 1996; Bäckman et al. 1999). Nurr1 is required for the acquisition of neurotransmitter identity (terminal differentiation) and the survival of post-mitotic mDA neurons (Saucedo-Cardenas et al. 1998; Sakurada et al. 1999; Kadkhodaei et al. 2009). In addition, the persistent expression in postnatal mDA neurons is involved in the vulnerability to a parkinsonian toxin, MPTP (1-methyl-4-phenyl-1,2,3,6-tetrahydropyridine) (Le et al. 1999).

10.4.3 Pitx3

Pitx3 is a homeodomain-containing transcription factor and expressed in post-mitotic mDA neurons (Smidt et al. 1997). Like Nurr1, Pitx3 is required for the survival of mDA neurons (Hwang et al. 2003; Nunes et al. 2003; van den Munckhof et al. 2003; Smidt et al. 2004). Accumulated evidence shows that Nurr1 and Pitx3 may cooperate to promote the terminal differentiation of and survival of mDA neurons via similar mechanism by regulating the expression of neurotrophic factors, such as brain-derived neurotrophic factor (BDNF) and glial cell line-derived neurotrophic factor (GDNF) (Martinat et al. 2006; Peng et al. 2011).

10.4.4 Engrailed (En) 1 and 2

Transcription factors En1 and En2 are required for the formation of the isthmic organizer. mDA neurons initiate En expression around E11.5 and maintain its expression through adulthood. mDA neurons are lost in En1 and En2 double mutants (Simon et al. 2001; Albéri et al. 2004). Although a cell-autonomous role of En in the survival of mDA neurons has been claimed, non-cell-autonomous contribution by the loss of En-controlled isthmic inductive activity cannot be fully excluded. En heterozygous mice display a progressive degeneration of mDA neurons in postnatal period (Sgadò et al. 2006; Sonnier et al. 2007). This resembles key pathological features of PD and supports the idea that En genes are important in promoting the survival of mDA neurons during embryonic and postnatal stages.

10.5 Generation of mDA Neurons for Cell-Based Therapies for PD

Current main clinical treatment for PD is dopamine replacement using dopamine precursor (L-Dopa) and/or dopamine receptor agonists. Although this treatment can improve PD symptoms at initial stages, the efficacy is gradually lost and unacceptable side effects develop with prolonged treatment. As motor impairments in PD are largely due to loss of mDA neurons in SNc, PD has long been considered to be one of the most promising diseases suitable for cell-based therapy.

Indeed, proof of concept for this therapy strategy has been provided using human fetal ventral mesencephalic tissue (hFVM), which contains mDA neuron progenitors (Kordower et al. 1995; Freed et al. 2001; Olanow et al. 2003; Kordower et al. 2008; Barker et al. 2013). Although hFVM transplantation has shown to be effective in some PD patients, the issues of tissue availability and ethical implications will limit its widespread clinical application. Thus, a number of alternative cell sources such as neural stem cells (NSCs) and pluripotent stem cells (PSCs), have been extensively investigated for the generation of mDA neurons suitable for transplantation in PD patients during the past years (☐ Fig. 10.2). Here we mainly focus on the application of human PSCs, including embryonic stem cells (ESCs) and induced pluripotent stem cells (iPSCs).

10.5.1 Human PSCs as a Source of mDA Neurons

In 1998, the first hESC lines were isolated and established from the inner cell mass of blastocyst, which can be differentiated into any cell types of the body, opening a new era of PSC-based cellular therapy (Thomson et al. 1998). Because of the potential use as a regenerative therapy for PD, the generation of mDA neurons from human PSCs has been the subject of extensive research. The ability to produce human mDA neurons *in vitro* has built in large part on the growing knowledge of how these cells develop *in vivo* (☐ Fig. 10.1). Several protocols for efficient differentiation of mDA neurons from PSCs have been devel-

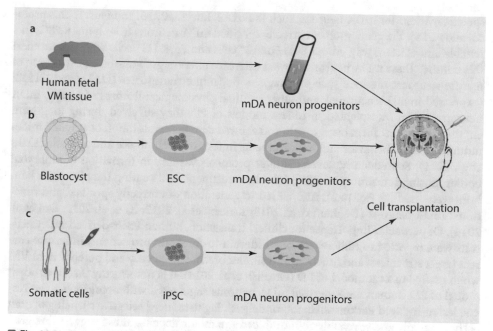

a

Human fetal
VM tissue

mDA neuron progenitors

b

Blastocyst   ESC   mDA neuron progenitors

Cell transplantation

c

Somatic cells   iPSC   mDA neuron progenitors

🔲 **Fig. 10.2** Various cell sources for mDA neuron-based therapy for Parkinson's disease. **a** Human fetal ventral mesencephalon (VM) tissues from aborted embryos consist of a mixed neuronal population including mDA progenitors, which can be used for neural grafts. The pluripotent stem cells, **b** ESC derived from blastocyst or **c** iPSC derived from patient's somatic cells such as fibroblasts, can be differentiated to mDA neuron progenitors *in vitro*, providing cell sources for transplantation

oped during the last two decades. These protocols use stromal co-culture method or embryoid body formation for neutralization. The morphogens and growth factors (such as Fgf8, Shh, BDNF, GDNF, TGFβ3) essential for mDA neuron development were utilized. Although some of these initial differentiation protocols generated a relatively high number of TH+ neurons, many issues remained, including the highly mixed cell composition, inconsistent and incomplete differentiation. In addition, these TH+ neurons did not co-express two transcription factors required for proper mDA neuron specification (e.g., FoxA2 and Lmx1a), indicating that they failed to give rise to authentic mDA neurons. The differentiated TH+ neurons also showed poor *in vivo* performance after grafting in animal models of PD. In some cases, teratoma formation or neural overgrowth was observed after transplantation into animal brains.

Recently, several studies on the developmental and cellular ontogeny of mDA neurons reported that these neurons were derived from ventral mesencephalic FP cells of expressing Corin, FoxA2, and Lmx1a, but not from the neuroepithelial cells expressing Pax6 (Ono et al. 2007; Bonilla et al. 2008). Thereafter, the novel FP-based strategy to generate human mDA neurons from PSCs was developed, guided by their unique developmental origin. During this protocol, human PSCs were first differentiated into FP cells through the dual inhibition of SMADs (inhibiting BMP, Nodal, activin and TGFβ signaling to improve neural induction), dual activation of Wnt and Shh pathways, together with the patterning factor Fgf8 (Chambers et al. 2009; Fasano et al. 2010; Kriks et al. 2011). The engraftable mDA neurons were then obtained from these midbrain FP precursors. In contrast to the DA neurons generated via neuroepithelial cells, the FP-derived cells expressed specific markers for mDA neurons, such as FoxA2, Lmx1a, Pitx3, and Nurr1. This proto-col allowed for the generation of correctly specified mDA neurons from human PSCs in a reliable and efficient way. About 80% FoxA2$^+$ cells and 75% TH$^+$ cells were produced in this culture. These mDA neurons showed extensive fiber outgrowth, robust expression of mature neuronal markers such as synapsin, dopamine transporter (DAT), and GIRK2 (expressed in mDA neurons), and other cardinal physiological features of mature mDA neurons. When transplanted in rodent models of PD, they survived during long-term engraftment, didn't form overgrowth, and showed complete restoration of amphetamine-induced rotation behavior and improvements in tests of forelimb use and akinesia (Kriks et al. 2011). Subsequently, several similar protocols varying in formation of embryoid bodies, dosing and timing of GSK3 inhibitor treatment, and coating the plate with lam-inin, were used (◘ Fig. 10.2). And all led to generation of correctly specified and func-tional mDA neurons (Denham et al. 2012; Kirkeby et al. 2012; Xi et al. 2012; Doi et al. 2014). Of particular importance for clinical translation, human PSCs-derived mDA neu-rons were reported to fully regenerate midbrain-to-forebrain projections, innervate cor-rect target structures, and function with similar preclinical efficacy and potency to hFVM when grafted in a rat model of PD (Grealish et al. 2014). It is noteworthy that in a rodent model of PD, human ESCs-derived mDA neurons engineered with optogenetics can res-cue lesion-induced Parkinsonian motor deficits. Light-induced selective silencing of graft activity rapidly and reversibly re-introduced the motor deficits. These suggest that graft neuronal activity and dopamine release have been successfully established *in vivo* (Steinbeck et al. 2015).

As with the prior hFVM grafts, the mDA neurons derived from human ESCs will also be allogenic, and their transplantation into patients might require a period of immuno-suppression. Besides, there are also ethical issues with the use of embryonic tissue in some cultures. In contrast, human iPSCs derived from various somatic cell sources such as skin, adipose, hair follicles, and peripheral blood cells, provide a means to generate patient-specific autologous graft. This approach will circumvent the requirement of immunosup-pression and avoid the ethical issues. For human iPSCs-based cell therapy for PD, the protocols for generation of mDA neurons from human iPSCs are almost the same as those of human ESCs (Cooper et al. 2010; Hargus et al. 2010; Kikuchi et al. 2011, 2017; Doi et al. 2014). Recently, the autologous iPSCs-derived mDA neurons have been shown to survive in large numbers for 2 years after transplantation into the striatum of nonhuman primate model of PD (Hallett et al. 2015). They could give rise to extensive re-innervation and improve the motor function. Another study showed that human iPSCs-derived mDA neurons functioned in a primate MPTP-induced PD model. Cells sorted by the FP marker Corin did not form any tumors after transplantation into the brains for at least 2 years (Kikuchi et al. 2017).

10.5.2 Direct Reprogramming of Somatic Cells into mDA Neurons

Recently, developments have been made in the direct conversion of human fibroblasts or other cell types to induced DA (iDA) neurons. These iDA neurons are generated directly from somatic cells in a short period of time, bypassing lengthy differentiation from human PSCs and the concern of potentially tumorigenic mitotic cells. Many groups have inde-pendently identified various combinations of transcription factors, capable to convert fibroblasts into iDA neurons. As expected, these transcription factors are involved in the specification of DA neurons during development, such as FoxA2, Lmx1a, Nurr1, Pitx3, En1, Lmx1b, Otx2 and Ascl1 (Wernig et al. 2008; Caiazzo et al. 2011; Pfisterer et al. 2011; Anderegg et al. 2015; Xu et al. 2017). These iDA neurons all express neuronal markers (such as Tuj1, Map 2, NeuN), dopaminergic markers (such as TH, AADC, VMAT2, DAT), and midbrain markers (such as FoxA1, En1). They can release and reuptake dopamine, and show electrophysiological activities. Some studies even reported that transplantation of iDA neurons can partially alleviate motor deficits in 6-OHDA-lesioned mice (Kim et al. 2011). It has been reported that using Lmx1b instead of Lmx1a could also reprogram mouse primary cortex astrocytes into functional iDA cells (Addis et al. 2011). Interestingly, in adult mouse brain, the striatal neurons can be reprogrammed into dopaminergic neuron-like cells (iDALs) with the aid of a stem cell factor (Sox2), three dopaminergic neuron-enriched transcription regulators (Nurr1, Lmx1a, and FoxA2), and a chemical compound (valproic acid) (Niu et al. 2018).

10.5.3 Considerations for Cell-Based Therapies for PD

During the 1980s, transplantation of hFVM tissue to the striatum of PD patients pioneered at Lund University has provided proof-of-principle that cell replacement therapy for PD can show long-term clinical benefits. Since previous hFVM tissue transplantation trials showed heterogeneity of clinical responses and even generated adverse effects, an EU-funded multicenter study/trial called TRANSEURO has been initiated. This clinical trial is aiming to provide standardized protocols, including PD patient selection, hFVM tissue preparation and implantation, immunosuppression regime, and patient assessment (Barker et al. 2015; Kirkeby et al. 2017b). Although TRANSEURO trial will be highly informative for cell-based therapies for PD, the human fetal tissue is problematic. Human PSCs with the recently established protocols to robustly differentiate into authentic mDA neurons become a very attractive cell source. Several first-in-human clinical trials using human PSCs-derived mDA neurons for PD have been initiated, such as EUROPEAN STEM-PD, NTSTEM-PD, CiRA Trial, and Summit for PD Trial (the first two trials use human ESCs as cell source, and the latter two with human iPSCs) (Barker et al. 2017). CiRA Trial, the world first clinical trial to investigate human iPSC-derived cells for PD has gained green light from Pharmaceuticals and Medical Devices Agency (PMDA) in Japan, and started recruiting patients in August 2018.

Due to the biological complexity, human PSCs-derived cell-based therapies for PD result in specific technical, logistical, and regulatory challenges. These include the choice of starting material (e.g., human ESCs vs iPSCs), reproducible and consistent manufacturing, and preclinical safety and efficacy assessment. Specifically, the derived mDA cell products for clinical use need to be of sufficient quality and quantity. High percentage of mDA neurons, no contaminating pluripotent cells, and no or very limited forebrain and hindbrain progenitors (i.e. $Pax6^+$, $FoxG1^+$, $Gbx2^+$) should be detected (Parmar 2018). FACS based on the FP marker Corin has been employed to enrich mDA neurons derived from human iPSCs (Kikuchi et al. 2017). Meanwhile, the patient selection is not straightforward. Whether patients younger and at early disease stage or patients older and at late stages are recruited remains debatable, but they need to show a clear response to oral DA medications. It is important to recognize that the mDA neuron grafts need many months and even years for their terminal maturation and integration into host neural circuit after transplantation. Thus, the primary end point for the early clinical trials will be tolerability and feasibility, instead of the restoration of impaired motor function. Another aspect to be mentioned is that PD patients also show non-motor symptoms such as neuropsychiatric symptoms and cognitive decline, which will unlikely be improved by these mDA neuron-based therapy. Since the human PSCs-derived mDA neuron treatment for PD patients is still in its infancy, the patients should ideally be followed for the rest of their lives. More correlations among clinical benefits, post-mortem pathology, and mDA transplant profiling should be analyzed to guide future clinical trial design.

10.6 Conclusion and Future Perspectives

Although our knowledge of the mDA neuron developmental mechanisms has gained great progress, some new approaches have been used to further define mDA neuron diversity (Poulin et al. 2014; Anderegg et al. 2015; Bodea and Blaess 2015; La Manno et al. 2016; Kee et al. 2017; Kirkeby et al. 2017a). Single cell sequencing of the midbrain Lmx1a-expressing progenitors during development showed that FoxA2 and Lmx1a, used as the markers for mDA neurons derived from ventral mesencephalic FP cells, were also expressed in early diencephalic progenitors giving rise to subthalamic nucleus. Similarly through single-cell gene expression profiling, a subtype of mDA neurons located in SNc positive for Aldh1a1 has been characterized to be especially vulnerable in the MPTP model of PD (Poulin et al. 2014). These studies will not only identify molecular heterogeneity of mDA neurons, but also enable further refinement and standardization of mDA neuron differentiation protocols for application in the clinic.

Most of the approaches under development for stem cell-based therapies for PD involve the transplantation of immature mDA neurons, which will require months *in vivo* for their functional maturation and integration into neural circuit. It is of critical importance to establish certain markers in these immature mDA neurons, which can predict the long-term graft outcome. Besides, these markers will also be used for quality control of stem cell-derived mDA neurons for clinical transplantation for PD patients. Indeed, recent research showed that some markers (such as FoxA2, Lmx1a) traditionally used to monitor mDA neuron differentiation failed to predict graft outcome, but a specific set of markers (such as En1/2, Pax8, Fgf8) associated with the caudal midbrain correlated with high dopaminergic yield after transplantation *in vivo* (Kirkeby et al. 2017a).

Although human iPSCs-derived mDA neurons can potentially provide autologous grafts and may avoid the need for immunosuppression, one disadvantage of this approach is that the grafted tissue will contain any genetic PD-susceptibility factors that contributed to the PD pathogenesis in the host. Thus, the transplanted mDA neurons from human iPSCs may degenerate more rapidly comparing to those derived from human ESCs. Theoretically, for certain forms of PD, for example monogenic early-onset PD, the mutations in these genes can be corrected with gene editing technology, including CRISPR/Cas9, ZFNs and TALENs sequence-specific nucleases (Calatayud et al. 2017). Nonetheless, the high degree of variability among individuals in the reprogramming of somatic cells and the differentiation into mDA neurons will pose challenges for regulatory approval and extensive preclinical testing. One alternative strategy is to use haplobanks and accept a degree of human leukocyte antigen (HLA).

The field of stem cell-based therapies for PD has gained remarkable progress through the years. Several first-in-human clinical trials using hPSCs-derived mDA neurons are starting to recruit patients or planned to initiate shortly (Barker et al. 2017). Experience and data from these clinical trials will help further optimize the differentiation protocols and manufacture of mDA neuron, and better design stem cell-based treatments for PD.

Take Home Message

1. mDA neurons are generated in the floor plate region of the midbrain during embryonic development.
2. mDA neuron generation is controlled by induction signals arising from the floor plate and isthmic organizer in the midbrain, including Shh, Fgf8, and Wnt1.
3. Lmx1a functions as a key transcription factor in the determination of mDA progenitors in the midbrain FP region.
4. Some key genes are involved in the development of post-mitotic mDA neurons, including Nurr1, Pitx3, En1 and En2.
5. Knowledge about molecular mechanisms underlying mDA neurons morphogenesis has instructed generation of mDA neurons from pluripotent stem cells or from somatic cells via direct reprogramming.
6. Several first-in-human clinical trials using human PSCs-derived mDA neurons for PD have been initiated.

Acknowledgements   We are grateful for editing help from Dr. Ning-Ning Song. This work was supported by the National Natural Science Foundation of China (81571332, 91232724) and the National key R&D program of China (2017YFA0104002).

References

Addis, R. C., Hsu, F. C., Wright, R. L., Dichter, M. A., Coulter, D. A., & Gearhart, J. D. (2011). Efficient conversion of astrocytes to functional midbrain dopaminergic neurons using a single polycistronic vector. *PLoS One, 6*(12), e28719.

Albéri, L., Sgadò, P., & Simon, H. H. (2004). Engrailed genes are cell-autonomously required to prevent apoptosis in mesencephalic dopaminergic neurons. *Development, 131*(13), 3229–3236.

Anderegg, A., Poulin, J. F., & Awatramani, R. (2015). Molecular heterogeneity of midbrain dopaminergic neurons–Moving toward single cell resolution. *FEBS Letters, 589*(24 Pt A), 3714–3726.

Andersson, E., Tryggvason, U., Deng, Q., Friling, S., Alekseenko, Z., Robert, B., Perlmann, T., & Ericson, J. (2006). Identification of intrinsic determinants of midbrain dopamine neurons. *Cell, 124*(2), 393–405.

Bäckman, C., Perlmann, T., Wallén, A., Hoffer, B. J., & Morales, M. (1999). A selective group of dopaminergic neurons express Nurr1 in the adult mouse brain. *Brain Research, 851*(1–2), 125–132.

Barker, R. A., Barrett, J., Mason, S. L., & Bjorklund, A. (2013). Fetal dopaminergic transplantation trials and the future of neural grafting in Parkinson's disease. *Lancet Neurology, 12*(1), 84–91.

Barker, R. A., Drouin-Ouellet, J., & Parmar, M. (2015). Cell-based therapies for Parkinson disease-past insights and future potential. *Nature Reviews. Neurology, 11*(9), 492–503.

Barker, R. A., Parmar, M., Studer, L., & Takahashi, J. (2017). Human trials of stem cell-derived dopamine neurons for Parkinson's disease: Dawn of a new era. *Cell Stem Cell, 21*(5), 569–573.

Bayly, R. D., Brown, C. Y., & Agarwala, S. (2012). A novel role for FOXA2 and SHH in organizing midbrain signaling centers. *Developmental Biology, 369*(1), 32–42.

Bodea, G. O., & Blaess, S. (2015). Establishing diversity in the dopaminergic system. *FEBS Letters, 589*(24 Pt A), 3773–3785.

Bonilla, S., Hall, A. C., Pinto, L., Attardo, A., Gotz, M., Huttner, W. B., & Arenas, E. (2008). Identification of midbrain floor plate radial glia-like cells as dopaminergic progenitors. *Glia, 56*(8), 809–820.

Brodski, C., Weisenhorn, D. M. V., Signore, M., Sillaber, I., Oesterheld, M., Broccoli, V., Acampora, D., Simeone, A., & Wurst, W. (2003). Location and size of dopaminergic and serotonergic cell populations are controlled by the position of the midbrain – Hindbrain organizer. *The Journal of Neuroscience, 23*(10), 4199–4207.

Caiazzo, M., Dell'Anno, M. T., Dvoretskova, E., Lazarevic, D., Taverna, S., Leo, D., Sotnikova, T. D., Menegon, A., Roncaglia, P., Colciago, G., Russo, G., Carninci, P., Pezzoli, G., Gainetdinov, R. R., Gustincich, S., Dityatev, A., & Broccoli, V. (2011). Direct generation of functional dopaminergic neurons from mouse and human fibroblasts. *Nature, 476*(7359), 224–227.

Calatayud, C., Carola, G., Consiglio, A., & Raya, A. (2017). Modeling the genetic complexity of Parkinson's disease by targeted genome edition in iPS cells. *Current Opinion in Genetics & Development, 46*, 123–131.

Chambers, S. M., Fasano, C. A., Papapetrou, E. P., Tomishima, M., Sadelain, M., & Studer, L. (2009). Highly efficient neural conversion of human ES and iPS cells by dual inhibition of SMAD signaling. *Nature Biotechnology, 27*(3), 275–280.

Chung, S., Leung, A., Han, B. S., Chang, M. Y., Moon, J. I., Kim, C. H., Hong, S., Pruszak, J., Isacson, O., & Kim, K. S. (2009). Wnt1-lmx1a forms a novel autoregulatory loop and controls midbrain dopaminergic differentiation synergistically with the SHH-FoxA2 pathway. *Cell Stem Cell, 5*(6), 646–658.

Cooper, O., Hargus, G., Deleidi, M., Blak, A., Osborn, T., Marlow, E., Lee, K., Levy, A., Perez-Torres, E., Yow, A., & Isacson, O. (2010). Differentiation of human ES and Parkinson's disease iPS cells into ventral midbrain dopaminergic neurons requires a high activity form of SHH, FGF8a and specific regionalization by retinoic acid. *Molecular and Cellular Neurosciences, 45*(3), 258–266.

Crossley, P. H., & Martin, G. R. (1995). The mouse Fgf8 gene encodes a family of polypeptides and is expressed in regions that direct outgrowth and patterning in the developing embryo. *Development, 121*(2), 439–451.

Dai, J.-X., Hu, Z.-L., Shi, M., Guo, C., & Ding, Y.-Q. (2008). Postnatal ontogeny of the transcription factor Lmx1b in the mouse central nervous system. *The Journal of Comparative Neurology, 509*(4), 341–355.

de Lau, L. M., & Breteler, M. M. (2006). Epidemiology of Parkinson's disease. *Lancet Neurology, 5*(6), 525–535.

Denham, M., Bye, C., Leung, J., Conley, B. J., Thompson, L. H., & Dottori, M. (2012). Glycogen synthase kinase 3beta and activin/nodal inhibition in human embryonic stem cells induces a pre-neuroepithelial state that is required for specification to a floor plate cell lineage. *Stem Cells, 30*(11), 2400–2411.

Doi, D., Samata, B., Katsukawa, M., Kikuchi, T., Morizane, A., Ono, Y., Sekiguchi, K., Nakagawa, M., Parmar, M., & Takahashi, J. (2014). Isolation of human induced pluripotent stem cell-derived dopaminergic progenitors by cell sorting for successful transplantation. *Stem Cell Reports, 2*(3), 337–350.

Fasano, C. A., Chambers, S. M., Lee, G., Tomishima, M. J., & Studer, L. (2010). Efficient derivation of functional floor plate tissue from human embryonic stem cells. *Cell Stem Cell, 6*(4), 336–347.

Freed, C. R., Greene, P. E., Breeze, R. E., Tsai, W. Y., DuMouchel, W., Kao, R., Dillon, S., Winfield, H., Culver, S., Trojanowski, J. Q., Eidelberg, D., & Fahn, S. (2001). Transplantation of embryonic dopamine neurons for severe Parkinson's disease. *The New England Journal of Medicine, 344*(10), 710–719.

Grealish, S., Diguet, E., Kirkeby, A., Mattsson, B., Heuer, A., Bramoulle, Y., Van Camp, N., Perrier, A. L., Hantraye, P., Bjorklund, A., & Parmar, M. (2014). Human ESC-derived dopamine neurons show similar preclinical efficacy and potency to fetal neurons when grafted in a rat model of Parkinson's disease. *Cell Stem Cell, 15*(5), 653–665.

Guo, C., Qiu, H.-Y., Huang, Y., Chen, H., Yang, R.-Q., Chen, S.-D., Chen, Z.-F., & Ding, Y.-Q. (2007). Lmx1b is essential for Fgf8 and Wnt1 expression in the isthmic organizer during tectum and cerebellum development in mice. *Development, 134*, 317–325.

Hallett, P. J., Deleidi, M., Astradsson, A., Smith, G. A., Cooper, O., Osborn, T. M., Sundberg, M., Moore, M. A., Perez-Torres, E., Brownell, A. L., Schumacher, J. M., Spealman, R. D., & Isacson, O. (2015). Successful function of autologous iPSC-derived dopamine neurons following transplantation in a non-human primate model of Parkinson's disease. *Cell Stem Cell, 16*(3), 269–274.

Hargus, G., Cooper, O., Deleidi, M., Levy, A., Lee, K., Marlow, E., Yow, A., Soldner, F., Hockemeyer, D., Hallett, P. J., Osborn, T., Jaenisch, R., & Isacson, O. (2010). Differentiated Parkinson patient-derived induced pluripotent stem cells grow in the adult rodent brain and reduce motor asymmetry in Parkinsonian rats. *Proceedings of the National Academy of Sciences of the United States of America, 107*(36), 15921–15926.

Hebsgaard, J. B., Nelander, J., Sabelstrom, H., Jonsson, M. E., Stott, S., & Parmar, M. (2009). Dopamine neuron precursors within the developing human mesencephalon show radial glial characteristics. *Glia, 57*(15), 1648–1658.

Hegarty, S. V., Sullivan, A. M., & O'Keeffe, G. W. (2013). Midbrain dopaminergic neurons: A review of the molecular circuitry that regulates their development. *Developmental Biology, 379*(2), 123–138.

Hwang, D. Y., Ardayfio, P., Kang, U. J., Semina, E. V., & Kim, K. S. (2003). Selective loss of dopaminergic neurons in the substantia nigra of Pitx3-deficient aphakia mice. *Molecular Brain Research, 114*(2), 123–131.

Hynes, M., Porter, J. A., Chiang, C., Chang, D., Tessier-Lavigne, M., Beachy, P. A., & Rosenthal, A. (1995a). Induction of midbrain dopaminergic neurons by Sonic hedgehog. *Neuron, 15*(1), 35–44.

Hynes, M., Poulsen, K., Tessier-Lavigne, M., & Rosenthal, A. (1995b). Control of neuronal diversity by the floor plate: Contact-mediated induction of midbrain dopaminergic neurons. *Cell, 80*(1), 95–101.

Hynes, M., Stone, D. M., Dowd, M., Pitts-Meek, S., Goddard, A., Gurney, A., & Rosenthal, A. (1997). Control of cell pattern in the neural tube by the zinc finger transcription factor and oncogene Gli-1. *Neuron, 19*(1), 15–26.

Jessell, T. M. (2000). Neuronal specification in the spinal cord: Inductive signals and transcriptional codes. *Nature reviews. Genetics, 1*(1), 20–29.

Kadkhodaei, B., Ito, T., Joodmardi, E., Mattsson, B., Rouillard, C., Carta, M., Muramatsu, S.-I., Sumi-Ichinose, C., Nomura, T., Metzger, D., Chambon, P., Lindqvist, E., Larsson, N.-G., Olson, L., Björklund, A., Ichinose, H., & Perlmann, T. (2009). Nurr1 is required for maintenance of maturing and adult midbrain dopamine neurons. *The Journal of Neuroscience, 29*(50), 15923–15932.

Kee, N., Volakakis, N., Kirkeby, A., Dahl, L., Storvall, H., Nolbrant, S., Lahti, L., Bjorklund, A. K., Gillberg, L., Joodmardi, E., Sandberg, R., Parmar, M., & Perlmann, T. (2017). Single-cell analysis reveals a close relationship between differentiating dopamine and subthalamic nucleus neuronal lineages. *Cell Stem Cell, 20*(1), 29–40.

Kikuchi, T., Morizane, A., Doi, D., Onoe, H., Hayashi, T., Kawasaki, T., Saiki, H., Miyamoto, S., & Takahashi, J. (2011). Survival of human induced pluripotent stem cell-derived midbrain dopaminergic neurons in the brain of a primate model of Parkinson's disease. *Journal of Parkinson's Disease, 1*(4), 395–412.

Kikuchi, T., Morizane, A., Doi, D., Magotani, H., Onoe, H., Hayashi, T., Mizuma, H., Takara, S., Takahashi, R., Inoue, H., Morita, S., Yamamoto, M., Okita, K., Nakagawa, M., Parmar, M., & Takahashi, J. (2017). Human iPS cell-derived dopaminergic neurons function in a primate Parkinson's disease model. *Nature, 548*(7669), 592–596.

Kim, J., Su, S. C., Wang, H., Cheng, A. W., Cassady, J. P., Lodato, M. A., Lengner, C. J., Chung, C. Y., Dawlaty, M. M., Tsai, L. H., & Jaenisch, R. (2011). Functional integration of dopaminergic neurons directly converted from mouse fibroblasts. *Cell Stem Cell, 9*(5), 413–419.

Kirkeby, A., Grealish, S., Wolf, D. A., Nelander, J., Wood, J., Lundblad, M., Lindvall, O., & Parmar, M. (2012). Generation of regionally specified neural progenitors and functional neurons from human embryonic stem cells under defined conditions. *Cell Reports, 1*(6), 703–714.

Kirkeby, A., Nolbrant, S., Tiklova, K., Heuer, A., Kee, N., Cardoso, T., Ottosson, D. R., Lelos, M. J., Rifes, P., Dunnett, S. B., Grealish, S., Perlmann, T., & Parmar, M. (2017a). Predictive markers guide differentiation to improve graft outcome in clinical translation of hESC-based therapy for Parkinson's disease. *Cell Stem Cell, 20*(1), 135–148.

Kirkeby, A., Parmar, M., & Barker, R. A. (2017b). Strategies for bringing stem cell-derived dopamine neurons to the clinic: A European approach (STEM-PD). *Progress in Brain Research, 230*, 165–190.

Kordower, J. H., Freeman, T. B., Snow, B. J., Vingerhoets, F. J., Mufson, E. J., Sanberg, P. R., Hauser, R. A., Smith, D. A., Nauert, G. M., Perl, D. P., et al. (1995). Neuropathological evidence of graft survival and striatal reinnervation after the transplantation of fetal mesencephalic tissue in a patient with Parkinson's disease. *The New England Journal of Medicine, 332*(17), 1118–1124.

Kordower, J. H., Chu, Y., Hauser, R. A., Freeman, T. B., & Olanow, C. W. (2008). Lewy body-like pathology in long-term embryonic nigral transplants in Parkinson's disease. *Nature Medicine, 14*(5), 504–506.

Kriks, S., Shim, J. W., Piao, J., Ganat, Y. M., Wakeman, D. R., Xie, Z., Carrillo-Reid, L., Auyeung, G., Antonacci, C., Buch, A., Yang, L., Beal, M. F., Surmeier, D. J., Kordower, J. H., Tabar, V., & Studer, L. (2011). Dopamine neurons derived from human ES cells efficiently engraft in animal models of Parkinson's disease. *Nature, 480*(7378), 547–551.

La Manno, G., Gyllborg, D., Codeluppi, S., Nishimura, K., Salto, C., Zeisel, A., Borm, L. E., Stott, S. R. W., Toledo, E. M., Villaescusa, J. C., Lonnerberg, P., Ryge, J., Barker, R. A., Arenas, E., & Linnarsson, S. (2016). Molecular diversity of midbrain development in mouse, human, and stem cells. *Cell, 167*(2), 566–580. e519.

Laguna, A., Schintu, N., Nobre, A., Alvarsson, A., Volakakis, N., Jacobsen, J. K., Gómez-Galán, M., Sopova, E., Joodmardi, E., Yoshitake, T., Deng, Q., Kehr, J., Ericson, J., Svenningsson, P., Shupliakov, O., & Perlmann, T. (2015). Dopaminergic control of autophagic-lysosomal function implicates Lmx1b in Parkinson's disease. *Nature Neuroscience, 18*, 826.

Lahti, L., Peltopuro, P., Piepponen, T. P., & Partanen, J. (2012). Cell-autonomous FGF signaling regulates anteroposterior patterning and neuronal differentiation in the mesodiencephalic dopaminergic progenitor domain. *Development, 139*(5), 894–905.

Le, W. D., Conneely, O. M., He, Y., Jankovic, J., & Appel, S. H. (1999). Reduced Nurr1 expression increases the vulnerability of mesencephalic dopamine neurons to MPTP-induced injury. *Journal of Neurochemistry, 73*(5), 2218–2221.

Lees, A. J., Hardy, J., & Revesz, T. (2009). Parkinson's disease. *The Lancet, 373*(9680), 2055–2066.

Lin, W., Metzakopian, E., Mavromatakis, Y. E., Gao, N., Balaskas, N., Sasaki, H., Briscoe, J., Whitsett, J. A., Goulding, M., Kaestner, K. H., & Ang, S.-L. (2009). Foxa1 and Foxa2 function both upstream of and cooperatively with Lmx1a and Lmx1b in a feedforward loop promoting mesodiencephalic dopaminergic neuron development. *Developmental Biology, 333*(2), 386–396.

Liu, A., & Joyner, A. L. (2001). EN and GBX2 play essential roles downstream of FGF8 in patterning the mouse mid/hindbrain region. *Development, 128*(2), 181–191.

Martinat, C., Bacci, J.-J., Leete, T., Kim, J., Vanti, W. B., Newman, A. H., Cha, J. H., Gether, U., Wang, H., & Abeliovich, A. (2006). Cooperative transcription activation by Nurr1 and Pitx3 induces embryonic stem cell maturation to the midbrain dopamine neuron phenotype. *Proceedings of the National Academy of Sciences of the United States of America, 103*(8), 2874–2879.

Nissim-Eliraz, E., Zisman, S., Schatz, O., & Ben-Arie, N. (2013). Nato3 integrates with the Shh-Foxa2 transcriptional network regulating the differentiation of midbrain dopaminergic neurons. *Journal of Molecular Neuroscience, 51*(1), 13–27.

Niu, W., Zang, T., Wang, L. L., Zou, Y., & Zhang, C. L. (2018). Phenotypic reprogramming of striatal neurons into dopaminergic neuron-like cells in the adult mouse brain. *Stem Cell Reports, 11*(5), 1156–1170.

Nunes, I., Tovmasian, L. T., Silva, R. M., Burke, R. E., & Goff, S. P. (2003). Pitx3 is required for development of substantia nigra dopaminergic neurons. *Proceedings of the National Academy of Sciences, 100*(7), 4245–4250.

Olanow, C. W., Goetz, C. G., Kordower, J. H., Stoessl, A. J., Sossi, V., Brin, M. F., Shannon, K. M., Nauert, G. M., Perl, D. P., Godbold, J., & Freeman, T. B. (2003). A double-blind controlled trial of bilateral fetal nigral transplantation in Parkinson's disease. *Annals of Neurology, 54*(3), 403–414.

Ono, Y., Nakatani, T., Sakamoto, Y., Mizuhara, E., Minaki, Y., Kumai, M., Hamaguchi, A., Nishimura, M., Inoue, Y., Hayashi, H., Takahashi, J., & Imai, T. (2007). Differences in neurogenic potential in floor plate cells along an anteroposterior location: Midbrain dopaminergic neurons originate from mesencephalic floor plate cells. *Development, 134*(17), 3213–3225.

Ono, Y., Nakatani, T., Minaki, Y., & Kumai, M. (2010). The basic helix-loop-helix transcription factor Nato3 controls neurogenic activity in mesencephalic floor plate cells. *Development, 137*(11), 1897–1906.

Park, H. L., Bai, C., Platt, K. A., Matise, M. P., Beeghly, A., Hui, C. C., Nakashima, M., & Joyner, A. L. (2000). Mouse Gli1 mutants are viable but have defects in SHH signaling in combination with a Gli2 mutation. *Development, 127*(8), 1593–1605.

Parmar, M. (2018). Towards stem cell based therapies for Parkinson's disease. *Development, 145*(1), dev156117.

Peng, C., Aron, L., Klein, R., Li, M., Wurst, W., Prakash, N., & Le, W. (2011). Pitx3 is a critical mediator of GDNF-induced BDNF expression in nigrostriatal dopaminergic neurons. *The Journal of Neuroscience, 31*(36), 12802–12815.

Pfisterer, U., Kirkeby, A., Torper, O., Wood, J., Nelander, J., Dufour, A., Bjorklund, A., Lindvall, O., Jakobsson, J., & Parmar, M. (2011). Direct conversion of human fibroblasts to dopaminergic neurons. *Proceedings of the National Academy of Sciences of the United States of America, 108*(25), 10343–10348.

Placzek, M., & Briscoe, J. (2005). The floor plate: Multiple cells, multiple signals. *Nature Reviews. Neuroscience, 6*(3), 230–240.

Poulin, J. F., Zou, J., Drouin-Ouellet, J., Kim, K. Y., Cicchetti, F., & Awatramani, R. B. (2014). Defining midbrain dopaminergic neuron diversity by single-cell gene expression profiling. *Cell Reports, 9*(3), 930–943.

Prakash, N., Brodski, C., Naserke, T., Puelles, E., Gogoi, R., Hall, A., Panhuysen, M., Echevarria, D., Sussel, L., Weisenhorn, D. M. V., Martinez, S., Arenas, E., Simeone, A., & Wurst, W. (2006). A Wnt1-regulated genetic network controls the identity and fate of midbrain-dopaminergic progenitors in vivo. *Development, 133*(1), 89–98.

Saarimäki-Vire, J., Peltopuro, P., Lahti, L., Naserke, T., Blak, A. A., Vogt Weisenhorn, D. M., Yu, K., Ornitz, D. M., Wurst, W., & Partanen, J. (2007). Fibroblast growth factor receptors cooperate to regulate neural progenitor properties in the developing midbrain and hindbrain. *The Journal of Neuroscience, 27*(32), 8581–8592.

Sakurada, K., Ohshima-Sakurada, M., Palmer, T. D., & Gage, F. H. (1999). Nurr1, an orphan nuclear receptor, is a transcriptional activator of endogenous tyrosine hydroxylase in neural progenitor cells derived from the adult brain. *Development, 126*(18), 4017–4026.

Saucedo-Cardenas, O., Quintana-Hau, J. D., Le, W.-D., Smidt, M. P., Cox, J. J., De Mayo, F., Burbach, J. P. H., & Conneely, O. M. (1998). Nurr1 is essential for the induction of the dopaminergic phenotype and the survival of ventral mesencephalic late dopaminergic precursor neurons. *Proceedings of the National Academy of Sciences, 95*(7), 4013–4018.

Sgadò, P., Albéri, L., Gherbassi, D., Galasso, S. L., Ramakers, G. M. J., Alavian, K. N., Smidt, M. P., Dyck, R. H., & Simon, H. H. (2006). Slow progressive degeneration of nigral dopaminergic neurons in postnatal engrailed mutant mice. *Proceedings of the National Academy of Sciences, 103*(41), 15242–15247.

Shi, M., Hu, Z.-L., Zheng, M.-H., Song, N.-N., Huang, Y., Zhao, G., Han, H., & Ding, Y.-Q. (2012). Notch–Rbpj signaling is required for the development of noradrenergic neurons in the mouse locus coeruleus. *Journal of Cell Science, 125*(18), 4320–4332.

Simon, H. H., Saueressig, H., Wurst, W., Goulding, M. D., & O'Leary, D. D. M. (2001). Fate of midbrain dopaminergic neurons controlled by the engrailed genes. *The Journal of Neuroscience, 21*(9), 3126–3134.

Smidt, M. P., van Schaick, H. S. A., Lanctôt, C., Tremblay, J. J., Cox, J. J., van der Kleij, A. A. M., Wolterink, G., Drouin, J., & Burbach, J. P. H. (1997). A homeodomain gene Ptx3 has highly restricted brain expression in mesencephalic dopaminergic neurons. *Proceedings of the National Academy of Sciences, 94*(24), 13305–13310.

Smidt, M. P., Asbreuk, C. H., Cox, J. J., Chen, H., Johnson, R. L., & Burbach, J. P. (2000). A second independent pathway for development of mesencephalic dopaminergic neurons requires Lmx1b. *Nature Neuroscience, 3*(4), 337–341.

Smidt, M. P., Smits, S. M., Bouwmeester, H., Hamers, F. P. T., van der Linden, A. J. A., Hellemons, A. J. C. G. M., Graw, J., & Burbach, J. P. H. (2004). Early developmental failure of substantia nigra dopamine neurons in mice lacking the homeodomain gene Pitx3. *Development, 131*(5), 1145–1155.

Sonnier, L., Le Pen, G., Hartmann, A., Bizot, J. C., Trovero, F., Krebs, M. O., & Prochiantz, A. (2007). Progressive loss of dopaminergic neurons in the ventral midbrain of adult mice heterozygote for Engrailed1. *Journal of Neuroscience, 27*(5), 1063–1071.

Steinbeck, J. A., Choi, S. J., Mrejeru, A., Ganat, Y., Deisseroth, K., Sulzer, D., Mosharov, E. V., & Studer, L. (2015). Optogenetics enables functional analysis of human embryonic stem cell-derived grafts in a Parkinson's disease model. *Nature Biotechnology, 33*(2), 204–209.

Thomson, J. A., Itskovitz-Eldor, J., Shapiro, S. S., Waknitz, M. A., Swiergiel, J. J., Marshall, V. S., & Jones, J. M. (1998). Embryonic stem cell lines derived from human blastocysts. *Science, 282*(5391), 1145–1147.

van den Munckhof, P., Luk, K. C., Ste-Marie, L., Montgomery, J., Blanchet, P. J., Sadikot, A. F., & Drouin, J. (2003). Pitx3 is required for motor activity and for survival of a subset of midbrain dopaminergic neurons. *Development, 130*(11), 2535–2542.

Wernig, M., Zhao, J. P., Pruszak, J., Hedlund, E., Fu, D., Soldner, F., Broccoli, V., Constantine-Paton, M., Isacson, O., & Jaenisch, R. (2008). Neurons derived from reprogrammed fibroblasts functionally integrate into the fetal brain and improve symptoms of rats with Parkinson's disease. *Proceedings of the National Academy of Sciences of the United States of America, 105*(15), 5856–5861.

Xi, J., Liu, Y., Liu, H., Chen, H., Emborg, M. E., & Zhang, S. C. (2012). Specification of midbrain dopamine neurons from primate pluripotent stem cells. *Stem Cells, 30*(8), 1655–1663.

Xu, Z., Chu, X., Jiang, H., Schilling, H., Chen, S., & Feng, J. (2017). Induced dopaminergic neurons: A new promise for Parkinson's disease. *Redox Biology, 11*, 606–612.

Yan, C. H., Levesque, M., Claxton, S., Johnson, R. L., & Ang, S.-L. (2011). Lmx1a and Lmx1b function cooperatively to regulate proliferation, specification, and differentiation of midbrain dopaminergic progenitors. *The Journal of Neuroscience, 31*(35), 12413–12425.

Ye, W., Shimamura, K., Rubenstein, J. L., Hynes, M. A., & Rosenthal, A. (1998). FGF and Shh signals control dopaminergic and serotonergic cell fate in the anterior neural plate. *Cell, 93*(5), 755–766.

Zetterström, R. H., Williams, R., Perlmann, T., & Olson, L. (1996). Cellular expression of the immediate early transcription factors Nurr1 and NGFI-B suggests a gene regulatory role in several brain regions including the nigrostriatal dopamine system. *Molecular Brain Research, 41*(1–2), 111–120.

Zou, H.-L., Su, C.-J., Shi, M., Zhao, G.-Y., Li, Z.-Y., Guo, C., & Ding, Y.-Q. (2009). Expression of the LIM-homeodomain gene Lmx1a in the postnatal mouse central nervous system. *Brain Research Bulletin, 78*(6), 306–312.

Liver Disease Modelling

Nina Graffmann, Lucas-Sebastian Spitzhorn, Audrey Ncube,
Wasco Wruck, and James Adjaye

© Springer Nature Switzerland AG 2020
B. Brand-Saberi (ed.), *Essential Current Concepts in Stem Cell Biology*,
Learning Materials in Biosciences, https://doi.org/10.1007/978-3-030-33923-4_11

What You Will Learn in This Chapter

This chapter describes how *in vitro* generated hepatic cells that are derived from induced pluripotent stem cells (iPSCs) or embryonic stem cells (ESCs) can be used for modelling liver disease.

We will introduce you to the essential liver functions and the different cell types in the liver. You will learn how the liver develops *in vivo* and how this is reflected in *in vitro* differentiation protocols. We will present you a widely used protocol for *in vitro* differentiation of iPSCs into hepatocyte-like cells (HLCs) and will then discuss advantages and disadvantages of iPSC derived hepatic cells for disease modelling. Next, we will introduce *in vitro* models for several diseases affecting hepatic cells or manifesting primarily in distant organs. You will learn how these models can be used to improve our knowledge of these diseases and to test novel treatment strategies.

Finally, we will demonstrate the use of bioinformatic tools to systematically analyse transcriptome data related to these disease models.

11.1   The Functions of the Liver

The liver is a versatile organ involved in many essential metabolic processes in the body and its high secretory activity classifies it as the biggest gland. One of the liver's main functions is the storage and release of micro- and macro-nutrients according to the needs of the body. It is capable of synthesizing, storing and breaking down the main energy sources for the body, namely glucose and fatty acids. Under the control of the insulin counterpart glucagon -a hormone which signals low blood glucose levels to the liver- it produces glucose either out of stored glycogen or by gluconeogenesis, thus providing a constant energy supply to the body and especially the brain (Rui 2014). Fatty acids which are derived either from the diet or from lipolysis taking place in adipocytes are taken up by the liver and stored as triacylglycerides in lipid droplets. From there, they can be released into the blood or they are used directly for energy generation via β-oxidation (Rui 2014).

In addition, the liver synthesizes and secretes many proteins, e.g. albumin -the major blood protein –, clotting factors, and factors related to immune reactions (Barle et al. 1997). It also produces bile acids which are stored in the gall bladder and released into the gut after food uptake (Chiang 2013). Furthermore, the liver plays a major role in detoxification of natural and synthetic compounds. Protein catabolism takes place in the liver and toxic ammonia originating from this process is transformed into urea which can be safely excreted (Mian and Lee 2002). In addition, the large family of cytochrome P-450 (CYP) enzymes is responsible for detoxifying a plethora of molecules, including drugs (Danielson 2002). In the latter case, CYP enzymes are not only needed for detoxification but also to generate the metabolically active form of the drug in the first place (Danielson 2002).

11.2   Cell Types in the Liver

The liver performs versatile functions and has an intricate architecture essential for this role. The adult human liver is composed of four lobes which consist of many small functional units, the so-called liver-lobules (Kruepunga et al. 2019).

About 70% of the cells in the liver are hepatocytes which are responsible for the metabolic and de-toxifying functions of the liver. Due to the anatomy of the liver lobule, the oxygen level of the blood that passes the hepatocytes decreases from the periphery to the centre of the liver lobules and this is accompanied by functional alterations of the hepatocytes. Periportal hepatocytes are more active in oxygen consuming tasks like β-oxidation of fatty acids and they are involved in gluconeogenesis and glucose delivery, while perivenous hepatocytes perform glycolysis, and metabolize drugs with their active CYP enzymes (Hijmans et al. 2014; Kietzmann 2017).

Cholangiocytes are the second characteristic cell type in the liver (Tabibian et al. 2013). They line the intra-hepatic bile ducts and modify primary bile as it flows towards the gall bladder by transporting ions, solutes, and water across their membranes (Tabibian et al. 2013).

In addition to these so-called parenchymal cells, the liver consists of several non-parenchymal cells as well. These are sinusoidal endothelial cells which line the small blood vessels inside the liver (Wisse et al. 1996), Kupffer cells which belong to the macrophage lineage (Bilzer et al. 2006), and hepatic stellate cells, a population of mesenchymal stem cells (MSCs) which store retinoic acid and fat soluble vitamins and play an important role during liver regeneration and fibrosis (Bansal 2016; Kordes et al. 2014).

As the liver performs many distinct functions, it is also affected by a wide variety of diseases, ranging from genetic disorders affecting either hepatocytes or cholangiocytes themselves or peripheral organs which lack proteins synthesized by the liver to metabolically induced diseases as for example steatosis, fibrosis, cirrhosis or hepatocellular carcinoma. Currently, there are approximately 29 million people in the European Union afflicted with chronic liver diseases (Blachier et al. 2013). In order to further increase our meagre knowledge on the development and treatment options of these diseases, *in vitro* models based on hepatocyte-like cells (HLCs) derived from induced pluripotent stem cells (iPSCs) have emerged as tools for studying the aetiology of these diseases.

11.3 Liver Development *in vivo*

All *in vitro* differentiation protocols are based on insights from development *in vivo*. In human, the onset of liver development is within the third week of gestation, when the endoderm emerges from the anterior primitive streak (Gordillo et al. 2015; Lemaigre 2009; Zorn 2008). Driven by WNT3A and high levels of Activin/Nodal signalling, cells specify towards definitive endoderm (DE) and migrate towards the anterior of the embryo. Nodal signalling induces the expression of the endoderm-associated transcription factors SOX17 and FOXA1–3 (Tsankov et al. 2015). Afterwards, the DE cells form a tube comprising the foregut, hindgut and midgut, which become specified according to gradients of BMP2/4, FGF, and WNT secreted by the adjacent mesoderm cells (Dessimoz et al. 2006; McLin et al. 2007; Tiso et al. 2002). Only low to intermediate concentrations of these factors induce foregut development. Foregut cells then develop into liver, whilst also giving rise to the ventral pancreas, stomach, lungs and thyroid (Gordillo et al. 2015; Lemaigre 2009; Zorn 2008).

Between day 23 and 26 of human development, the liver diverticulum arises from the ventral foregut adjacent to the cardiac mesoderm (Gordillo et al. 2015; Lemaigre 2009; Zorn 2008). This process is guided by FGF signalling from the cardiac mesoderm and BMP2/4 signalling from the septum transversum while low levels of WNT signalling are required. At this point the cells specify into hepatic endoderm (HE). Next, the monolayer of HE cells develops into a multilayer of hepatoblasts -bipotential cells that can develop into hepatocytes or cholangiocytes. They form the so called liver bud by invading the septum transversum. Hepatoblast proliferation and migration is promoted by hepatocyte growth factor (HGF) signalling (Bladt et al. 1995; Michalopoulos et al. 2003; Schmidt et al. 1995).

Hepatoblasts express beside the fetal liver characteristic protein α-fetoprotein (AFP) and the more general factors hepatic nuclear factor (HNF) 4α and cytokeratin (CK) 18 also albumin (ALB) which is characteristic for mature adult hepatocytes. In addition, they express markers characteristic for cholangiocytes e.g. CK19. They start differentiating into hepatocytes and cholangiocytes between day 56–58 of gestation. A gradient of TGFβ, Notch, WNT, BMP, and FGF signalling from the portal vein area to the periphery is responsible for driving the cells into one of the two fates (Antoniou et al. 2009; Raynaud et al. 2011). Cholangiocyte precursors reside at the so-called ductal plate which is located next to the portal vein. Under the influence of the above mentioned cytokines, especially Notch2, they mature into cholangiocytes, bud off from the ductal plate (which eventually regresses) and form tubules (Antoniou et al. 2009; Raynaud et al. 2011). Cells at the periphery of this signalling gradient develop into hepatocytes (Raynaud et al. 2011). This process is guided by glucocorticoid hormones in combination with the liver specific cytokines oncostatin M (OSM) and HGF as well as WNT (Kamiya et al. 2001; Michalopoulos et al. 2003).

11.4 *In vitro* Hepatocyte Differentiation

Obtaining liver biopsies of healthy adult primary human hepatocytes (PHH) or cholangiocytes for *in vitro* studies is complicated due to ethical issues and the fact that healthy individuals rarely undergo liver biopsies. Thus, in almost all cases the biopsies will be from a diseased individual. An even stronger obstacle for using these cells for *in vitro* models is the fact that they immediately start to dedifferentiate in cell culture which precludes any long term experiments (Godoy et al. 2016).

Therefore, many research groups have developed protocols for the differentiation of ESCs and iPSCs into hepatocyte-like cells (HLCs) or cholangiocyte-like cells (CLCs) based on our knowledge of signalling pathways active during liver development (Agarwal et al. 2008; Graffmann et al. 2016; Hannan et al. 2013; Jozefczuk et al. 2011; Matz et al. 2017; Sauer et al. 2014; Sgodda et al. 2017; Sgodda et al. 2013; Si-Tayeb et al. 2010; Siller et al. 2015; Wang et al. 2017).

There are many laboratory-specific variations of a common core protocol for HLC differentiation such that we cannot explain all in detail in this book chapter. We will focus on the common scheme used in one of these protocol types. Besides HLC differentiation, numerous laboratories have also established protocols for CLC differentiation, however, this is not the subject of this chapter (Dianat et al. 2014; Ogawa et al. 2015; Sampaziotis et al. 2017).

Most HLC differentiation schemes are based on the three major steps of differentiation DE, HE and HLC stage (☐ Fig. 11.1).

Step 1: From iPSCs to DE

In the first step, iPSCs are directed towards DE by Activin A and WNT signalling. The latter can be either induced by directly adding WNT3A to the medium of by inhibiting GSK3β with CHIR99021 and thus activating β-catenin which enters the nucleus and initiates a WNT-specific transcriptional program.

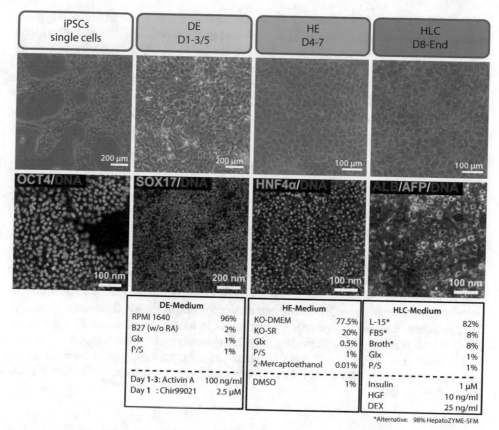

| iPSCs single cells | DE D1-3/5 | HE D4-7 | HLC D8-End |

| OCT4/DNA | SOX17/DNA | HNF4α/DNA | ALB/AFP/DNA |

DE-Medium		HE-Medium		HLC-Medium	
RPMI 1640	96%	KO-DMEM	77.5%	L-15*	82%
B27 (w/o RA)	2%	KO-SR	20%	FBS*	8%
Glx	1%	Glx	0.5%	Broth*	8%
P/S	1%	P/S	1%	Glx	1%
		2-Mercaptoethanol	0.01%	P/S	1%
Day 1-3: Activin A	100 ng/ml	DMSO	1%	Insulin	1 µM
Day 1 : Chir99021	2.5 µM			HGF	10 ng/ml
				DEX	25 ng/ml

*Alternative: 98% HepatoZYME-SFM

Fig. 11.1 *In vitro* differentiation of iPSCs into HLCs. Upper lane: Bright-field images demonstrating the changes in morphology during the differentiation process. Middle lane: Immunocytochemistry confirming the expression of characteristic markers for the respective state. Lower lane: Medium composition for each step (Graffmann et al. 2016)

At this stage, some groups additionally modulate FGF2 and/or BMP4 signalling to reduce self-renewal and increase differentiation. After 3–5 days of high proliferation accompanied by substantial cell death, the cells develop the typical petal like shape of DE. Successful accomplishment of this step can be confirmed by staining for the endoderm specific transcription factor SOX17 or for the surface marker CXCR4.

Step 2: From DE to HE

In the next step, HE is induced by adding DMSO to the medium for 4 days. In addition, it is possible to also include BMP2 or 4 and/or FGF4 or 10. The morphology of the cells further changes and they acquire a polygonal shape which is characteristic of hepatocytes. At this stage they express the hepatic transcription factor HNF4α and sometimes even the early hepatocyte marker AFP.

Step 3: From HE to HLCs

Differentiation into HLCs is promoted by HGF, dexamethasone, and insulin. At this stage most protocols include also OSM, although this has been strongly questioned recently. While early data suggest that OSM is needed for differentiation and especially albumin secretion (Kamiya et al. 1999, 2001), recent studies indicate that differentiation efficiency is not reduced in the absence of OSM (Sgodda et al. 2017).

Overall, HLCs usually appear 12–15 days after the start of differentiation. However, maturation increases till about day 21 and in most instances cells start to de-differentiate beyond this period. Mature HLCs are positive for CK18, ALB, E-Cadherin, as well as many more hepatocyte markers (Agarwal et al. 2008; Graffmann et al. 2016; Matz et al. 2017). In addition, they express AFP which is not detectable in adult primary hepatocytes and thus classifies the *in vitro* cells as more fetal. HLCs also have functional activities. They synthesize urea, uptake and release indocyanine dye and store glycogen. In addition, some of the CYP enzymes are active, especially the more fetal ones (Cameron et al. 2015; Graffmann et al. 2016; Hay et al. 2011; Matz et al. 2017).

Numerous factors besides the medium have an influence on the success of the differentiation. In earlier protocols, cells were seeded onto matrigel coated dishes for the differentiation process. However, these days a 1:3 mixture of Laminin 521 and 111 is preferred as this increases differentiation efficiency and is a step towards a xeno-free protocol (Cameron et al. 2015). In all cases it is important to maintain the cells at a high density, in order to prevent them from developing into large and granular endoderm-derived epithelial cells of unknown function (Graffmann et al. 2018).

Overall, HLCs in 2D cultures lack maturity and therefore it is the "Holy Grail" in this field of research to direct the cells towards a more adult phenotype similar to liver-biopsy derived mature hepatocytes. 3D cultures are very promising to achieve this stage of maturity. They comprise methods where cells are embedded in a collagen sandwich (Gieseck et al. 2014) as well as techniques with floating hepatic organoids under constant agitation or stirring (Sgodda et al. 2017). Finally, it is possible to include other cell types in the organoid in order to mimic the *in vivo* structures with sinusoids and mesenchymal cells (Camp et al. 2017; Nie et al. 2018).

11.5   Advantages and Disadvantages of Modelling Liver Diseases with iPSC Derived Cells

Disease modelling with *in vitro* derived HLCs or CLCs is still in a relatively early phase and there is no consensus model that can be used for all diseases. Rather, every laboratory has optimized their models in order to fit best for the disease under investigation. Despite this heterogeneity, there are many advantages of iPSC derived liver cells for disease modelling over traditional ways, which include animal models, transformed cancer cell lines, and primary human hepatocytes (PHH) (◘ Table 11.1).

◘ **Table 11.1**   Advantages and disadvantages of iPSC derived hepatic cells for disease modelling

Advantages	Disadvantages
iPSCs are available from all genetic backgrounds	Differentiated cells resemble the fetal instead of the adult state
Minor risks for the patient	
Limited ethical considerations	
Unlimited expansion of cells, unrestricted availability of cells	Ageing effects are diluted out during reprogramming
Differentiation into distinct tissues/cell types from one donor	
Increased lifetime of *in vitro* derived HLCs over primary hepatocytes	Epigenetic memory influences differentiation efficiency
Multifactorial diseases can be analysed	

11.6 Examples of Liver Diseases Modelled with iPSC Derived Hepatic Cells

11.6.1 Liver Diseases That Affect Hepatocytes

In recent years, several diseases have been modelled with iPSC derived hepatic cells, however, we can only present a selection of these in this chapter.

11.6.1.1 Steatosis

Steatosis is a metabolic disease which is associated with the accumulation of fat within hepatocytes. Steatosis or fatty liver generally has two distinct aetiologies: Alcoholic liver disease is triggered by high and regular intakes of alcohol, while nonalcoholic fatty liver disease (NAFLD) depends on imbalances between calorie intake and expenditure. Since obesity as well as insulin resistance lead to chronic inflamed adipocyte tissue, they are the major risk factors which promote NAFLD. In later stages this disease can lead to steatohepatitis (NASH), liver fibrosis, cirrhosis and cancer (Basaranoglu et al. 2013; Perry et al. 2015). Besides nutrition and alcohol drugs (e.g. valproic acid, tetracycline, amiodarone) are also crucial factors for steatosis development (Szalowska et al. 2014). Although there are some mice models, they fail to recapitulate all aspects of the disease since the murine metabolism reacts differently upon diet changes compared to the human setting which does not allow to analyse all levels of the disease in a single mouse model (Machado et al. 2015; Takahashi et al. 2012).

In vitro modulation of this condition was done by generation of iPSC derived HLCs. Upon addition of oleic acid these HLCs incorporated fat leading to changes in the expression of metabolic associated genes and upregulation of lipid associated proteins (Graffmann et al. 2016).

Being a multifactorial disease, NAFLD is not caused by a single mutation. In this setting it is of particular interest to study the mechanisms underlying the disease in HLCs derived from a variety of patients in order to identify common disease associated pathways (Jozefczuk et al. 2012).

11.6.1.2 Hepatitis B/C

Even-though vaccination and therapy against hepatitis B virus (HBV) have been developed, approximately 260 million people are affected with chronic infection leading to a variety of outcomes. In some cases, the patient is symptom-free in other cases liver cirrhosis or hepatocellular carcinoma ensue (Revill et al. 2016; Zeng et al. 2008).

Apart from man only chimpanzees are fully susceptible for this infection which has a high tropism towards the liver. This fact makes it very challenging to study the disease. Since the research on great apes is limited other experimental models are needed (Allweiss and Dandri 2016). Efforts have been made to generate transgenic mice which are susceptible for the virus infection due to genetic modifications (Chisari et al. 1985; Yang et al. 2014). A more advanced approach is the generation of chimeric mice harbouring human hepatocytes as well as human immune cells and this has made it possible to study the interplay between human hepatocytes and the adaptive immune response (Bility et al. 2014).

Successful HBV models with iPSC-derived HLCs have been described (Shlomai et al. 2014). Shlomai et al. demonstrated that HLCs can be infected with HBV and observed changes over a time course of 24 days. In 2018, HLCs derived from human iPSCs in combination with human umbilical vein endothelial cells (HUVECs) and MSCs were used to derive liver organoids for modelling HBV infection. These 3D structures were more permissive for HBV and the virus propagation maintained for a longer time span than the 2D cultured HLCs. This personalized model enabled the modulation of HBV infection and mirrored virus induced hepatic dysfunction as well as the life cycle of the HBV (Nie et al. 2018).

Hepatitis C virus (HCV) is also a major health burden all over the world affecting approximately 550 million people and leading to secondary liver malfunctions. Established medication can suppress the replication of the virus and recently even drugs that cure the disease have been developed, although they are not affordable for the majority of the patients (Hayes and Chayama 2017). HCV infection is particularly hard to deal with because symptoms frequently only occur 15–20 years post-infection which limits treatment options. Many patients eventually need a liver transplant because of high levels of liver cirrhosis. However, the reinfection rate of transplanted liver is 100% (Garcia-Retortillo et al. 2002).

As for HBV, it could be shown that HLCs from iPSCs are a relevant tool for mimicking host-HCV interplay, allowing insights into innate immune response as well as the synthesis of lipoproteins (Schobel et al. 2018). In contrast to the Huh7.5 cell line, HLCs have been be infected with various HCV strains isolated from patients, thus enabling a more detailed study of infection and replication modes (Wu et al. 2012). It has also been shown that shRNA mediated repression of receptors important for the infection with HCV only prevented the infection with wild-type virus strains and not with mutated ones (Wu et al. 2012).

Another level of HLC mediated disease modelling for both viruses is the combination of the HLC approach with the chimeric mice approach. Chimeras harbouring human HLCs were created for investigating the pathogenesis of HBV and HCV and as a platform to test anti-viral components (Carpentier et al. 2014; Yuan et al. 2018).

11.6.1.3 Wilsons Disease

Wilson's disease (WD) is caused by accumulation of copper in the liver and the brain. It is a rare autosomal recessive disease manifesting in a mutated *ATP7B* gene. This gene is crucial within the biliary excretion pathway of copper. Due to this impaired excretion, copper accumulates over time and primarily manifests in the brain and the liver of patients leading to hepatitis, cirrhosis, liver failure or neurological malfunctions (Brewer and Askari 2005).

To date, four established rodent models of WD exist, but they do not carry the most frequent missense mutation of the *ATP7B* gene (de Bie et al. 2007). Furthermore, the metabolic pathways in these animals are distinct to that observed in human (Ranucci et al. 2017).

Zhang et al. differentiated iPSCs from a Chinese patient suffering from WD with a Chinese hotspot mutation into HLCs. These HLCs mimicked the disease phenotype *in vitro* by showing impaired ATP7B proteins in the cytoplasm and thus resulting in reduced levels of copper transport. Using a lentivirus expressing the functional version of ATP7B it was possible to cure the disease *in vitro*. Furthermore, addition of the chaperone drug curcumin in the cell culture medium could rescue the disease phenotype (Zhang et al. 2011).

Yi et al. derived iPSCs from a patient carrying a Caucasian hotspot mutation. These cells were differentiated into neurons and HLCs expressing the gene defect thereby modelling both major affected cell types in WD (Yi et al. 2012).

11.6.1.4 Drug Induced Hepatic Toxicity

Although drug-induced liver injury (DILI) is a small factor for acute liver injury it causes most of the acute liver failures (Larrey and Pageaux 2005). DILI is also one of the main reasons for non-approval of drugs, terminations of clinical trials as well as withdrawing drugs from the market (Stevens and Baker 2009). During DILI, metabolites of the drugs and patients' proteins form adducts. These are presented on the cell surface as neo-antigens leading either to an immune-allergic reaction or to non-allergic toxic reactions (Kaplowitz 2005). The subsequent hepatic reaction cascades include impairment of transporters, elevated levels of oxidative stress, and pro-inflammatory cytokines (Russmann et al. 2010). Especially anti-cancer drugs such as protein kinase inhibitors, immune checkpoint inhibitors or the epidermal growth factor receptor tyrosine kinase inhibitor can cause severe hepatic damages (Takeda et al. 2015), although the commonly used pain relieve medication paracetamol (acetaminophen) is the leading causative drug in the UK and the US (Larrey and Pageaux 2005).

In vitro cell models including PHH, hepatic cell lines, and HLCs in 2D and 3D composition have been used to promote DILI research (Gomez-Lechon et al. 2010; Lu et al. 2015). With these cell-based models it has been possible to gain insights into DILI-associated mechanisms such as endoplasmic and oxidative stress and transporter inhibition (Bell et al. 2017; Godoy et al. 2013; Schadt et al. 2015). In addition, it has been shown that HLCs derived from patients with alpha-1-antitrypsin deficiency (see ▸ Sect. 11.6.2.1.) react much more sensitive to several common drugs than those derived from healthy controls (Wilson et al. 2015). In line with the need for in depth knowledge about pharmacokinetics of certain drugs in a specific environment, iPSCs have been derived from donors with distinct known cytochrome P450 genotypes in order to generate HLCs which can be used for drug testing (Bohndorf et al. 2017). To enable high throughput screening of drugs, HLC differentiation has been achieved in 384 wells (Carpentier et al. 2016).

11.6.1.5 Tangier Disease

Tangier disease (TD) is a rare disease first found on Tangier island in Virginia where an isolated community of descendants from settlers live. TD patients have no high density lipoprotein (HDL) and thus cholesterol efflux is disturbed. TD symptoms due to cholesterol ester deposits range from orange tonsils over hepatosplenomegaly and neuropathy to cardiovascular disease. TD is caused by mutation in the gene *ABCA1* which regulates cholesterol efflux via HDL (Brooks-Wilson et al. 1999; Scott 1999) and an iPSC-based model has been reported (Bi et al. 2017). The authors of the study generated iPSCs from TD patients and matched healthy individuals and subsequently differentiated them into HLCs. TD iPSC-derived HLCs had impaired HDL production and cholesterol efflux and increased triacylglycerol secretion. This iPSC-derived HLC model of TD has been instrumental in confirming the involvement of *ABCA1* in the molecular mechanisms controlling cholesterol efflux.

11.6.2 Diseases That Affect Hepatocytes, But Manifest Primarily in Other Organs

Some liver diseases have extrahepatic manifestations, meaning that other organs besides the liver are affected.

11.6.2.1 Alpha-1-Antitrypsin Deficiency

Alpha-1-antitrypsin deficiency (A1AD) is a genetic disorder with a worldwide prevalence of 1:2500. It is caused by mutations in the *SERPINA1* gene. This gene codes for the protein alpha-1-anti-trpysin (A1AT) mainly synthesized in hepatocytes. A1AT is responsible for the protection against the enzyme neutrophil elastase, which is released by neutrophils during immune response, but also damages normal tissue if not carefully regulated by A1AT (Brantly et al. 1988). Lack of A1AT's anti-protease activity results in uncontrolled circulating neutrophil elastase that destroys lung air sacs leading to emphysema or chronic obstructive pulmonary disease (COPD). In addition, mutated A1AT protein forms insoluble precipitates in hepatocytes which cause damage, thus resulting in scaring and chronic liver disease.

Using multiple donors to model the disease, Wilson et al. showed differential expression of 135 genes between HLCs derived from patients carrying the mutation and healthy controls (Wilson et al. 2015). They and others could show that A1AT forms insoluble aggregates in patient derived HLCs and is not exported efficiently. They could improve the phenotype by treating the cells with a proteasome inhibitor or by increasing autophagy (Rashid et al. 2010; Wilson et al. 2015). In addition, they also showed that HLCs derived from A1AD patients are more sensitive for common drugs like paracetamol (acetaminophen) than those from healthy controls (Wilson et al. 2015).

11.6.2.2 Transthyretin-Related Hereditary Amyloidosis

Transthyretin (TTR) amyloidosis (or familial amyloidotic polyneuropathy (FAP)) is a systemic familiarly hereditary disorder caused by mutations that destabilise the TTR protein in the liver (Ando et al. 2013). Misfolded and aggregated amyloid fibrils consisting of TTR – a plasma protein produced in the liver to transport thyroxine and vitamin A– are deposited and accumulate in various organs and tissues, such as in the nervous system and cardiac tissue, leading to inherent dysfunction. It is characterised by a slowly progressive peripheral sensory motor and autonomic neuropathy, and later involves visceral organs such as the kidney, hence becoming fatal (Ando et al. 2013).

Leung et al. generated TTR patient-specific iPSCs which were further directly differentiated into hepatic, cardiac and neuronal lineages, to model three tissue types involved in the disease. The HLCs produced and secreted mutant TTR protein, which caused cell death of neurons and cardiac cells incubated with the TTR containing supernatant. Treatment with stabilizers of TTR improved the phenotype. In essence, Leung et al. proved that this iPSC-based system is an ideal platform for further investigating the role of the liver in TTR (Leung et al. 2013).

11.6.3 Diseases That Affect Cholangiocytes

11.6.3.1 Cystic Fibrosis

Cystic fibrosis (CF) is a hereditary disease caused by mutations in the gene encoding the CF transmembrane conductance regulator (CFTR). It involves several organ systems, with manifestations mostly as lung disease but it also affects the liver (Accurso 2006). In the liver, the *CFTR* gene expression occurs only at the level of the epithelium of the bile ducts and the CFTR protein anchors on the apical membrane of the cholangiocytes and regulates electrolyte and water content in the bile. Mutations in CFTR result in impaired secretion and thick viscous bile causing damage to both the cholangiocytes and hepatocytes – periportal fibrosis, which over the years leads to cirrhosis (van Mourik 2017). Obstruction of the intra-hepatic bile ductules with thick mucus, leads to the development of conditions such as progressive cholestasis or multi-lobular biliary cirrhosis (O'Brien et al. 1992).

In vitro differentiation of iPSCs into cholangiocyte-like cells (CLCs) in a 3D system resulted in spheres and cysts which had functional activity measured by rhodamine uptake and swelling of the cysts (Ogawa et al. 2015; Sampaziotis et al. 2015). Patient derived CLCs did not only show reduced cyst formation, but also had impaired functions related to dye uptake and swelling due to the defects in CFTR function. In the *in vitro* assays, the function could be partly restored by treating the cysts with small molecules promoting proper folding of CFTR (Ogawa et al. 2015; Sampaziotis et al. 2015). Thus, this system is suitable for *in vitro* drug testing.

11.6.3.2 Alagille Syndrome

Alagille syndrome (ALGS) is a rare genetic disorder with varying severity, where bile duct formation is impaired. It is considered a multi-system disorder as it also affects the liver, skeleton, eyes and heart. It is predominantly caused by mutations in the gene *Jagged 1* (*JAG1*) which is a ligand in the Notch signalling pathway and thus essential for cholangiocyte development (Kamath et al. 2010). Typical symptoms such as cholestasis, jaundice, itching, as well as heart and skeletal problems manifest during the first 3 months of life with the severe form, with cholestasis being a direct consequence due to insufficient bile ducts.

To improve our understanding of the mechanisms that cause ALGS liver pathology, studies using hepatic organoids from iPSCs were recently carried out by Guan et al. (2017). The organoids comprised a mixture of HLCs and CLCs. Organoid formation and function was massively impaired in ALGS patient derived iPSCs. Inducing a disease causing mutation in wildtype iPSCs by CRISPR (Clustered Regularly Interspaced Short Palindromic Repeats) lead to the same phenotype as observed in ALGS patient derived organoids. In addition, it was also possible to correct the mutation in ALGS patient derived cells which reverted the phenotype and thus proved that the mutation was responsible for the disease (Guan et al. 2017).

11.7 Bioinformatic Analysis

11.7.1 Datasets from Public Repositories

Bioinformatic methods enable meta-analysis of a plethora of datasets available at public repositories such as the ones shown in ❑ Table 11.2.

For a more comprehensive compilation of public repositories we refer to the list recommended by the journal Scientific Data: ► https://www.nature.com/sdata/policies/repositories. For liver disease modeling there exist datasets generated from liver biopsies or from iPSC-derived HLCs.

In this example, we focus on a dataset based on HLCs derived from Tangier disease (TD, see ► Sect. 11.6.1.5) patients for which no liver biopsies or primary hepatocytes existed (Bi et al. 2017). We compare the transcriptome data of these cells with transcriptomes of liver biopsies from NAFLD and NASH patients with the intention of gaining in

◘ Table 11.2 Selection of repositories for transcriptome and genome data

Repository	URL	Comments
National Center for biotechnology information (NCBI) gene expression omnibus (GEO)	► https://www.ncbi.nlm.nih.gov/geo	Microarrays and sequencing data
NCBI sequence read archive (SRA)	► https://www.ncbi.nlm.nih.gov/sra	Sequencing data
European Bioinformatic institute (EMBL-EBI) ArrayExpress	► https://www.ebi.ac.uk/arrayexpress/	Microarrays and sequencing data
1000 genomes project	► http://www.internationalgenome.org/	Sequencing data

depth insights on cholesterol reverse transport which is disabled in TD patients due to a lack of HDL.

Previous bioinformatic analyses have unveiled a gene signature which correlates with the progression of NAFLD to NASH (Wruck et al. 2017). This gene signature was functionally annotated with cholesterol-related processes. Similar, although not as dramatic as in TD, high plasma triacylglyceride levels and low HDL cholesterol often can be observed in NAFLD patients (Arguello et al. 2015). This indicates that both diseases have a common phenotype and that bioinformatic comparisons of genome-wide gene expression can increase our understanding of both diseases. The role of *ABCA1* in reverse cholesterol transport suggests that its decreased expression in NAFLD/NASH may result in increased liver triacylglycerides (Arguello et al. 2015; Vega-Badillo et al. 2016).

11.7.2 Comparison of Gene Expression Datasets

Employing a venn diagram analysis, we identified common genes that correlated to the progression of NAFLD and were also differentially expressed in the iPSC-derived HLC model of TD (◘ Fig. 11.2a). Each fraction in this diagram represents the number of genes that are either uniquely expressed in one sample or shared by the indicated samples. The overlap common to the TD model and all NAFLD datasets consists only of two genes: *LRRC31* and *C3orf58* (◘ Fig. 11.2b).

The classical way to gain a deeper insight into the functions of these genes is to review publications, however, for these two factors the number of publications is rather limited. Another way is to check which gene ontologies (GOs) are assigned to the gene of interest. GOs specify the biological role of genes (Ashburner et al. 2000) and starting with the top categories biological process (BP), cellular component (CC) and molecular function (MF) stepwise refine these roles to very detailed terms, e.g. the BP "somatotropin secreting cell differentiation".

C3orf58, one of the genes present in all datasets related to the progression of NAFLD to NASH, is associated with GO terms which can be roughly grouped to Golgi, coat protein (COP) I coated vesicles and phosphatidylinositol 3-kinase signaling. COPI and COPII coated vesicles perform the retrograde and anterograde transport of proteins between the Golgi

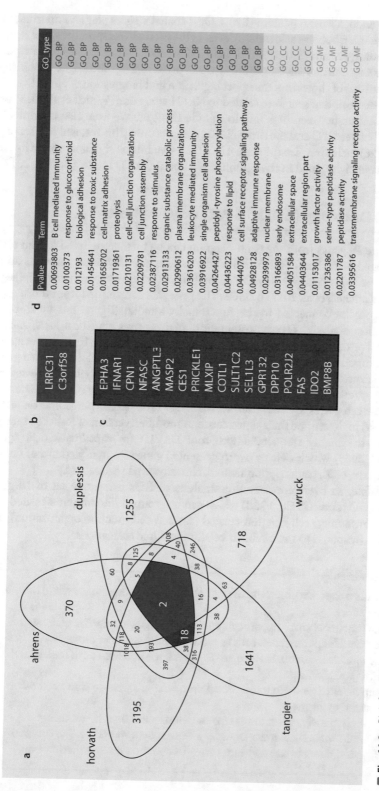

Fig. 11.2 Bioinformatic analysis of genes regulated in TD and NAFLD. **a** Liver transcriptome datasets related to the progression of NAFLD to NASH (Ahrens et al. 2013; du Plessis et al. 2015; Horvath et al. 2014; Wruck et al. 2015) and TD (Bi et al. 2017) were compared in a venn diagram. **b** The genes *C3orf58* and *LRRC31* were found in the intersection of all datasets. **c** 18 genes were found in the intersection of all NASH-related datasets and the TD dataset but not the high- vs low-grade NAFLD dataset. **d** Gene ontologies (GOs) overrepresented in the 18-gene-set in **c**

complex and the endoplasmic reticulum (ER). COPII coated vesicles play an important role in cholesterol regulation because at low sterol levels in the ER they transport the sterol regulatory element binding protein (SREBP) to the Golgi for proteolytic release as transcription factor (Ikonen 2008). Further investigation of these proteins may shed light on hitherto unknown mechanisms of cholesterol transport regulation and progression of NAFLD.

Three of the steatosis datasets were related to NASH while one (Wruck et al. 2015) was related to a distinction between high and low grade steatosis. We were also interested in the comparison between TD and the three NASH datasets only. This resulted in a set of 18 genes (◻ Fig. 11.2c) including also the gene *ANGPTL3* – an inhibitor of lipoprotein lipase - which Bi et al. have highlighted in their publication about the TD model (Bi et al. 2017).

11.7.3 Gene Ontology (GO) and Pathway Analysis

Gene sets like the 18 genes from the venn diagram can be further analyzed to find functionality associated with them. This functionality can be given by other gene sets annotated with specific functions, such as KEGG-pathways (▶ https://www.genome.jp/kegg/pathway.html) (Kanehisa et al. 2017) or GOs. The gene sets determined from the analysis of experiments can be connected to the functionally annotated gene sets by so-called "overrepresentation analysis". Overrepresentation analysis uses statistical tests, such as Fisher's exact test or the hypergeometric test, to determine if the genes from the set appear more often in a functional group of genes than in the background of all annotated genes. Usually, in gene expression analysis genes are categorized as (differentially) expressed/not (differentially) expressed and in the pathway (GO)/not in the pathway (GO). The number of genes in these categories are subjected to statistical tests (e.g. hypergeometric test or Fisher's exact test) to determine their significance. A modified version of Fisher's exact test is employed in the highly frequented web tool DAVID (▶ https://david.ncifcrf.gov/) (Huang da et al. 2009) which detects overrepresented gene sets in a plethora of dataset including pathways, GOs, transcription factors, chromosomal regions, etc.

In our example, an over-representation analysis of GOs using this set of 18 genes revealed several GO terms (◻ Fig. 11.2d) which can be grouped into immune-related (e.g. B cell mediated immunity) cell junction-related (e.g. cell-cell-junction organization), and response-related (response to toxic substance/stimulus/lipid) categories.

> **Take Home Message**
> 1. The liver is a complex organ with essential functions for metabolism and drug detoxification
> 2. Its main functional cells –hepatocytes and cholangiocytes- can be generated *in vitro* from pluripotent stem cells to model a wide variety of diseases
> 3. Disease models recapitulate the disease phenotype and help us to understand underlying mechanisms
> 4. They can be used for (high throughput) drug testing and are a step towards personalized medicine
> 5. 3D models generally reflect the *in vivo* situation better than 2D models
> 6. Bioinformatic tools allow the analysis of a wide spectrum of data. The comparison of related diseases can give hints towards common disease mechanisms

References

Accurso, F. J. (2006). Update in cystic fibrosis 2005. *American Journal of Respiratory and Critical Care Medicine, 173*, 944–947.

Agarwal, S., Holton, K. L., & Lanza, R. (2008). Efficient differentiation of functional hepatocytes from human embryonic stem cells. *Stem Cells, 26*, 1117–1127.

Ahrens, M., Ammerpohl, O., von Schonfels, W., Kolarova, J., Bens, S., Itzel, T., Teufel, A., Herrmann, A., Brosch, M., Hinrichsen, H., et al. (2013). DNA methylation analysis in nonalcoholic fatty liver disease suggests distinct disease-specific and remodeling signatures after bariatric surgery. *Cell Metabolism, 18*, 296–302.

Allweiss, L., & Dandri, M. (2016). Experimental in vitro and in vivo models for the study of human hepatitis B virus infection. *Journal of Hepatology, 64*, S17–S31.

Ando, Y., Coelho, T., Berk, J. L., Cruz, M. W., Ericzon, B. G., Ikeda, S., Lewis, W. D., Obici, L., Plante-Bordeneuve, V., Rapezzi, C., et al. (2013). Guideline of transthyretin-related hereditary amyloidosis for clinicians. *Orphanet Journal of Rare Diseases, 8*, 31.

Antoniou, A., Raynaud, P., Cordi, S., Zong, Y., Tronche, F., Stanger, B. Z., Jacquemin, P., Pierreux, C. E., Clotman, F., & Lemaigre, F. P. (2009). Intrahepatic bile ducts develop according to a new mode of tubulogenesis regulated by the transcription factor SOX9. *Gastroenterology, 136*, 2325–2333.

Arguello, G., Balboa, E., Arrese, M., & Zanlungo, S. (2015). Recent insights on the role of cholesterol in nonalcoholic fatty liver disease. *Biochimica et Biophysica Acta, 1852*, 1765–1778.

Ashburner, M., Ball, C. A., Blake, J. A., Botstein, D., Butler, H., Cherry, J. M., Davis, A. P., Dolinski, K., Dwight, S. S., Eppig, J. T., et al. (2000). Gene ontology: Tool for the unification of biology. The Gene Ontology Consortium. *Nature Genetics, 25*, 25–29.

Bansal, M. B. (2016). Hepatic stellate cells: Fibrogenic, regenerative or both? Heterogeneity and context are key. *Hepatology International, 10*, 902–908.

Barle, H., Nyberg, B., Essen, P., Andersson, K., McNurlan, M. A., Wernerman, J., & Garlick, P. J. (1997). The synthesis rates of total liver protein and plasma albumin determined simultaneously in vivo in humans. *Hepatology, 25*, 154–158.

Basaranoglu, M., Basaranoglu, G., & Senturk, H. (2013). From fatty liver to fibrosis: A tale of "second hit". *World Journal of Gastroenterology: WJG, 19*, 1158–1165.

Bell, C. C., Lauschke, V. M., Vorrink, S. U., Palmgren, H., Duffin, R., Andersson, T. B., & Ingelman-Sundberg, M. (2017). Transcriptional, functional, and mechanistic comparisons of stem cell-derived hepatocytes, HepaRG cells, and three-dimensional human hepatocyte spheroids as predictive in vitro systems for drug-induced liver injury. *Drug Metabolism and Disposition: The Biological Fate of Chemicals, 45*, 419–429.

Bi, X., Pashos, E. E., Cuchel, M., Lyssenko, N. N., Hernandez, M., Picataggi, A., McParland, J., Yang, W., Liu, Y., Yan, R., et al. (2017). ATP-binding cassette transporter A1 deficiency in human induced pluripotent stem cell-derived hepatocytes abrogates HDL biogenesis and enhances triglyceride secretion. *eBioMedicine, 18*, 139–145.

Bility, M. T., Cheng, L., Zhang, Z., Luan, Y., Li, F., Chi, L., Zhang, L., Tu, Z., Gao, Y., Fu, Y., et al. (2014). Hepatitis B virus infection and immunopathogenesis in a humanized mouse model: Induction of human-specific liver fibrosis and M2-like macrophages. *PLoS Pathogens, 10*, e1004032.

Bilzer, M., Roggel, F., & Gerbes, A. L. (2006). Role of Kupffer cells in host defense and liver disease. *Liver International: Official Journal of the International Association for the Study of the Liver, 26*, 1175–1186.

Blachier, M., Leleu, H., Peck-Radosavljevic, M., Valla, D. C., & Roudot-Thoraval, F. (2013). The burden of liver disease in Europe: A review of available epidemiological data. *Journal of Hepatology, 58*, 593–608.

Bladt, F., Riethmacher, D., Isenmann, S., Aguzzi, A., & Birchmeier, C. (1995). Essential role for the C-met receptor in the migration of myogenic precursor cells into the limb bud. *Nature, 376*, 768–771.

Bohndorf, M., Ncube, A., Spitzhorn, L. S., Enczmann, J., Wruck, W., & Adjaye, J. (2017). Derivation and characterization of integration-free iPSC line ISRM-UM51 derived from SIX2-positive renal cells isolated from urine of an African male expressing the CYP2D6 *4/*17 variant which confers intermediate drug metabolizing activity. *Stem Cell Research, 25*, 18–21.

Brantly, M., Nukiwa, T., & Crystal, R. G. (1988). Molecular basis of alpha-1-antitrypsin deficiency. *The American Journal of Medicine, 84*, 13–31.

Brewer, G. J., & Askari, F. K. (2005). Wilson's disease: Clinical management and therapy. *Journal of Hepatology, 42*(Suppl), S13–S21.

Brooks-Wilson, A., Marcil, M., Clee, S. M., Zhang, L. H., Roomp, K., van Dam, M., Yu, L., Brewer, C., Collins, J. A., Molhuizen, H. O., et al. (1999). Mutations in ABC1 in Tangier disease and familial high-density lipoprotein deficiency. *Nature Genetics, 22*, 336–345.

Cameron, K., Tan, R., Schmidt-Heck, W., Campos, G., Lyall, M. J., Wang, Y., Lucendo-Villarin, B., Szkolnicka, D., Bates, N., Kimber, S. J., et al. (2015). Recombinant laminins drive the differentiation and self-organization of hESC-derived hepatocytes. *Stem Cell Reports, 5*, 1250–1262.

Camp, J. G., Sekine, K., Gerber, T., Loeffler-Wirth, H., Binder, H., Gac, M., Kanton, S., Kageyama, J., Damm, G., Seehofer, D., et al. (2017). Multilineage communication regulates human liver bud development from pluripotency. *Nature, 546*(7659), 533.

Carpentier, A., Tesfaye, A., Chu, V., Nimgaonkar, I., Zhang, F., Lee, S. B., Thorgeirsson, S. S., Feinstone, S. M., & Liang, T. J. (2014). Engrafted human stem cell-derived hepatocytes establish an infectious HCV murine model. *The Journal of Clinical Investigation, 124*, 4953–4964.

Carpentier, A., Nimgaonkar, I., Chu, V., Xia, Y., Hu, Z., & Liang, T. J. (2016). Hepatic differentiation of human pluripotent stem cells in miniaturized format suitable for high-throughput screen. *Stem Cell Research, 16*, 640–650.

Chiang, J. Y. (2013). Bile acid metabolism and signaling. *Comprehensive Physiology, 3*, 1191–1212.

Chisari, F. V., Pinkert, C. A., Milich, D. R., Filippi, P., McLachlan, A., Palmiter, R. D., & Brinster, R. L. (1985). A transgenic mouse model of the chronic hepatitis B surface antigen carrier state. *Science, 230*, 1157–1160.

Danielson, P. B. (2002). The cytochrome P450 superfamily: Biochemistry, evolution and drug metabolism in humans. *Current Drug Metabolism, 3*, 561–597.

de Bie, P., Muller, P., Wijmenga, C., & Klomp, L. W. (2007). Molecular pathogenesis of Wilson and Menkes disease: Correlation of mutations with molecular defects and disease phenotypes. *Journal of Medical Genetics, 44*, 673–688.

Dessimoz, J., Opoka, R., Kordich, J. J., Grapin-Botton, A., & Wells, J. M. (2006). FGF signaling is necessary for establishing gut tube domains along the anterior-posterior axis in vivo. *Mechanisms of Development, 123*, 42–55.

Dianat, N., Dubois-Pot-Schneider, H., Steichen, C., Desterke, C., Leclerc, P., Raveux, A., Combettes, L., Weber, A., Corlu, A., & Dubart-Kupperschmitt, A. (2014). Generation of functional cholangiocyte-like cells from human pluripotent stem cells and HepaRG cells. *Hepatology, 60*, 700–714.

du Plessis, J., van Pelt, J., Korf, H., Mathieu, C., van der Schueren, B., Lannoo, M., Oyen, T., Topal, B., Fetter, G., Nayler, S., et al. (2015). Association of adipose tissue inflammation with histologic severity of nonalcoholic fatty liver disease. *Gastroenterology, 149*, 635–648.e614.

Garcia-Retortillo, M., Forns, X., Feliu, A., Moitinho, E., Costa, J., Navasa, M., Rimola, A., & Rodes, J. (2002). Hepatitis C virus kinetics during and immediately after liver transplantation. *Hepatology, 35*, 680–687.

Gieseck, R. L., 3rd, Hannan, N. R., Bort, R., Hanley, N. A., Drake, R. A., Cameron, G. W., Wynn, T. A., & Vallier, L. (2014). Maturation of induced pluripotent stem cell derived hepatocytes by 3D-culture. *PLoS One, 9*, e86372.

Godoy, P., Hewitt, N. J., Albrecht, U., Andersen, M. E., Ansari, N., Bhattacharya, S., Bode, J. G., Bolleyn, J., Borner, C., Bottger, J., et al. (2013). Recent advances in 2D and 3D in vitro systems using primary hepatocytes, alternative hepatocyte sources and non-parenchymal liver cells and their use in investigating mechanisms of hepatotoxicity, cell signaling and ADME. *Archives of Toxicology, 87*, 1315–1530.

Godoy, P., Widera, A., Schmidt-Heck, W., Campos, G., Meyer, C., Cadenas, C., Reif, R., Stober, R., Hammad, S., Putter, L., et al. (2016). Gene network activity in cultivated primary hepatocytes is highly similar to diseased mammalian liver tissue. *Archives of Toxicology, 90*(10), 2513–2529.

Gomez-Lechon, M. J., Lahoz, A., Gombau, L., Castell, J. V., & Donato, M. T. (2010). In vitro evaluation of potential hepatotoxicity induced by drugs. *Current Pharmaceutical Design, 16*, 1963–1977.

Gordillo, M., Evans, T., & Gouon-Evans, V. (2015). Orchestrating liver development. *Development, 142*, 2094–2108.

Graffmann, N., Ring, S., Kawala, M. A., Wruck, W., Ncube, A., Trompeter, H. I., & Adjaye, J. (2016). Modeling nonalcoholic fatty liver disease with human pluripotent stem cell-derived immature hepatocyte-like cells reveals activation of PLIN2 and confirms regulatory functions of peroxisome proliferator-activated receptor alpha. *Stem Cells and Development, 25*, 1119–1133.

Graffmann, N., Ncube, A., Wruck, W., & Adjaye, J. (2018). Cell fate decisions of human iPSC-derived bipotential hepatoblasts depend on cell density. *PLoS One, 13*, e0200416.

Guan, Y., Xu, D., Garfin, P. M., Ehmer, U., Hurwitz, M., Enns, G., Michie, S., Wu, M., Zheng, M., Nishimura, T., et al. (2017). Human hepatic organoids for the analysis of human genetic diseases. *JCI Insight, 2*, 94954.

Hannan, N. R., Segeritz, C. P., Touboul, T., & Vallier, L. (2013). Production of hepatocyte-like cells from human pluripotent stem cells. *Nature Protocols, 8*, 430–437.

Hay, D. C., Pernagallo, S., Diaz-Mochon, J. J., Medine, C. N., Greenhough, S., Hannoun, Z., Schrader, J., Black, J. R., Fletcher, J., Dalgetty, D., et al. (2011). Unbiased screening of polymer libraries to define novel substrates for functional hepatocytes with inducible drug metabolism. *Stem Cell Research, 6*, 92–102.

Hayes, C. N., & Chayama, K. (2017). Why highly effective drugs are not enough: The need for an affordable solution to eliminating HCV. *Expert Review of Clinical Pharmacology, 10*, 583–594.

Hijmans, B. S., Grefhorst, A., Oosterveer, M. H., & Groen, A. K. (2014). Zonation of glucose and fatty acid metabolism in the liver: Mechanism and metabolic consequences. *Biochimie, 96*, 121–129.

Horvath, S., Erhart, W., Brosch, M., Ammerpohl, O., von Schonfels, W., Ahrens, M., Heits, N., Bell, J. T., Tsai, P. C., Spector, T. D., et al. (2014). Obesity accelerates epigenetic aging of human liver. *Proceedings of the National Academy of Sciences of the United States of America, 111*, 15538–15543.

Huang da, W., Sherman, B. T., & Lempicki, R. A. (2009). Systematic and integrative analysis of large gene lists using DAVID bioinformatics resources. *Nature Protocols, 4*, 44–57.

Ikonen, E. (2008). Cellular cholesterol trafficking and compartmentalization. *Nature Reviews Molecular Cell Biology, 9*, 125–138.

Jozefczuk, J., Prigione, A., Chavez, L., & Adjaye, J. (2011). Comparative analysis of human embryonic stem cell and induced pluripotent stem cell-derived hepatocyte-like cells reveals current drawbacks and possible strategies for improved differentiation. *Stem Cells and Development, 20*, 1259–1275.

Jozefczuk, J., Kashofer, K., Ummanni, R., Henjes, F., Rehman, S., Geenen, S., Wruck, W., Regenbrecht, C., Daskalaki, A., Wierling, C., et al. (2012). A systems biology approach to deciphering the etiology of steatosis employing patient-derived dermal fibroblasts and iPS cells. *Frontiers in Physiology, 3*, 339.

Kamath, B. M., Schwarz, K. B., & Hadzic, N. (2010). Alagille syndrome and liver transplantation. *Journal of Pediatric Gastroenterology and Nutrition, 50*, 11–15.

Kamiya, A., Kinoshita, T., Ito, Y., Matsui, T., Morikawa, Y., Senba, E., Nakashima, K., Taga, T., Yoshida, K., Kishimoto, T., et al. (1999). Fetal liver development requires a paracrine action of oncostatin M through the gp130 signal transducer. *EMBO Journal, 18*, 2127–2136.

Kamiya, A., Kinoshita, T., & Miyajima, A. (2001). Oncostatin M and hepatocyte growth factor induce hepatic maturation via distinct signaling pathways. *FEBS Letters, 492*, 90–94.

Kanehisa, M., Furumichi, M., Tanabe, M., Sato, Y., & Morishima, K. (2017). KEGG: New perspectives on genomes, pathways, diseases and drugs. *Nucleic Acids Research, 45*, D353–D361.

Kaplowitz, N. (2005). Idiosyncratic drug hepatotoxicity. *Nature Reviews. Drug Discovery, 4*, 489–499.

Kietzmann, T. (2017). Metabolic zonation of the liver: The oxygen gradient revisited. *Redox Biology, 11*, 622–630.

Kordes, C., Sawitza, I., Gotze, S., Herebian, D., & Haussinger, D. (2014). Hepatic stellate cells contribute to progenitor cells and liver regeneration. *The Journal of Clinical Investigation, 124*, 5503–5515.

Kruepunga, N., Hakvoort, T. B. M., Hikspoors, J., Kohler, S. E., & Lamers, W. H. (2019). Anatomy of rodent and human livers: What are the differences? *Biochemistry Biophysics Acta Molecular Basis of Disease, 1865*(5), 869–878. Amsterdam, Netherlands.

Larrey, D., & Pageaux, G. P. (2005). Drug-induced acute liver failure. *European Journal of Gastroenterology & Hepatology, 17*, 141–143.

Lemaigre, F. P. (2009). Mechanisms of liver development: Concepts for understanding liver disorders and design of novel therapies. *Gastroenterology, 137*, 62–79.

Leung, A., Nah, S. K., Reid, W., Ebata, A., Koch, C. M., Monti, S., Genereux, J. C., Wiseman, R. L., Wolozin, B., Connors, L. H., et al. (2013). Induced pluripotent stem cell modeling of multisystemic, hereditary transthyretin amyloidosis. *Stem Cell Reports, 1*, 451–463.

Lu, J., Einhorn, S., Venkatarangan, L., Miller, M., Mann, D. A., Watkins, P. B., & LeCluyse, E. (2015). Morphological and functional characterization and assessment of iPSC-derived hepatocytes for in vitro toxicity testing. *Toxicological Sciences: An Official Journal of the Society of Toxicology, 147*, 39–54.

Machado, M. V., Michelotti, G. A., Xie, G., de Almeida, T. P., Boursier, J., Bohnic, B., Guy, C. D., & Diehl, A. M. (2015). Mouse models of diet-induced nonalcoholic steatohepatitis reproduce the heterogeneity of the human disease. *PLoS One, 10*, e0127991.

Matz, P., Wruck, W., Fauler, B., Herebian, D., Mielke, T., & Adjaye, J. (2017). Footprint-free human fetal fore-skin derived iPSCs: A tool for modeling hepatogenesis associated gene regulatory networks. *Scientific Reports, 7*, 6294.

McLin, V. A., Rankin, S. A., & Zorn, A. M. (2007). Repression of Wnt/beta-catenin signaling in the anterior endoderm is essential for liver and pancreas development. *Development, 134,* 2207–2217.

Mian, A., & Lee, B. (2002). Urea-cycle disorders as a paradigm for inborn errors of hepatocyte metabolism. *Trends in Molecular Medicine, 8,* 583–589.

Michalopoulos, G. K., Bowen, W. C., Mule, K., & Luo, J. (2003). HGF-, EGF-, and dexamethasone-induced gene expression patterns during formation of tissue in hepatic organoid cultures. *Gene Expression, 11,* 55–75.

Nie, Y. Z., Zheng, Y. W., Miyakawa, K., Murata, S., Zhang, R. R., Sekine, K., Ueno, Y., Takebe, T., Wakita, T., Ryo, A., et al. (2018). Recapitulation of hepatitis B virus-host interactions in liver organoids from human induced pluripotent stem cells. *eBioMedicine, 35,* 114–123.

O'Brien, S., Keogan, M., Casey, M., Duffy, G., McErlean, D., Fitzgerald, M. X., & Hegarty, J. E. (1992). Biliary complications of cystic fibrosis. *Gut, 33,* 387–391.

Ogawa, M., Ogawa, S., Bear, C. E., Ahmadi, S., Chin, S., Li, B., Grompe, M., Keller, G., Kamath, B. M., & Ghanekar, A. (2015). Directed differentiation of cholangiocytes from human pluripotent stem cells. *Nature Biotechnology, 33,* 853–861.

Perry, R. J., Camporez, J. P., Kursawe, R., Titchenell, P. M., Zhang, D., Perry, C. J., Jurczak, M. J., Abudukadier, A., Han, M. S., Zhang, X. M., et al. (2015). Hepatic acetyl CoA links adipose tissue inflammation to hepatic insulin resistance and type 2 diabetes. *Cell, 160,* 745–758.

Ranucci, G., Polishchuck, R., & Iorio, R. (2017). Wilson's disease: Prospective developments towards new therapies. *World journal of gastroenterology: WJG, 23,* 5451–5456.

Rashid, S. T., Corbineau, S., Hannan, N., Marciniak, S. J., Miranda, E., Alexander, G., Huang-Doran, I., Griffin, J., Ahrlund-Richter, L., Skepper, J., et al. (2010). Modeling inherited metabolic disorders of the liver using human induced pluripotent stem cells. *The Journal of Clinical Investigation, 120,* 3127–3136.

Raynaud, P., Carpentier, R., Antoniou, A., & Lemaigre, F. P. (2011). Biliary differentiation and bile duct morphogenesis in development and disease. *The International Journal of Biochemistry & Cell Biology, 43,* 245–256.

Revill, P., Testoni, B., Locarnini, S., & Zoulim, F. (2016). Global strategies are required to cure and eliminate HBV infection. *Nature Reviews. Gastroenterology & Hepatology, 13,* 239–248.

Rui, L. (2014). Energy metabolism in the liver. *Comprehensive Physiology, 4,* 177–197.

Russmann, S., Jetter, A., & Kullak-Ublick, G. A. (2010). Pharmacogenetics of drug-induced liver injury. *Hepatology, 52,* 748–761.

Sampaziotis, F., Cardoso de Brito, M., Madrigal, P., Bertero, A., Saeb-Parsy, K., Soares, F. A., Schrumpf, E., Melum, E., Karlsen, T. H., Bradley, J. A., et al. (2015). Cholangiocytes derived from human induced pluripotent stem cells for disease modeling and drug validation. *Nature Biotechnology, 33*(8), 845–852.

Sampaziotis, F., de Brito, M. C., Geti, I., Bertero, A., Hannan, N. R., & Vallier, L. (2017). Directed differentiation of human induced pluripotent stem cells into functional cholangiocyte-like cells. *Nature Protocols, 12,* 814–827.

Sauer, V., Roy-Chowdhury, N., Guha, C., & Roy-Chowdhury, J. (2014). Induced pluripotent stem cells as a source of hepatocytes. *Current Pathobiology Reports, 2,* 11–20.

Schadt, S., Simon, S., Kustermann, S., Boess, F., McGinnis, C., Brink, A., Lieven, R., Fowler, S., Youdim, K., Ullah, M., et al. (2015). Minimizing DILI risk in drug discovery – A screening tool for drug candidates. *Toxicology In Vitro An International Journal Published in Association with BIBRA, 30,* 429–437.

Schmidt, C., Bladt, F., Goedecke, S., Brinkmann, V., Zschiesche, W., Sharpe, M., Gherardi, E., & Birchmeier, C. (1995). Scatter factor/hepatocyte growth-factor is essential for liver development. *Nature, 373,* 699–702.

Schobel, A., Rosch, K., & Herker, E. (2018). Functional innate immunity restricts hepatitis C virus infection in induced pluripotent stem cell-derived hepatocytes. *Scientific Reports, 8,* 3893.

Scott, J. (1999). Heart disease – Good cholesterol news. *Nature, 400,* 816.

Sgodda, M., Mobus, S., Hoepfner, J., Sharma, A. D., Schambach, A., Greber, B., Ott, M., & Cantz, T. (2013). Improved hepatic differentiation strategies for human induced pluripotent stem cells. *Current Molecular Medicine, 13,* 842–855.

Sgodda, M., Dai, Z., Zweigerdt, R., Sharma, A.D., Ott, M., & Cantz, T. (2017). *A scalable approach for the generation of human pluripotent stem cell-derived hepatic organoids with sensitive hepatotoxicity features. Stem cells and development.* New York, USA.

Shlomai, A., Schwartz, R. E., Ramanan, V., Bhatta, A., de Jong, Y. P., Bhatia, S. N., & Rice, C. M. (2014). Modeling host interactions with hepatitis B virus using primary and induced pluripotent stem cell-derived hepatocellular systems. *Proceedings of the National Academy of Sciences of the United States of America, 111,* 12193–12198.

Siller, R., Greenhough, S., Naumovska, E., & Sullivan, G.J. (2015). *Small-molecule-driven hepatocyte differentiation of human pluripotent stem cells. Stem cell reports.* Bar Harbor, USA.

Si-Tayeb, K., Noto, F. K., Nagaoka, M., Li, J., Battle, M. A., Duris, C., North, P. E., Dalton, S., & Duncan, S. A. (2010). Highly efficient generation of human hepatocyte-like cells from induced pluripotent stem cells. *Hepatology, 51*, 297–305.

Stevens, J. L., & Baker, T. K. (2009). The future of drug safety testing: Expanding the view and narrowing the focus. *Drug Discovery Today, 14*, 162–167.

Szalowska, E., van der Burg, B., Man, H. Y., Hendriksen, P. J., & Peijnenburg, A. A. (2014). Model steatogenic compounds (amiodarone, valproic acid, and tetracycline) alter lipid metabolism by different mechanisms In mouse liver slices. *PLoS One, 9*, e86795.

Tabibian, J. H., Masyuk, A. I., Masyuk, T. V., O'Hara, S. P., & LaRusso, N. F. (2013). Physiology of cholangiocytes. *Comprehensive Physiology, 3*, 541–565.

Takahashi, Y., Soejima, Y., & Fukusato, T. (2012). Animal models of nonalcoholic fatty liver disease/nonalcoholic steatohepatitis. *World journal of gastroenterology: WJG, 18*, 2300–2308.

Takeda, M., Okamoto, I., & Nakagawa, K. (2015). Pooled safety analysis of EGFR-TKI treatment for EGFR mutation-positive non-small cell lung cancer. *Lung Cancer, 88*, 74–79.

Tiso, N., Filippi, A., Pauls, S., Bortolussi, M., & Argenton, F. (2002). BMP signalling regulates anteroposterior endoderm patterning in zebrafish. *Mechanisms of Development, 118*, 29–37.

Tsankov, A. M., Gu, H., Akopian, V., Ziller, M. J., Donaghey, J., Amit, I., Gnirke, A., & Meissner, A. (2015). Transcription factor binding dynamics during human ES cell differentiation. *Nature, 518*, 344–349.

van Mourik, I. D. M. (2017). Liver disease in cystic fibrosis. *Paediatrics and Child Health, 27*, 552–555.

Vega-Badillo, J., Gutierrez-Vidal, R., Hernandez-Perez, H. A., Villamil-Ramirez, H., Leon-Mimila, P., Sanchez-Munoz, F., Moran-Ramos, S., Larrieta-Carrasco, E., Fernandez-Silva, I., Mendez-Sanchez, N., et al. (2016). Hepatic miR-33a/miR-144 and their target gene ABCA1 are associated with steatohepatitis in morbidly obese subjects. *Liver International, 36*, 1383–1391.

Wang, Y., Alhaque, S., Cameron, K., Meseguer-Ripolles, J., Lucendo-Villarin, B., Rashidi, H., and Hay, D.C. (2017). Defined and scalable generation of hepatocyte-like cells from human pluripotent stem cells. *Journal of Visualized Experiments: JoVE.*

Wilson, A.A., Ying, L., Liesa, M., Segeritz, C.P., Mills, J.A., Shen, S.S., Jean, J., Lonza, G.C., Liberti, D.C., Lang, A.H., et al. (2015). *Emergence of a stage-dependent human liver sisease signature with directed differentiation of alpha-1 antitrypsin-deficient iPS cells. Stem cell reports.* Bar Harbor, USA.

Wisse, E., Braet, F., Luo, D., De Zanger, R., Jans, D., Crabbe, E., & Vermoesen, A. (1996). Structure and function of sinusoidal lining cells in the liver. *Toxicologic Pathology, 24*, 100–111.

Wruck, W., Kashofer, K., Rehman, S., Daskalaki, A., Berg, D., Gralka, E., Jozefczuk, J., Drews, K., Pandey, V., Regenbrecht, C., et al. (2015). Multi-omic profiles of human non-alcoholic fatty liver disease tissue highlight heterogenic phenotypes. *Scientific Data, 2*, 150068.

Wruck, W., Graffmann, N., Kawala, M. A., & Adjaye, J. (2017). Concise review: Current status and future directions on research related to nonalcoholic fatty liver disease. *Stem Cells, 35*, 89–96.

Wu, X., Robotham, J. M., Lee, E., Dalton, S., Kneteman, N. M., Gilbert, D. M., & Tang, H. (2012). Productive hepatitis C virus infection of stem cell-derived hepatocytes reveals a critical transition to viral permissiveness during differentiation. *PLoS Pathogens, 8*, e1002617.

Yang, D., Liu, L., Zhu, D., Peng, H., Su, L., Fu, Y. X., & Zhang, L. (2014). A mouse model for HBV immunotolerance and immunotherapy. *Cellular and molecular immunology, 11*, 71–78.

Yi, F., Qu, J., Li, M., Suzuki, K., Kim, N. Y., Liu, G. H., & Belmonte, J. C. (2012). Establishment of hepatic and neural differentiation platforms of Wilson's disease specific induced pluripotent stem cells. *Protein & Cell, 3*, 855–863.

Yuan, L., Liu, X., Zhang, L., Li, X., Zhang, Y., Wu, K., Chen, Y., Cao, J., Hou, W., Zhang, J., et al. (2018). A chimeric humanized mouse model by engrafting the human induced pluripotent stem cell-derived hepatocyte-like cell for the chronic hepatitis B virus infection. *Frontiers in Microbiology, 9*, 908.

Zeng, Z., Guan, L., An, P., Sun, S., O'Brien, S. J., Winkler, C. A., & consortium H.B.V.s. (2008). A population-based study to investigate host genetic factors associated with hepatitis B infection and pathogenesis in the Chinese population. *BMC Infectious Diseases, 8*, 1.

Zhang, S., Chen, S., Li, W., Guo, X., Zhao, P., Xu, J., Chen, Y., Pan, Q., Liu, X., Zychlinski, D., et al. (2011). Rescue of ATP7B function in hepatocyte-like cells from Wilson's disease induced pluripotent stem cells using gene therapy or the chaperone drug curcumin. *Human Molecular Genetics, 20*, 3176–3187.

Zorn, A.M. (2008). *Liver development.* StemBook (Ed) The Stem Cell Research Community, StemBook.

Organoids in Developmental Biology Research and Application

Tobias Cantz

© Springer Nature Switzerland AG 2020
B. Brand-Saberi (ed.), *Essential Current Concepts in Stem Cell Biology*,
Learning Materials in Biosciences, https://doi.org/10.1007/978-3-030-33923-4_12

What You Will Learn in This Chapter

As introduction into the topic of Organoids in Research and Application you will learn about the relevant cornerstones of two-dimensional (2D) and three-dimensional (3D) culture technologies, with a focus on the propagation and differentiation of stem cells in monolayer and aggregate systems. In the subsequent section the most immanent hallmarks of organoid cultivation systems were discussed and a subset of prominent findings is elaborated to gain insights into organoids derived from pluripotent stem cells or their derivatives on the one hand, and from adult tissue-related stem or progenitor cells on the other hand. In the following section, critical developmental cues and mechanisms of liver homeostasis in hepatic tissues were revealed as one example for applied research exploiting the organoid culture technology. Similarly, another section discusses the current state of studying metabolism in hepatic organoids. Finally, you will learn about the use of organoid research in infection medicine, where models were applied to study the pathophysiology of Zika virus, Helicobacter pylori, and Noro Virus infections in suitable organoid culture assays.

12.1 Relevant Cornerstones of 2D and 3D Culture Technologies

Since its infancy, cell culture technology of primary human cells was fostered by innovations that aimed at the expression of a more authentic cellular phenotype in comparison with the respective cell type in the organism. The conventional two-dimensional (2D) cell culture systems offered scalability from microwells to larger tissue culture flasks and provided uniform supply of nutrients and medium components to all cells. 2D culture conditions also allowed more complex conditions, where different cell types were seeded in the same dish, such as embryonic stem cells growing on feeder cells, which was the state-the-art-condition to cultivate pluripotent stem cells before more defined media compositions were established (Silva et al. 2008; Ludwig et al. 2006; Chen et al. 2011). Interestingly, the early protocols for differentiating murine embryonic stem cells took advantage of spontaneous self-organization and specifications events, when a given number of stem cells were aggregated as so-called embryoid bodies, often by forcing their annealing in hanging drops (Rohwedel et al. 1994). Putatively, gradients of nutrients and media factors varying between outer cells to inner cells supported the differentiation into progenitor cells of a given germ-layer. Not only for stem cells, but also for somatic cells, such as primary hepatocytes, three-dimensional cell culture systems were elaborated about 30 years ago aiming at cells with higher molecular or physiological similarity to the liver tissue (Barcellos-Hoff et al. 1989; Petersen et al. 1992). In these days, the terms "organoid" was coined. Literally this term describes multicellular aggregates that should represent a given tissue-specific architecture and functionality. In contrast, the term "spheroids" usually describes (cystic) aggregates that are grown in three-dimensional (3D)-cultivation systems from primary cells such as intestinal stem cells or other organs' progenitor cells. A recent review article by Simian and Bissel (2016) discusses this controversy in terminology and concludes that a variety of cell aggregates are covered by the term organoids and that some authors uses the terms "spheroids" and "organoids" even interchangeably but not distinctly.

12.2 Hallmarks of Organoid Cultivation Systems

In the context of stem cell biology, the term organoid is nowadays defined as three-dimensional aggregate that consists of various tissue-specific cell types in a spatial composition that allows organotypic functional interaction of the cells. Such pluripotent stem cell derived organoids were described by the generation of structures that resemble optical cups (Eiraku et al. 2011), pituitary epithelium (Suga et al. 2011), the intestine (Sato and Clevers 2013) and the cerebrum (Lancaster and Knoblich 2014). These findings clearly demonstrated the potential of the organoid cell culture technology, but the more complex the described structures became, the more variable and less reproducible cellular composition of the organoids was noted. In the field of brain organoids, Quadrato et al. studied cellular diversity in organoids by analyzing single cell gene expression profiles from 80.000 cells obtained from 31 human organoids. In that study, the authors described a considerable organoid-to-organoid variability in the generation of neural cell types and identified cell types that were reproducibly generated and others that appeared more sporadically (Quadrato et al. 2017). Clearly, such findings demonstrate the need for advanced protocols to grow and propagate the organoids in a more defined (bioreactor) setting. Such protocols for the generation of brain organoids were recently elaborated in the work of Velasco et al., where four approaches to generate brain spheroids and organoids, respectively, were refined. In particular, the dorsally patterned forebrain organoid protocol reliably resulted in a rich, but reproducible diversity of cell types appropriate for the human cerebral cortex. Single-cell RNA-sequencing analysis of 21 individual organoids demonstrated that the vast majority of these organoids exhibited an indistinguishable compendium of cell types that follow similar developmental trajectories. More importantly, the cellular composition displayed a very low organoid-to-organoid variability that was in a comparable range as seen for differences amongst normal human brains. Strikingly, when the authors investigated organoids derived from different stem cell lines a consistent reproducibility in the cellular composition and respective gene expression profiles was observed (Velasco et al. 2019). Similar protocols need to be established for other tissue types, in order to pave the way for valuable organoid model to study developmental abnormalities and alterations associated with human diseases.

Complimentary to this approach, i.e. the generation of tissue-resembling organoids from undifferentiated pluripotent stem cells in one step, Takebe demonstrated the feasibility to generate organoids, like mini-livers, by assembling pre-differentiated stem cell derivatives such as pluripotent stem cell-derived hepatic cells, mesenchymal stroma cells and endothelial cells in a self-condensation system (Takebe et al. 2013). In subsequent studies, he was able to adapt this concept for the generation of brain, lung, heart, kidney, and intestinal organ buds, which show remarkable tissue organization features in comparison with their respective organs' counterparts (Takebe et al. 2015). During embryonic development, cross-talk between endoderm and mesoderm is essential for early germ layer patterning, and this self-condensation approach allows for such epithelial-mesenchymal interactions that are considered to be key for cellular differentiation processes and tissue morphogenesis. With respect to mesodermal derivatives, Little and colleagues described the generation of kidney organoids from human pluripotent stem cells that recapitulate nephrogenesis by transferring monolayer cultures of pluripotent stem cell derived ureteric epithelium and metanephric mesenchyme to organoid cultures (Takasato et al. 2015). At this stage one could conclude that several concepts in organoid formation from pluripotent stem cells merge in a way, that controlling physical parameters (flow, contraction, gas) as well as biological parameters (matrices, growth factors, nutrients, supporting cells) is implemented in a narrative engineering approach, where defined combination of un- and pre-differentiated stem cell derivatives were self-assembled (Takebe and Wells 2019).

This process has been investigated since several decades and a couple of principles were elaborated in the past. The basis of this organ self-assembly seems to arise from segregation of cells with similar adhesive properties into domains that achieve the most thermodynamically stable pattern as suggested by Steinberg's differential adhesion hypothesis (Steinberg and Roth 1964; Steinberg 2007). A second mechanism that can influence tissue morphogenesis is related to proper spatially restricted progenitor fate decisions. The combination of both, sorting out and fate specification in governing self-organization, is particularly evident in tumors called teratomas, where the spontaneous development various tissues from the inoculated pluripotent stem cells presumably represents the recapitulation of both cell segregation and fate specification (Lancaster and Knoblich 2014). As discussed above, organoid cultivation of pluripotent stem cell derivatives can be understood as similar technique like the embryoid body culture that provides the respective patterning factors driving particular cell identities and to eventually derive 3D self-organized tissue-like entities.

Providing an alternative to conventional two-dimensional settings, several studies reported the use of aggregates for scalable expansion of human pluripotent stem cells (Zweigerdt et al. 2011; Kropp et al. 2016), lineage-specific differentiation (Kempf et al. 2014, 2015, 2016) and enhanced maturation of pluripotent cells to their specific somatic lineage (Jo et al. 2016; Volkner et al. 2016; Dekkers et al. 2016; Sgodda et al. 2017). Notably, the in vitro propagation of adult tissue-derived stem cells is less established than the growth of pluripotent stem cells, but recent advances in 3D cultivation systems and more detailed insights into the interplay of extracelluar matrices and signaling pathways paved the way for the maintenance and expansion of adult stem and progenitor cells from various tissues. While the formation of three-dimensional structures depends on several factors of the cell itself, like stiffness, adhesion and cohesion molecules, paracrine signals seem to be responsible for the maturation of three-dimensional cell aggregates (Asai et al. 2017). The seminal work of Hans Clevers' lab elucidated the characteristics of Lgr5-positive cells in the intestine and other organs. His lab described such a 3D in vitro culture system that allowed to grow intestinal epithelial organoids not only derived from dissociated intestinal cells as described before (Ootani et al. 2009), but also from a single adult Lgr5+ intestinal cell (Sato et al. 2009). The growth of these organoids was supported by inoculation into Matrigel as extracellular matrix and supplementation of epidermal growth factor (EGF), Noggin, and the Wnt agonist R-spondin. Interestingly, these culture conditions allowed the propagation of the intestinal organoids beyond the Hayflick limit, which predicts that primary non-transformed cells only divide 40–60 times (~2–3 months) before undergoing senescence (Huch and Koo 2015). Meanwhile, similar cultivation conditions have been elucidated for other organoids derived from further endodermal tissues like stomach, colon, pancreas, and liver (Sato and Clevers 2015).

12.3 Revealing Developmental Cues and Liver Homeostasis in Hepatic Organoids

The adult liver is primarily composed of hepatocytes and cholangiocytes that interact with non-parenchymal endothelial and mesenchymal cell types. During fetal development, both hepatocytes and cholangiocytes derive from endodermal fetal liver progenitor cells, the hepatoblasts. However, in normal homeostasis, the adult liver is mostly maintained by the self-replication of existing adult mature hepatocytes and cholangiocytes, while the contribution of progenitor cells to the normal homeostasis is negligible (Huch and Koo 2015). However, the role of putative liver stem cells, so-called oval cells, remained controversial. Very recent analyses of Stuart Forbes lab indeed demonstrate that such oval cells contribute to liver regeneration, when hepatocyte proliferation is blocked (Raven et al. 2017). However, most likely, these progenitor cells do not originate from resident liver stem cells, but rather from (de-differentiated) ductular cells such as cholangiocytes. On the other hand, a seemingly bi-potent ductal population, either from a damaged or an undamaged liver, is considered to be the source for adult liver organoid cultures as described in several other studies (Huch et al. 2013; Huch et al. 2015; Dorrell et al. 2014). In contrast to previous work, which probably reflects the oval cell progenitor cell type in biliary duct-derived organoid cultures, recent work from Hans Clevers lab now describes the long-term expansion of murine and human hepatocytes in organoid culture systems, that recapitulates the hepatocyte proliferative response upon hepatic injury (Hu et al. 2018). Murine and human organoids can be established from single hepatocytes and further propagated for multiple months, while their key morphological, functional and gene expression features were maintained. Importantly, hepatocytes grown in 3D organoids with similar features were described by Roel Nusse's lab, which demonstrated the necessity of tumor necrosis factor alpha (TNFα), an injury-induced inflammatory cytokine, to promote the expansion of hepatocytes in 3D culture, which enabled serial passaging and long-term culture of hepatic cells for more than 6 months (Peng et al. 2018).

12.4 Studying Metabolism in Hepatic Organoids

Pluripotent stem cell derived liver organoids also represent a new type of in vitro liver model for understanding disease mechanism and drug testing. In particular, disease-specific liver organoids can be investigated, if they were differentiated from affected patients-derived induced pluripotent stem cells. Usually, the pluripotent stem cells were initially differentiated into hepatic endoderm or progenitor cells under monolayer culture conditions, before they were further propagated in a 3D culture system for further differentiation (Sgodda et al. 2017; Wu et al. 2019). However, the application of the organoid technologies for the development of high throughput approaches remains a big challenge and has not yet reached its full potential, as it adds another layer of complexity on top of the challenges of conventional 2D-based screening approaches (Horman et al. 2015; Arlotta 2018). But, if organoid culture systems are to be integrated in screening workflows, they need to become robust enough to self-organize into well-defined, homogeneous, reproducible 3D tissues to enable screening in quantity (Friese et al. 2019). Till date, such systems are not yet fully applicable, but first attempts were made for instance to grow hepatic organoids on a perfusable micropillar chip system that exhibited a marked enhancement of liver-specific functions, including Cytochrome P450 enzymes-related metabolic capabilities (Wang et al. 2018). Moreover, the on-a-chip-differentiated liver organoids exhibited a sensitive hepatotoxic response after exposure to acetaminophen (APAP) in a dose- and time-dependent manner. The latter assay was also used by Sgodda and colleagues, when small and large organoids were assembled from pluripotent stem cell-derived hepatic endodermal monolayer cultures. Here, clearly the more homogenous smaller hepatic organoids exhibited the more sensitive toxicity features, when the hepatic cells were exposed to APAP (Sgodda et al. 2017). Propagation of the pre-differentiated monolayer cells as hepatic organoids resulted in an increased level of hepatic gene expression, cytochrome activity, and albumin secretion, which is in line with further reports on the propagation of hepatic cells in 3D aggregates (Gieseck 3rd et al. 2014; Takebe et al. 2013).

12.5 Organoid Research in Infection Medicine

Since organoids are composed of several, if not all tissue-related cell types of a particular organ they represent also an interesting tool for the study of infectious diseases, especially of pathogens that lack a suitable animal model or conventional cell culture system. The 2016 outbreak of Zika virus (ZIKV) infections in Brazil was closely correlated to an increased number of newborns suffering from microencephaly. In fact, ZIKV was detected by electron microscopy and RT-qPCR in brains and amniotic fluid of microcephalic fetuses, strengthening the causal link between ZIKV and increased incidence of microcephaly (Mlakar et al. 2016). In order to investigate how Zika virus infection may lead to microcephaly, the above described pluripotent stem cell-derived cerebral organoids were used to recapitulate early stage, first trimester fetal brain development. Interestingly, the Zika virus strain MR766 efficiently infected organoids and caused a decrease in overall organoid size that correlated with the kinetics of viral copy number (Dang et al. 2016). The authors could further demonstrate that the innate immune receptor Toll-Like-Receptor 3 (TLR3) was upregulated after Zika virus infection of human organoids and mouse neurospheres, while TLR3 inhibition reduced the phenotypic effects of Zika virus infection. Pathway analysis of gene expression changes during TLR3 activation highlighted 41 genes also related to neuronal development, suggesting a mechanistic connection to the disrupted neurogenesis. Based on these findings, the author could demonstrate a strong link between ZIKV-mediated TLR3 activation, perturbed cell fate and a reduction in organoid volume reminiscent of microcephaly, and thus provide convincing evidence that the Zika virus infection of pregnant women would be indeed the most probable reason for the reported microcephaly in their newborn children, which was further supported by an independent study demonstrating that Zika virus abrogates neurogenesis during human brain development in various culture systems, including cerebral organoid cultures (Garcez et al. 2016).

As another example, human gastric organoids represent a new model of self-renewing gastric epithelium grown from stem cells that can be directed into the different lineages of the stomach, which suggest the superiority of this culture model to currently used cell lines. Further differentiation of such gastric organoids also allowed mechanistic studies on Helicobacter pylori infection, that depends on the cell types present in the organoids. The three-dimensional aggregation of the various cell types is of particular importance, because Helicobacter pylori colonizes the lumen of the stomach and has there contact with the apical side of the epithelium. In the organoids, the apical side of the polarized epithelium faces the lumen of the 3D structure, which is accessible by microinjection of the organoids and, thus, infectibility and bacterial growth could be studied inside the organoids (Bartfeld et al. 2015). Furthermore, intestinal organoids have also been applied to study norovirus infection, which has previously been refractory to in vitro culture attempts, despite its capacity to consistently and repetitively cause outbreaks of severe gastroenteritis (Bartfeld and Clevers 2017).

Take Home Message

1. The term *organoid* is not unequivocally defined but might be roughly understood as three-dimensional cellular aggregate that consists of various tissue-related cell types that exhibits a functionally relevant spatial organization.
2. Organoids can be derived from undifferentiated pluripotent stem cells, assembled from (various) pre-differentiated pluripotent stem cell derivatives, or from adult (tissue-related) stem cells.
3. Organoids can be studied as surrogate for in vivo development as well as in applications such as metabolic assays or infection models.

References

Arlotta, P. (2018). Organoids required! A new path to understanding human brain development and disease. *Nature Methods, 15*(1), 27–29. https://doi.org/10.1038/nmeth.4557.

Asai, A., Aihara, E., Watson, C., Mourya, R., Mizuochi, T., Shivakumar, P., Phelan, K., Mayhew, C., Helmrath, M., Takebe, T., Wells, J., & Bezerra, J. A. (2017). Paracrine signals regulate human liver organoid maturation from induced pluripotent stem cells. *Development, 144*(6), 1056–1064. https://doi.org/10.1242/dev.142794.

Barcellos-Hoff, M. H., Aggeler, J., Ram, T. G., & Bissell, M. J. (1989). Functional differentiation and alveolar morphogenesis of primary mammary cultures on reconstituted basement membrane. *Development, 105*(2), 223–235.

Bartfeld, S., & Clevers, H. (2017). Stem cell-derived organoids and their application for medical research and patient treatment. *Journal of Molecular Medicine (Berlin, Germany), 95*(7), 729–738. https://doi.org/10.1007/s00109-017-1531-7.

Bartfeld, S., Bayram, T., van de Wetering, M., Huch, M., Begthel, H., Kujala, P., Vries, R., Peters, P. J., & Clevers, H. (2015). In vitro expansion of human gastric epithelial stem cells and their responses to bacterial infection. *Gastroenterology, 148*(1), 126–136. e126. https://doi.org/10.1053/j.gastro.2014.09.042.

Chen, G., Gulbranson, D. R., Hou, Z., Bolin, J. M., Ruotti, V., Probasco, M. D., Smuga-Otto, K., Howden, S. E., Diol, N. R., Propson, N. E., Wagner, R., Lee, G. O., Antosiewicz-Bourget, J., Teng, J. M., & Thomson, J. A. (2011). Chemically defined conditions for human iPSC derivation and culture. *Nature Methods, 8*(5), 424–429. https://doi.org/10.1038/nmeth.1593.

Dang, J., Tiwari, S. K., Lichinchi, G., Qin, Y., Patil, V. S., Eroshkin, A. M., & Rana, T. M. (2016). Zika virus depletes neural progenitors in human cerebral organoids through activation of the innate immune receptor TLR3. *Cell Stem Cell, 19*(2), 258–265. https://doi.org/10.1016/j.stem.2016.04.014.

Dekkers, J. F., Berkers, G., Kruisselbrink, E., Vonk, A., de Jonge, H. R., Janssens, H. M., Bronsveld, I., van de Graaf, E. A., Nieuwenhuis, E. E., Houwen, R. H., Vleggaar, F. P., Escher, J. C., de Rijke, Y. B., Majoor, C. J., Heijerman, H. G., de Winter-de Groot, K. M., Clevers, H., van der Ent, C. K., & Beekman, J. M. (2016). Characterizing responses to CFTR-modulating drugs using rectal organoids derived from subjects with cystic fibrosis. *Science Translational Medicine, 8*(344), 344ra384. https://doi.org/10.1126/scitranslmed.aad8278.

Dorrell, C., Tarlow, B., Wang, Y., Canaday, P. S., Haft, A., Schug, J., Streeter, P. R., Finegold, M. J., Shenje, L. T., Kaestner, K. H., & Grompe, M. (2014). The organoid-initiating cells in mouse pancreas and liver are phenotypically and functionally similar. *Stem Cell Research, 13*(2), 275–283. https://doi.org/10.1016/j.scr.2014.07.006.

Eiraku, M., Takata, N., Ishibashi, H., Kawada, M., Sakakura, E., Okuda, S., Sekiguchi, K., Adachi, T., & Sasai, Y. (2011). Self-organizing optic-cup morphogenesis in three-dimensional culture. *Nature, 472*(7341), 51–56. https://doi.org/10.1038/nature09941.

Friese, A., Ursu, A., Hochheimer, A., Scholer, H. R., Waldmann, H., & Bruder, J. M. (2019). The convergence of stem cell technologies and phenotypic drug discovery. *Cell Chemical Biology, 26*(8), 1050–1066. https://doi.org/10.1016/j.chembiol.2019.05.007.

Garcez, P. P., Loiola, E. C., Madeiro da Costa, R., Higa, L. M., Trindade, P., Delvecchio, R., Nascimento, J. M., Brindeiro, R., Tanuri, A., & Rehen, S. K. (2016). Zika virus impairs growth in human neurospheres and brain organoids. *Science, 352*(6287), 816–818. https://doi.org/10.1126/science.aaf6116.

Gieseck, R. L., 3rd, Hannan, N. R., Bort, R., Hanley, N. A., Drake, R. A., Cameron, G. W., Wynn, T. A., & Vallier, L. (2014). Maturation of induced pluripotent stem cell derived hepatocytes by 3D-culture. *PLoS One, 9*(1), e86372. https://doi.org/10.1371/journal.pone.0086372.

Horman, S. R., Hogan, C., Delos Reyes, K., Lo, F., & Antczak, C. (2015). Challenges and opportunities toward enabling phenotypic screening of complex and 3D cell models. *Future Medicinal Chemistry, 7*(4), 513–525. https://doi.org/10.4155/fmc.14.163.

Hu, H., Gehart, H., Artegiani, B., LO-I, C., Dekkers, F., Basak, O., van Es, J., Chuva de Sousa Lopes, S. M., Begthel, H., Korving, J., van den Born, M., Zou, C., Quirk, C., Chiriboga, L., Rice, C. M., Ma, S., Rios, A., Peters, P. J., de Jong, Y. P., & Clevers, H. (2018). Long-term expansion of functional mouse and human hepatocytes as 3D organoids. *Cell, 175*(6), 1591–1606. e1519. https://doi.org/10.1016/j.cell.2018.11.013.

Huch, M., & Koo, B. K. (2015). Modeling mouse and human development using organoid cultures. *Development, 142*(18), 3113–3125. https://doi.org/10.1242/dev.118570.

Huch, M., Dorrell, C., Boj, S. F., van Es, J. H., Li, V. S., van de Wetering, M., Sato, T., Hamer, K., Sasaki, N., Finegold, M. J., Haft, A., Vries, R. G., Grompe, M., & Clevers, H. (2013). In vitro expansion of single Lgr5+ liver stem cells induced by Wnt-driven regeneration. *Nature, 494*(7436), 247–250. https://doi.org/10.1038/nature11826.

Huch, M., Gehart, H., van Boxtel, R., Hamer, K., Blokzijl, F., Verstegen, M. M., Ellis, E., van Wenum, M., Fuchs, S. A., de Ligt, J., van de Wetering, M., Sasaki, N., Boers, S. J., Kemperman, H., de Jonge, J., Ijzermans, J. N., Nieuwenhuis, E. E., Hoekstra, R., Strom, S., Vries, R. R., van der Laan, L. J., Cuppen, E., & Clevers, H. (2015). Long-term culture of genome-stable bipotent stem cells from adult human liver. *Cell, 160*(1–2), 299–312. https://doi.org/10.1016/j.cell.2014.11.050.

Jo, J., Xiao, Y., Sun, A. X., Cukuroglu, E., Tran, H. D., Goke, J., Tan, Z. Y., Saw, T. Y., Tan, C. P., Lokman, H., Lee, Y., Kim, D., Ko, H. S., Kim, S. O., Park, J. H., Cho, N. J., Hyde, T. M., Kleinman, J. E., Shin, J. H., Weinberger, D. R., Tan, E. K., Je, H. S., & Ng, H. H. (2016). Midbrain-like organoids from human pluripotent stem cells contain functional dopaminergic and neuromelanin-producing neurons. *Cell Stem Cell, 19*(2), 248–257. https://doi.org/10.1016/j.stem.2016.07.005.

Kempf, H., Olmer, R., Kropp, C., Ruckert, M., Jara-Avaca, M., Robles-Diaz, D., Franke, A., Elliott, D. A., Wojciechowski, D., Fischer, M., Roa Lara, A., Kensah, G., Gruh, I., Haverich, A., Martin, U., & Zweigerdt, R. (2014). Controlling expansion and cardiomyogenic differentiation of human pluripotent stem cells in scalable suspension culture. *Stem Cell Reports, 3*(6), 1132–1146. https://doi.org/10.1016/j.stemcr.2014.09.017.

Kempf, H., Kropp, C., Olmer, R., Martin, U., & Zweigerdt, R. (2015). Cardiac differentiation of human pluripotent stem cells in scalable suspension culture. *Nature Protocols, 10*(9), 1345–1361. https://doi.org/10.1038/nprot.2015.089.

Kempf, H., Olmer, R., Haase, A., Franke, A., Bolesani, E., Schwanke, K., Robles-Diaz, D., Coffee, M., Gohring, G., Drager, G., Potz, O., Joos, T., Martinez-Hackert, E., Haverich, A., Buettner, F. F., Martin, U., & Zweigerdt, R. (2016). Bulk cell density and Wnt/TGFbeta signalling regulate mesendodermal patterning of human pluripotent stem cells. *Nature Communications, 7*, 13602. https://doi.org/10.1038/ncomms13602.

Kropp, C., Kempf, H., Halloin, C., Robles-Diaz, D., Franke, A., Scheper, T., Kinast, K., Knorpp, T., Joos, T. O., Haverich, A., Martin, U., Zweigerdt, R., & Olmer, R. (2016). Impact of feeding strategies on the scalable expansion of human pluripotent stem cells in single-use stirred tank bioreactors. *Stem Cells Translational Medicine, 5*(10), 1289–1301. https://doi.org/10.5966/sctm.2015-0253.

Lancaster, M. A., & Knoblich, J. A. (2014). Organogenesis in a dish: Modeling development and disease using organoid technologies. *Science, 345*(6194), 1247125. https://doi.org/10.1126/science.1247125.

Ludwig, T. E., Bergendahl, V., Levenstein, M. E., Yu, J., Probasco, M. D., & Thomson, J. A. (2006). Feeder-independent culture of human embryonic stem cells. *Nature Methods, 3*(8), 637–646. https://doi.org/10.1038/nmeth902.

Mlakar, J., Korva, M., Tul, N., Popovic, M., Poljsak-Prijatelj, M., Mraz, J., Kolenc, M., Resman Rus, K., Vesnaver Vipotnik, T., Fabjan Vodusek, V., Vizjak, A., Pizem, J., Petrovec, M., & Avsic Zupanc, T. (2016). Zika virus associated with microcephaly. *The New England Journal of Medicine, 374*(10), 951–958. https://doi.org/10.1056/NEJMoa1600651.

Ootani, A., Li, X., Sangiorgi, E., Ho, Q. T., Ueno, H., Toda, S., Sugihara, H., Fujimoto, K., Weissman, I. L., Capecchi, M. R., & Kuo, C. J. (2009). Sustained in vitro intestinal epithelial culture within a Wnt-dependent stem cell niche. *Nature Medicine, 15*(6), 701–706. https://doi.org/10.1038/nm.1951.

Peng, W. C., Logan, C. Y., Fish, M., Anbarchian, T., Aguisanda, F., Alvarez-Varela, A., Wu, P., Jin, Y., Zhu, J., Li, B., Grompe, M., Wang, B., & Nusse, R. (2018). Inflammatory cytokine TNFalpha promotes the long-term expansion of primary hepatocytes in 3D culture. *Cell, 175*(6), 1607–1619. e1615. https://doi.org/10.1016/j.cell.2018.11.012.

Petersen, O. W., Ronnov-Jessen, L., Howlett, A. R., & Bissell, M. J. (1992). Interaction with basement membrane serves to rapidly distinguish growth and differentiation pattern of normal and malignant human breast epithelial cells. *Proceedings of the National Academy of Sciences of the United States of America, 89*(19), 9064–9068.

Quadrato, G., Nguyen, T., Macosko, E. Z., Sherwood, J. L., Min Yang, S., Berger, D. R., Maria, N., Scholvin, J., Goldman, M., Kinney, J. P., Boyden, E. S., Lichtman, J. W., Williams, Z. M., McCarroll, S. A., & Arlotta, P. (2017). Cell diversity and network dynamics in photosensitive human brain organoids. *Nature, 545*(7652), 48–53. https://doi.org/10.1038/nature22047.

Raven, A., Lu, W. Y., Man, T. Y., Ferreira-Gonzalez, S., O'Duibhir, E., Dwyer, B. J., Thomson, J. P., Meehan, R. R., Bogorad, R., Koteliansky, V., Kotelevtsev, Y., Ffrench-Constant, C., Boulter, L., & Forbes, S. J. (2017). Cholangiocytes act as facultative liver stem cells during impaired hepatocyte regeneration. *Nature, 547*(7663), 350–354. https://doi.org/10.1038/nature23015.

Rohwedel, J., Maltsev, V., Bober, E., Arnold, H. H., Hescheler, J., & Wobus, A. M. (1994). Muscle cell differentiation of embryonic stem cells reflects myogenesis in vivo: Developmentally regulated expression of myogenic determination genes and functional expression of ionic currents. *Developmental Biology, 164*(1), 87–101. https://doi.org/10.1006/dbio.1994.1182.

Sato, T., & Clevers, H. (2013). Growing self-organizing mini-guts from a single intestinal stem cell: Mechanism and applications. *Science, 340*(6137), 1190–1194. https://doi.org/10.1126/science.1234852.

Sato, T., & Clevers, H. (2015). SnapShot: Growing organoids from stem cells. *Cell, 161*(7), 1700–1700. e1701. https://doi.org/10.1016/j.cell.2015.06.028.

Sato, T., Vries, R. G., Snippert, H. J., van de Wetering, M., Barker, N., Stange, D. E., van Es, J. H., Abo, A., Kujala, P., Peters, P. J., & Clevers, H. (2009). Single Lgr5 stem cells build crypt-villus structures in vitro without a mesenchymal niche. *Nature, 459*(7244), 262–265. https://doi.org/10.1038/nature07935.

Sgodda, M., Dai, Z., Zweigerdt, R., Sharma, A. D., Ott, M., & Cantz, T. (2017). A scalable approach for the generation of human pluripotent stem cell-derived hepatic organoids with sensitive hepatotoxicity features. *Stem Cells and Development, 26*(20), 1490–1504. https://doi.org/10.1089/scd.2017.0023.

Silva, J., Barrandon, O., Nichols, J., Kawaguchi, J., Theunissen, T. W., & Smith, A. (2008). Promotion of reprogramming to ground state pluripotency by signal inhibition. *PLoS Biology, 6*(10), e253. https://doi.org/10.1371/journal.pbio.0060253.

Simian, M., & Bissell, M. J. (2016). Organoids: A historical perspective of thinking in three dimensions. *The Journal of Cell Biology, 216*(1), 31–40. https://doi.org/10.1083/jcb.201610056.

Steinberg, M. S. (2007). Differential adhesion in morphogenesis: A modern view. *Current Opinion in Genetics & Development, 17*(4), 281–286. https://doi.org/10.1016/j.gde.2007.05.002.

Steinberg, M. S., & Roth, S. A. (1964). Phases in cell aggregation and tissue reconstruction an approach to the kinetics of cell aggregation. *The Journal of Experimental Zoology, 157*, 327–338.

Suga, H., Kadoshima, T., Minaguchi, M., Ohgushi, M., Soen, M., Nakano, T., Takata, N., Wataya, T., Muguruma, K., Miyoshi, H., Yonemura, S., Oiso, Y., & Sasai, Y. (2011). Self-formation of functional adenohypophysis in three-dimensional culture. *Nature, 480*(7375), 57–62. https://doi.org/10.1038/nature10637.

Takasato, M., Er, P. X., Chiu, H. S., Maier, B., Baillie, G. J., Ferguson, C., Parton, R. G., Wolvetang, E. J., Roost, M. S., Chuva de Sousa Lopes, S. M., & Little, M. H. (2015). Kidney organoids from human iPS cells contain multiple lineages and model human nephrogenesis. *Nature, 526*(7574), 564–568. https://doi.org/10.1038/nature15695.

Takebe, T., & Wells, J. M. (2019). Organoids by design. *Science, 364*(6444), 956–959. https://doi.org/10.1126/science.aaw7567.

Takebe, T., Sekine, K., Enomura, M., Koike, H., Kimura, M., Ogaeri, T., Zhang, R. R., Ueno, Y., Zheng, Y. W., Koike, N., Aoyama, S., Adachi, Y., & Taniguchi, H. (2013). Vascularized and functional human liver from an iPSC-derived organ bud transplant. *Nature, 499*(7459), 481–484. https://doi.org/10.1038/nature12271.

Takebe, T., Enomura, M., Yoshizawa, E., Kimura, M., Koike, H., Ueno, Y., Matsuzaki, T., Yamazaki, T., Toyohara, T., Osafune, K., Nakauchi, H., Yoshikawa, H. Y., & Taniguchi, H. (2015). Vascularized and complex organ buds from diverse tissues via mesenchymal cell-driven condensation. *Cell Stem Cell, 16*(5), 556–565. https://doi.org/10.1016/j.stem.2015.03.004.

Velasco, S., Kedaigle, A. J., Simmons, S. K., Nash, A., Rocha, M., Quadrato, G., Paulsen, B., Nguyen, L., Adiconis, X., Regev, A., Levin, J. Z., & Arlotta, P. (2019). Individual brain organoids reproducibly form

cell diversity of the human cerebral cortex. *Nature, 570*(7762), 523–527. https://doi.org/10.1038/
s41586-019-1289-x.

Volkner, M., Zschatzsch, M., Rostovskaya, M., Overall, R. W., Busskamp, V., Anastassiadis, K., & Karl, M. O.
(2016). Retinal organoids from pluripotent stem cells efficiently recapitulate Retinogenesis. *Stem Cell
Reports, 6*(4), 525–538. https://doi.org/10.1016/j.stemcr.2016.03.001.

Wang, Y., Wang, H., Deng, P., Chen, W., Guo, Y., Tao, T., & Qin, J. (2018). In situ differentiation and generation
of functional liver organoids from human iPSCs in a 3D perfusable chip system. *Lab on a Chip, 18*(23),
3606–3616. https://doi.org/10.1039/c8lc00869h.

Wu, F., Wu, D., Ren, Y., Huang, Y., Feng, B., Zhao, N., Zhang, T., Chen, X., Chen, S., & Xu, A. (2019). Generation
of hepatobiliary organoids from human induced pluripotent stem cells. *Journal of Hepatology, 70*(6),
1145–1158. https://doi.org/10.1016/j.jhep.2018.12.028.

Zweigerdt, R., Olmer, R., Singh, H., Haverich, A., & Martin, U. (2011). Scalable expansion of human pluripotent stem cells in suspension culture. *Nature Protocols, 6*(5), 689–700. https://doi.org/10.1038/
nprot.2011.318.

2

Extracellular Vesicles

André Görgens and Bernd Giebel

© Springer Nature Switzerland AG 2020
B. Brand-Saberi (ed.), *Essential Current Concepts in Stem Cell Biology*,
Learning Materials in Biosciences, https://doi.org/10.1007/978-3-030-33923-4_13

What You Will Learn in This Chapter

In this chapter, extracellular vesicles (EVs) will be introduced conceptually as still rather novel mediators of intercellular communication, and in particular in the context of regenerative medicine. You will learn, of how EVs were discovered historically and what we meanwhile have learned about different EV subtypes and their related functions. In the context of regenerative medicine, novel EV-based therapeutic approaches will be explained and the potential underlying mode of action will be discussed. Since the EV research field is rather young, and since their analysis is challenging due to their small size, we will explicitly mention inherent limitations and challenges throughout this chapter.

13.1   Concepts in Regenerative Medicine

Degenerative diseases are classically associated with the irreversible loss of tissue. In this context, it is broadly assumed that the potential of endogenous stem and progenitor cells, which normally control tissue homeostasis, is insufficient to promote tissue regeneration. Consequently, there have been many attempts to treat degenerative diseases with stem or progenitor cells with proposed developmental potentials comparable to endogenous stem cells.

In the early 2000's, with the interest in stem cell biology increasing exponentially, a number of observations implied plasticity in somatic stem cell compartments. Different studies suggested that immature brain cells, under appropriate environmental conditions, are able to create blood cells and *vice versa* (Bjornson et al. 1999; Mezey et al. 2000). Especially the interest in fibroblastoid cells, so-called *mesenchymal stem/stromal cells* (MSCs), increased. These cells were initially raised from adult bone marrow and showed multi-lineage differentiation capabilities (including bone, cartilage, fat, tendon, muscle, and bone marrow stroma) (Pittenger et al. 1999). In the following years MSCs were raised from different tissues, and their developmental potential was tested in various *in vitro* and *in vivo* assays. Several manuscripts reported developmental potentials far beyond those initially described by Pittenger and colleagues (1999), e.g. that MSCs may directly differentiate into neurons (Munoz-Elias et al. 2003).

MSCs quickly emerged as a potentially promising cell source in regenerative medicine and after hematopoietic stem cells are now the second most transplanted *stem cell-like* entity in NIH registered clinical trials. Although MSC-based therapeutic approaches and subsequent results are discussed controversially in the field, MSCs have been reported to improve the symptoms of several diseases, qualifying them as an important tool in regenerative medicine. Over the years, however, it became evident that MSCs – in contradiction to initial assumptions – are actually hardly integrated into disease affected tissues. Instead, resulting therapeutic effects have been associated in several disease models with their immunomodulatory properties that were first reported in 2002 (Di Nicola et al. 2002). Finally it turned out that MSCs act in a paracrine rather than in a cell-cell contact dependent manner (Caplan and Dennis 2006). Furthermore, the proposed stem cell-related features of MSCs have meanwhile been challenged. Accordingly, most scientists nowadays prefer to call them *mesenchymal stromal cells* rather than *mesenchymal stem cells*. Recently, Arnold Caplan, a pioneer in the MSC field, recommended to name them even more precisely *medicinal signaling cells* (Caplan 2017).

Regardless what the preferred terminus finally will be, it has become a major objective to identify the active components exerting the MSCs' pro-regenerative/immunomodulatory activities. Interestingly, so-called *extracellular vesicles* (EVs) have been identified to mediate related MSC functions in a variety of different disease models, including acute kidney injury and stroke (Börger et al. 2017; Bruno et al. 2009; Doeppner et al. 2015; Lai et al. 2010; Lener et al. 2015). In this chapter, we will subsequently summarize current concepts and elaborate the role of EVs in the field of regenerative medicine and highlight potential biomedical applications with a focus on MSC derived EVs.

13.2 Extracellular Vesicles

Extracellular vesicles (EVs) are submicron-sized biological vesicles which can be released by basically all cell types. Especially during the last decade, EVs have attracted lots of attention through the demonstration that they can transfer complex information or signals from releasing cells to other cells or tissues in a targeted manner, thereby influencing the biology and function of the recipient cell. Nowadays, it is well known that EVs are involved in a plethora of physiological and pathological processes (Yanez-Mo et al. 2015). Within recent years EVs have been connected to various therapeutic approaches including anti-tumor therapy, vaccination, modulation of the immune system and drug-delivery (Fais et al. 2016; Lener et al. 2015; Wiklander et al. 2019).

13.2.1 Pioneers of EV Research

EVs were first observed in different contexts without realizing that they actually represent a universal form of intercellular communication. Instead, they were initially described as platelet-derived particles in normal plasma and later referred to as "platelet dust" (Chargaff and West 1946; Wolf 1967). Subsequent reports in the 1970s/1980s describe the release of plasma membrane vesicles, virus-like particles in mammalian cell cultures, and the detection of biological vesicles in seminal plasma (Benz and Moses 1974; Dalton 1975; De Broe et al. 1975; Stegmayr and Ronquist 1982). In 1983, two groups performed ultrastructural studies of transferrin trafficking in reticulocytes. They observed that labelled transferrin is taken up by recipient cells through receptor-mediated endocytosis, and upon formation of late endosomes found on the membrane of intraluminal vesicles (ILVs) that are formed by the inward budding of the endosomal membrane, the so called limiting membrane. – Accordingly, these late endosomes are either named multivesicular bodies (MVBs) or multivesicular endosomes. Against the earlier opinion that the MVBs fuse with lysosomes to degrade the ILVs including their cargo, these groups demonstrated that the MVBs can fuse with the plasma membrane and release their ILVs into the extracellular environment (Harding et al. 1983; Pan and Johnstone 1983). Later they deciphered these MVB-derived vesicles as *exosomes* (Johnstone et al. 1987). The first report describing functional properties of exosomes in intercellular communication was finally published in 1996. Here, Graca Raposo and colleagues observed that B lymphoblastic cells release MHC class II carrying vesicles which are able to induce antigen-specific MHC class II-restricted T cell responses (Raposo et al. 1996). Subsequently, a number of studies reported functional relevance in tumor and immune biology. However, it took until 2006/2007 before finally "*exosomes*" were reported to contain different RNA species which can be transferred between cells and can modulate the gene expression in the target cell (Ratajczak et al. 2006; Valadi et al. 2007). Following those reports, exosome research started to become popular, and the field has been growing almost exponentially since then. In the following years the origin of different vesicle types was discussed since it turned out that aside of releasing vesicles (exosomes) via MVBs, cells can also shed vesicles from their plasma membrane, and again other vesicles can derive from apoptotic cells (so-called apoptotic bodies) or are formed via a multitude of other processes (Kim et al. 2015). Until today, there are no methods available which can separate the different vesicle types according to their origin. Traditionally, vesicles are processed by physical methods, commonly allowing enrichment of vesicles of comparable sizes and/or densities, but not regarding their origin. Aiming to emphasize this limitation, the *International Society of extracellular vesicles (ISEV)* which was founded in 2012 discussed the nomenclature intensively and agreed to define vesicles derived from MVBs as *exosomes* and vesicles deriving from the plasma membrane as *microvesicles*. According to the literature, exosomes have sizes of 50–150 nm and microvesicles from 100 to 1000 nm. Due to those overlapping size ranges and comparable densities, exosomes and smaller microvesicles as well as other smaller vesicle types cannot be separated from each other experimentally. Subsequently, most ISEV members agreed to name vesicles in an experimental context rather extracellular vesicles (EVs) than exosomes or microvesicles (Gould and Raposo 2013; Raposo and Stoorvogel 2013; Thery et al. 2018) (◻ Fig. 13.1). Unfortunately, not all scientists follow these recommendations and some EV researcher still decipher small EVs independent of their origin as *exosomes* and larger EVs as *microvesicles*. While nomenclature does not affect the biology itself, it is beneficial for the communication among scientists as well as for the reproducibility and comparability of experimental results if precise termini are used. This is especially of high

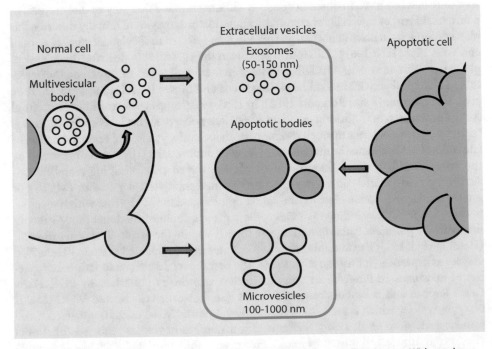

◻ **Fig. 13.1** Overview about the current nomenclature of the three main EV subtypes. With overlapping size ranges vesicles are defined according to their origin. Exosomes originate from the endosomal system and are released upon fusion of multivesicular bodies with the plasma membrane. Microvesicles are budded from the plasma membrane, and apoptotic bodies are larger vesicles formed by apoptotic cells

importance because most laboratories are still using a plethora of often very different procedures to enrich for EVs, resulting in experimental data often being hard to compare between different studies which in turn hampers their appropriate interpretation. In addition, due to their small size it is challenging to characterize and analyze EVs especially in the size-range of exosomes. While several rather established methods including nanoparticle tracking analysis (Dragovic et al. 2011; Sokolova et al. 2011), electron microscopy (Zabeo et al. 2017) and flow cytometry (Gorgens et al. 2019; Wiklander et al. 2018) meanwhile are used quite frequently in the field to quantify and analyze different properties of EVs, the field has just begun with methodical standardization and optimization, and also started to develop and explore novel methods to prepare and analyze EVs (Coumans et al. 2017; Giebel and Helmbrecht 2017; Shao et al. 2018; Welsh et al. 2017).

Upon discussing about experimental procedures, there is consensus in the field that certain details and information about the preparation and characterization should be reported in publications. Accordingly, guidelines have been formulated, due to the quick development of the EV field these guidelines have been just updated quite recently. We like to recommend that anybody intending to approach the EV field experimentally should study and consider the recommendations provided in these manuscripts (Consortium et al. 2017; Lotvall et al. 2014; Thery et al. 2018). Although, the guidelines are very well written, they are very condensed and the EV field still provides certain challenges. Thus, we highly recommend that anybody intending to approach the field experimentally, gets into contact with experts in the field. To promote the EV research, many national EV societies have been formed meanwhile in addition to ISEV. For example, the *German Society of Extracellular Vesicles* (GSEV) was founded in 2017. As one of their missions, it is a goal of the national societies to provide intellectual support to scientist stepping into the field.

At the functional level, EVs have been identified as particles being part of a newly discovered intercellular communication system (Ludwig and Giebel 2012; Yanez-Mo et al. 2015). Accordingly, they exert multiple different functions depending on the cell type of origin, both under physiological and pathological conditions (Yanez-Mo et al. 2015). EVs have been studied as markers of organ dysfunction and are under evaluation for diagnostic purposes (Fais et al. 2016; Gilani et al. 2016). Furthermore, EVs released by mesenchymal stem/stromal cells (MSCs) are key factors promoting tissue regeneration, counteracting apoptosis and promoting anti-inflammatory immune responses (Börger et al. 2017; Lener et al. 2015).

13.3 MSC-EVs in Regenerative Medicine

After realizing that MSCs in therapeutic contexts act rather in a paracrine than a cellular manner, several groups started to search for the underlying, therapeutically active components. Within the two landmark studies in the field, MSC conditioned cell culture media (CM) were fractioned by different protocols. Bruno and colleagues fractioned MSC-CM samples by differential ultracentrifugation and recovered the activity resembling the therapeutic effect of MSCs in an acute kidney tubular injury mouse model within the pellet resulting from a 100,000 × g ultracentrifugation step. Upon characterizing the pellet in more detail, vesicular structures with sizes between 80 nm and 1 μm (mean value of 135 nm) were discovered, which the authors deciphered as *microvesicles* (Bruno et al. 2009). Comparable to the MSCs themselves, the *microvesicle* fraction suppressed apoptosis rates and increased the proliferation of tubular epithelial cells *in vitro*. Lai and colleagues used a HPLC driven size-exclusion method and enriched a fraction containing particles with a hydrodynamic radius of 55–65 nm (Lai et al. 2010). Due to the presence of marker proteins being described as exosome-related, i.e. CD9, CD81 and Alix, the authors assigned the term *exosomes* to the recovered particles. Upon testing the obtained *exosome* fraction in a murine model for myocardial ischemia/reperfusion injury, a therapeutic effect, especially the reduction of the infarction size, was observed which resembled the effects the group had previously observed after treatment with MSCs and MSC-CM (Timmers et al. 2007).

In the following years, those findings were confirmed and further investigated by several other groups. At the example of an ischemic stroke mouse model, we demonstrated that systemically administered EVs derived from MSCs, similar to cellular treatments with the MSCs themselves, induced neurological recovery via mechanisms that involved long-term neuroprotection, promotion of neurogenesis and angiogenesis, as well as reversal of post-ischemic immune depression which is known to confer susceptibility to infection in the stroke recovery phase (Doeppner et al. 2015). Following application of EVs harvested from supernatants of human MSCs to a human GvHD patient, we showed that – similar as in the ischemic stroke model – the treatment with MSC EVs improved the GvHD symptoms and resulted in modulation of the immune responses (Kordelas et al. 2014). Indeed, immunomodulatory features of MSC-EVs have meanwhile been described in several disease models, implying that immunomodulation is an important part of the proposed EV-mediated mode of action that contributes to the pro-regenerative effects of MSC-EVs (Börger et al. 2017). Apparently, this mechanism is highly conserved during evolution, as we confirmed that human MSC-EVs produced with the same protocol (Ludwig et al. 2018) also exert therapeutic functions in different animal models including mouse, rat and sheep (Doeppner et al. 2015; Drommelschmidt et al. 2017; Ophelders et al. 2016).

As mentioned before at the example of ischemic stroke, in addition to their immunomodulatory activities, systemically administered EVs from MSCs induce neurological recovery by a combination of different mechanisms involving long-term neuroprotection, promotion of neurogenesis and angiogenesis. Accordingly, therapeutically active EVs also induce other pro-regenerative processes being required to promote successful tissue regeneration in addition to their immunomodulatory properties. For example, it has been demonstrated mechanistically that MSC-EVs can increase ATP levels in damaged cells, reduce oxidative stress and the severity of cell injury, and restore cellular metabolic activities (Arslan et al. 2013). For now, we do not know whether EVs from a certain stem cell type are better or more suitable for certain applications than others, and whether individual tissues or different cell types can provide EVs more suitable or potent to stimulate regeneration. However, due to the extensive use of MSCs in regenerative medicine in various animal models and due to the fact that MSC administration appears safe in a plethora of clinical studies in human recipients, we propose that focusing on the inherent function of MSCs-EVs for therapeutic purposes will be most likely both, safe and therapeutically effective.

Compared to cells, EVs have the huge advantage that they are not self-replicating and can be handled much easier than cells. Furthermore, they can be sterilized by filtration processes. Coupled with the experimental observation that MSC-EVs mediate comparable therapeutic effects than their parental cells, those first pre-clinical and clinical findings raised the overall interest in MSC-EVs intensively (Lener et al. 2015). This increase in attention and activity in the field is also reflected by the exponential increase of MSC-EV publications within the last decade (◻ Fig. 13.2).

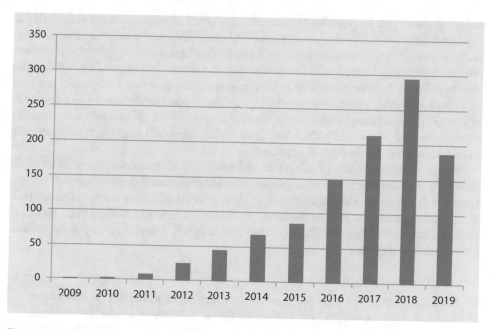

◘ Fig. 13.2 Number of publications per year found in PubMed (NIH) using the search string ("mesen-chymal stem cells" and "exosomes") or ("mesenchymal stem cells" and "microvesicles") or ("mesenchymal stem cells" and "extracellular vesicles") Search date: June 8th, 2019

13.4 Immunomodulatory Properties of (MSC-)EVs

Aside of EVs from MSCs, EVs from several other sources have been reported to contain immunomodulatory capacities: Starting with the fertilization process, immunomodulatory EVs, named prostasomes, have been identified in seminal plasma (Aalberts et al. 2014). Immunomodulatory EVs play essential roles during pregnancy (Nair and Salomon 2018) as well as in many developmental and regenerative processes and during tumor formation and expansion (Yanez-Mo et al. 2015). Upon comparing the different systems it becomes apparent that most related processes are connected to developmental and regen-erative processes and involve cell division. Suggestively, immunomodulatory properties mediated by EVs are part of somatic stemness programs. This implies tolerance and in turn regulatory immune responses are required to allow cell proliferation and successful tissue development or regeneration, respectively. In this context, it is worth mentioning that to our best knowledge all degenerative diseases including ischemia and allogeneic reactions are associated with prolonged acute inflammatory responses. Against current dogmas, we wonder whether many dividing cells and developing tissues are in principal immunogenic and may get attacked by the immune system in its acute inflammatory state. To allow development/regeneration the immune system might need to be switched from the acute inflammatory into the tolerance state. In such a scenario, endogenous somatic stem and progenitor cells may contribute to the immune modulation by releasing tolerance-inducing EVs which in turn induce an environment that is permissive for tissue development and regeneration. If biased by pathogenic mechanisms towards its acute inflammatory state, however, the EVs released by endogenous cells may not be sufficient to switch the immune response towards tolerance, resulting in a condition in which the inflamed tissue gets targeted by the immune system. Accordingly, tissue remodeling is impaired. Administration of somatic stem cells or their EVs may result in immunomodulation and create permissive environments for developmental and regenerative processes, effectively resulting in successful tissue remodeling. Of note and in good agreement with this hypothesis, tumors effectively induce tolerance and suppress anti-tumor immune responses. Currently, anti-tumor treatment with check-point inhibitors is a popular strategy to switch the immune system from the tolerance state back to the acute inflammatory state (Galon and Bruni 2019). Thus, EVs from proliferating cells might act more generally as tolerance inducing checkpoint activators.

Treatment of recipients of allogeneic tissues with immunomodulatory EVs should thus suppress allogenic immune responses against transplanted donor organs or tissues, respectively, similarly as we have previously observed in the GvHD patient (Kordelas et al. 2014). Indeed, as it was already discussed in 2014, EVs have successfully been shown to promote the alloantigen-specific tolerance and allograft acceptance in rodent models (Monguio-Tortajada et al. 2014).

Take Home Message

EVs are currently a hot topic and provide many promising aspects for the field of life sciences. As explained EVs from various cell types, especially stem and tumor cells, seem to mediate pro-regenerative effects amongst others by modulating the immune system in several diseases. Furthermore, as EVs seem to act in cell specific manners they are considered as novel drug delivery systems and provide promising biomarkers for many different diseases. Also in basic research we expect that they will largely affect our overall understanding of intercellular signaling. Despite these very promising aspects, however, the field is very young, and Biotech companies have just started to develop devices for appropriate EV analyses. We consider EVs as very heterogeneous entities and we have to learn how we can investigate and finally unravel this heterogeneity.

In analogy to the peripheral blood system one could say that we are at a stage in which we just would have discovered circulating blood cells without being able to discriminate different leukocyte subsets. Most of the currently available techniques enrich EVs, but depending on the methods, also a lot of byproducts, which for sure can interfere with the experimental results. Many findings in the field might be attributed to non-EV associated components co-purified with the EVs. Also, techniques being well accepted in the field might finally turn out to be not appropriate. Consequently, even though we are convinced of the huge impact EVs will have, several current findings may be challenged in the future. Thus, we like to end this chapter reminding readers to remain critically. Models including dogmas which today seem to be common knowledge remain challengeable, and upon scientific progress can turn out to be wrong after all. Thus, if your experiments reproducibly do not fit to the model you are working on, also start considering the correctness of the model. While we are aware that many ideas we have summarized and hypothesized in this chapter might turn out to be incorrect once improved techniques will be available for the EV research in the future, we still find this research field fascinating and exciting.

References

Aalberts, M., Stout, T. A., & Stoorvogel, W. (2014). Prostasomes: Extracellular vesicles from the prostate. *Reproduction, 147*, R1–R14.

Arslan, F., Lai, R. C., Smeets, M. B., Akeroyd, L., Choo, A., Aguor, E. N., Timmers, L., van Rijen, H. V., Doevendans, P. A., Pasterkamp, G., et al. (2013). Mesenchymal stem cell-derived exosomes increase ATP levels, decrease oxidative stress and activate PI3K/Akt pathway to enhance myocardial viability and prevent adverse remodeling after myocardial ischemia/reperfusion injury. *Stem Cell Research, 10*, 301–312.

Benz, E. W., Jr., & Moses, H. L. (1974). Small, virus-like particles detected in bovine sera by electron microscopy. *Journal of the National Cancer Institute, 52*, 1931–1934.

Bjornson, C. R., Rietze, R. L., Reynolds, B. A., Magli, M. C., & Vescovi, A. L. (1999). Turning brain into blood: A hematopoietic fate adopted by adult neural stem cells in vivo. *Science, 283*, 534–537.

Börger, V., Bremer, M., Ferrer-Tur, R., Gockeln, L., Stambouli, O., Becic, A., & Giebel, B. (2017). Mesenchymal stem/stromal cell-derived extracellular vesicles and their potential as novel immunomodulatory therapeutic agents. *International Journal of Molecular Sciences, 18*, E1450.

Bruno, S., Grange, C., Deregibus, M. C., Calogero, R. A., Saviozzi, S., Collino, F., Morando, L., Busca, A., Falda, M., Bussolati, B., et al. (2009). Mesenchymal stem cell-derived microvesicles protect against acute tubular injury. *Journal of the American Society of Nephrology: JASN, 20*, 1053–1067.

Caplan, A. I. (2017). Mesenchymal stem cells: Time to change the name! *Stem Cells Translational Medicine, 6*, 1445–1451.

Caplan, A. I., & Dennis, J. E. (2006). Mesenchymal stem cells as trophic mediators. *Journal of Cellular Biochemistry, 98*, 1076–1084.

Chargaff, E., & West, R. (1946). The biological significance of the thromboplastic protein of blood. *The Journal of Biological Chemistry, 166*, 189–197.

Consortium, E.-T., Van Deun, J., Mestdagh, P., Agostinis, P., Akay, O., Anand, S., Anckaert, J., Martinez, Z. A., Baetens, T., Beghein, E., et al. (2017). EV-TRACK: Transparent reporting and centralizing knowledge in extracellular vesicle research. *Nature Methods, 14*, 228–232.

Coumans, F. A. W., Brisson, A. R., Buzas, E. I., Dignat-George, F., Drees, E. E. E., El-Andaloussi, S., Emanueli, C., Gasecka, A., Hendrix, A., Hill, A. F., et al. (2017). Methodological guidelines to study extracellular vesicles. *Circulation Research, 120*, 1632–1648.

Dalton, A. J. (1975). Microvesicles and vesicles of multivesicular bodies versus "virus-like" particles. *Journal of the National Cancer Institute, 54*, 1137–1148.

De Broe, M., Wieme, R., & Roels, F. (1975). Letter: Membrane fragments with koinozymic properties released from villous adenoma of the rectum. *Lancet, 2*, 1214–1215.

Di Nicola, M., Carlo-Stella, C., Magni, M., Milanesi, M., Longoni, P. D., Matteucci, P., Grisanti, S., & Gianni, A. M. (2002). Human bone marrow stromal cells suppress T-lymphocyte proliferation induced by cellular or nonspecific mitogenic stimuli. *Blood, 99*, 3838–3843.

Doeppner, T. R., Herz, J., Gorgens, A., Schlechter, J., Ludwig, A. K., Radtke, S., de Miroschedji, K., Horn, P. A., Giebel, B., & Hermann, D. M. (2015). Extracellular vesicles improve post-stroke neuroregeneration and prevent postischemic immunosuppression. *Stem Cells Translational Medicine, 4*, 1131–1143.

Dragovic, R. A., Gardiner, C., Brooks, A. S., Tannetta, D. S., Ferguson, D. J., Hole, P., Carr, B., Redman, C. W., Harris, A. L., Dobson, P. J., et al. (2011). Sizing and phenotyping of cellular vesicles using nanoparticle tracking analysis. *Nanomedicine, 7*, 780–788.

Drommelschmidt, K., Serdar, M., Bendix, I., Herz, J., Bertling, F., Prager, S., Keller, M., Ludwig, A. K., Duhan, V., Radtke, S., et al. (2017). Mesenchymal stem cell-derived extracellular vesicles ameliorate inflammation-induced preterm brain injury. *Brain, Behavior, and Immunity, 60*, 220–232.

Fais, S., O'Driscoll, L., Borras, F. E., Buzas, E., Camussi, G., Cappello, F., Carvalho, J., Cordeiro da Silva, A., Del Portillo, H., El Andaloussi, S., et al. (2016). Evidence-based clinical use of nanoscale extracellular vesicles in nanomedicine. *ACS Nano, 10*, 3886–3899.

Galon, J., & Bruni, D. (2019). Approaches to treat immune hot, altered and cold tumours with combination immunotherapies. *Nature Reviews. Drug Discovery, 18*(3), 197.

Giebel, B., & Helmbrecht, C. (2017). Methods to analyze EVs. *Methods in Molecular Biology, 1545*, 1–20.

Gilani, S. I., Weissgerber, T. L., Garovic, V. D., & Jayachandran, M. (2016). Preeclampsia and extracellular vesicles. *Current Hypertension Reports, 18*, 68–68.

Gorgens, A., Bremer, M., Ferrer-Tur, R., Murke, F., Tertel, T., Horn, P. A., Thalmann, S., Welsh, J. A., Probst, C., Guerin, C., et al. (2019). Optimisation of imaging flow cytometry for the analysis of single extracellular vesicles by using fluorescence-tagged vesicles as biological reference material. *Journal of Extracellular Vesicles, 8*, 1587567.

Gould, S. J., & Raposo, G. (2013). As we wait: Coping with an imperfect nomenclature for extracellular vesicles. *Journal of Extracellular Vesicles, 2*.

Harding, C., Heuser, J., & Stahl, P. (1983). Receptor-mediated endocytosis of transferrin and recycling of the transferrin receptor in rat reticulocytes. *The Journal of Cell Biology, 97*, 329–339.

Johnstone, R. M., Adam, M., Hammond, J. R., Orr, L., & Turbide, C. (1987). Vesicle formation during reticulocyte maturation. Association of plasma membrane activities with released vesicles (exosomes). *The Journal of Biological Chemistry, 262*, 9412–9420.

Kim, D. K., Lee, J., Kim, S. R., Choi, D. S., Yoon, Y. J., Kim, J. H., Go, G., Nhung, D., Hong, K., Jang, S. C., et al. (2015). EVpedia: A community web portal for extracellular vesicles research. *Bioinformatics, 31*, 933–939.

Kordelas, L., Rebmann, V., Ludwig, A. K., Radtke, S., Ruesing, J., Doeppner, T. R., Epple, M., Horn, P. A., Beelen, D. W., & Giebel, B. (2014). MSC-derived exosomes: A novel tool to treat therapy-refractory graft-versus-host disease. *Leukemia, 28*, 970–973.

Lai, R. C., Arslan, F., Lee, M. M., Sze, N. S., Choo, A., Chen, T. S., Salto-Tellez, M., Timmers, L., Lee, C. N., El Oakley, R. M., et al. (2010). Exosome secreted by MSC reduces myocardial ischemia/reperfusion injury. *Stem Cell Research, 4*, 214–222.

Lener, T., Gimona, M., Aigner, L., Borger, V., Buzas, E., Camussi, G., Chaput, N., Chatterjee, D., Court, F. A., Del Portillo, H. A., et al. (2015). Applying extracellular vesicles based therapeutics in clinical trials – An ISEV position paper. *Journal of Extracellular Vesicles, 4*, 30087.

Lotvall, J., Hill, A. F., Hochberg, F., Buzas, E. I., Di Vizio, D., Gardiner, C., Gho, Y. S., Kurochkin, I. V., Mathivanan, S., Quesenberry, P., et al. (2014). Minimal experimental requirements for definition of extracellular vesicles and their functions: A position statement from the International Society for Extracellular Vesicles. *Journal of Extracellular Vesicles, 3*, 26913.

Ludwig, A. K., & Giebel, B. (2012). Exosomes: Small vesicles participating in intercellular communication. *The International Journal of Biochemistry & Cell Biology, 44*, 11–15.

Ludwig, A.-K., De Miroschedji, K., Doeppner, T. R., Börger, V., Ruesing, J., Rebmann, V., Durst, S., Jansen, S., Bremer, M., Behrmann, E., et al. (2018). Precipitation with polyethylene glycol followed by washing and pelleting by ultracentrifugation enriches extracellular vesicles from tissue culture supernatants in small and large scales. *Journal of Extracellular Vesicles, 7*, 1528109.

Mezey, E., Chandross, K. J., Harta, G., Maki, R. A., & McKercher, S. R. (2000). Turning blood into brain: Cells bearing neuronal antigens generated in vivo from bone marrow. *Science, 290*, 1779–1782.

Monguio-Tortajada, M., Lauzurica-Valdemoros, R., & Borras, F. E. (2014). Tolerance in organ transplantation: From conventional immunosuppression to extracellular vesicles. *Frontiers in Immunology, 5*, 416.

Munoz-Elias, G., Woodbury, D., & Black, I. B. (2003). Marrow stromal cells, mitosis, and neuronal differentiation: Stem cell and precursor functions. *Stem Cells, 21*, 437–448.

Nair, S., & Salomon, C. (2018). Extracellular vesicles and their immunomodulatory functions in pregnancy. *Seminars in Immunopathology, 40*, 425–437.

Ophelders, D. R., Wolfs, T. G., Jellema, R. K., Zwanenburg, A., Andriessen, P., Delhaas, T., Ludwig, A. K., Radtke, S., Peters, V., Janssen, L., et al. (2016). Mesenchymal stromal cell-derived extracellular vesicles protect the fetal brain after hypoxia-ischemia. *Stem Cells Translational Medicine, 5*, 754–763.

Pan, B. T., & Johnstone, R. M. (1983). Fate of the transferrin receptor during maturation of sheep reticulocytes in vitro: Selective externalization of the receptor. *Cell, 33*, 967–978.

Pittenger, M. F., Mackay, A. M., Beck, S. C., Jaiswal, R. K., Douglas, R., Mosca, J. D., Moorman, M. A., Simonetti, D. W., Craig, S., & Marshak, D. R. (1999). Multilineage potential of adult human mesenchymal stem cells. *Science, 284*, 143–147.

Raposo, G., & Stoorvogel, W. (2013). Extracellular vesicles: Exosomes, microvesicles, and friends. *The Journal of Cell Biology, 200*, 373–383.

Raposo, G., Nijman, H. W., Stoorvogel, W., Liejendekker, R., Harding, C. V., Melief, C. J., & Geuze, H. J. (1996). B lymphocytes secrete antigen-presenting vesicles. *The Journal of Experimental Medicine, 183*, 1161–1172.

Ratajczak, J., Miekus, K., Kucia, M., Zhang, J., Reca, R., Dvorak, P., & Ratajczak, M. Z. (2006). Embryonic stem cell-derived microvesicles reprogram hematopoietic progenitors: Evidence for horizontal transfer of mRNA and protein delivery. *Leukemia, 20*, 847–856.

Shao, H., Im, H., Castro, C. M., Breakefield, X., Weissleder, R., & Lee, H. (2018). New technologies for analysis of extracellular vesicles. *Chemical Reviews, 118*, 1917–1950.

3

Sokolova, V., Ludwig, A. K., Hornung, S., Rotan, O., Horn, P. A., Epple, M., & Giebel, B. (2011). Characterisation of exosomes derived from human cells by nanoparticle tracking analysis and scanning electron microscopy. *Colloids and Surfaces. B, Biointerfaces, 87*, 146–150.

Stegmayr, B., & Ronquist, G. (1982). Promotive effect on human sperm progressive motility by prostasomes. *Urological Research, 10*, 253–257.

Thery, C., Witwer, K. W., Aikawa, E., Alcaraz, M. J., Anderson, J. D., Andriantsitohaina, R., Antoniou, A., Arab, T., Archer, F., Atkin-Smith, G. K., et al. (2018). Minimal information for studies of extracellular vesicles 2018 (MISEV2018): A position statement of the International Society for Extracellular Vesicles and update of the MISEV2014 guidelines. *Journal of Extracellular Vesicles, 7*, 1535750.

Timmers, L., Lim, S. K., Arslan, F., Armstrong, J. S., Hoefer, I. E., Doevendans, P. A., Piek, J. J., El Oakley, R. M., Choo, A., Lee, C. N., et al. (2007). Reduction of myocardial infarct size by human mesenchymal stem cell conditioned medium. *Stem Cell Research, 1*, 129–137.

Valadi, H., Ekstrom, K., Bossios, A., Sjostrand, M., Lee, J. J., & Lotvall, J. O. (2007). Exosome-mediated transfer of mRNAs and microRNAs is a novel mechanism of genetic exchange between cells. *Nature Cell Biology, 9*, 654–659.

Welsh, J. A., Holloway, J. A., Wilkinson, J. S., & Englyst, N. A. (2017). Extracellular vesicle flow cytometry analysis and standardization. *Frontiers in Cell and Development Biology, 5*, 78.

Wiklander, O. P. B., Bostancioglu, R. B., Welsh, J. A., Zickler, A. M., Murke, F., Corso, G., Felldin, U., Hagey, D. W., Evertsson, B., Liang, X. M., et al. (2018). Systematic methodological evaluation of a multiplex bead-based flow cytometry assay for detection of extracellular vesicle surface signatures. *Frontiers in Immunology, 9*, 1326.

Wiklander, O. P. B., Brennan, M. A., Lotvall, J., Breakefield, X. O., & El Andaloussi, S. (2019). Advances in therapeutic applications of extracellular vesicles. *Science Translational Medicine, 11*, eaav8521.

Wolf, P. (1967). The nature and significance of platelet products in human plasma. *British Journal of Haematology, 13*, 269–288.

Yanez-Mo, M., Siljander, P. R., Andreu, Z., Zavec, A. B., Borras, F. E., Buzas, E. I., Buzas, K., Casal, E., Cappello, F., Carvalho, J., et al. (2015). Biological properties of extracellular vesicles and their physiological functions. *Journal of Extracellular Vesicles, 4*, 27066.

Zabeo, D., Cvjetkovic, A., Lässer, C., Schorb, M., Lötvall, J., & Höög, J. L. (2017). Exosomes purified from a single cell type have diverse morphology. *Journal of Extracellular Vesicles, 6*, 1329476–1329476.

Ethics in Stem Cell Applications

Michael Fuchs

© Springer Nature Switzerland AG 2020
B. Brand-Saberi (ed.), *Essential Current Concepts in Stem Cell Biology*,
Learning Materials in Biosciences, https://doi.org/10.1007/978-3-030-33923-4_14

What You Will Learn in This Chapter

In this chapter you will learn about the different sources to get embryonic cells for research or for therapeutic applications. Since getting ES cells from embryos in the blastocyst stage normally implies the destruction of the embryo the chapter deals with the moral status of the human embryo. You are informed about different positions and their background assumptions. In addition to the international ethical discussion the genesis of the legal regulation in Germany and Europe is described and explained. Finally the question of moral disagreement is addressed.

The ethics of medical research constitutes a part within the field of medical ethics and bioethics where a far-reaching consensus could be found, at least as far as the principles and necessary procedures are concerned. This consensus has been built upon the concept of informed consent, which is deemed a central aspect in the ethics of research on human beings and on human biological material (Faden and Beauchamp 1986). The idea behind this concept is to avoid any kind of instrumentalization of human persons and to respect human dignity.

If we look at stem cell research we have to be aware that several kinds of cells have to be distinguished. This distinction is based on differences in the biological potentials, the ontological status (what kind of thing is x?) and the moral status (what is the moral value of x?). Some cells do have the potential to differentiate and to create various tissues and even various types of tissue. This is the reason why they are of high interest for research and for future therapeutic applications. Some cells even have the potential to develop into an entire organism for instance an adult human being. In this regard the term "totipotency" stands for the potential to develop into an entire organism, "pluripotency" for the potential to develop into (theoretically) all cell types apart from extraembryonic tissues (Denker 2002).

Since cells differentiate and lose their potential during their live time, embryonic cells are of special interest. We have different sources to get embryonic cells for research or for therapeutic applications: EC cells (embryonic carcinoma cells) are taken from embryonic tumour cells, EG cells (embryonic germ cells) from fetal precursor cells of gametes and ES cells (embryonic stem cells) from early embryonic stages of development (blastocysts). ES cells (embryonic stem cells) may be subdivided into the following groups: ES cells generated from blastocysts created by in vitro fertilisation (IVF), ES cells generated from blastocysts created by somatic cell nuclear transfer (SCNT) (Tachibana et al. 2013; Meissner and Jaenisch 2005).

ES cells generated from blastocysts created by in vitro fertilisation (IVF) can be derived either from socalled surplus or supernumerary embryos or from embryos created for research purposes.

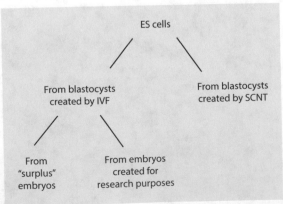

For many decades the discussion on stem cell research was focused on human embryonic stem cell (Hug and Hermerén 2011). And the central issue in the discussion on human embryonic stem cells was the moral status of the human embryo (Føllesdal 2006). Taking ES cells from embryos in the blastocyst stage normally implied and still implies the destruction of the embryo. Killing of an embryo or fetus is one of the most contentious moral issues (De Gracia 2012, 16–59).

14.1 Disagreements on the Question of the Status of Embryos

The spectrum of positions in the philosophical debate ranged from the positing of a moral imperative to pursue embryo research and therapeutic cloning (Merkel 2001), to the belief in a duty to ensure ungraduated protection of human dignity for all early stages of development, which biologically have be considered part of the "human family" (Spaemann 2001). A radical position was held by Peter Singer and Helga Kuhse: "We must recall however," they argued, "that when we kill a new-born infant there is no person whose life has begun. When I think of myself as the person I now am, I realize that I did not come into existence until sometime after my birth. At birth I had no sense of the future, and no experiences which I can now remember as ‚mine'. It is the beginning of the person, rather than of the physical organism, that is crucial so far as the right to life is concerned" (Singer and Kuhse 1985, 133). Singer and Kuhse agree with most of us, that adult human beings deserve respect and protection of their lives and even of their corporal integrity. But since this respect presupposes a specific degree of mental capacity Singer and Kuhse deny that already the newborn has to be kept alive in any case. Another position was marked by Michael Tooley who was looking for an analogy between brain death and the beginning of the brain: "Just as I shall live only as long as the relevant part of my brain remains essentially intact, so I came into existence only when the appropriate part or parts of my brain came into existence, or more precisely, reached the appropriate stage of development to sustain my identity as a human being, with the capacity for consciousness. When I came into existence is a matter of how far back the relevant neurophysiological continuity can be traced. Presumably, then, my life began somewhere between conception and birth" (Lockwood 1985, 23).

For the debate in the United Kingdom individuation was seen as a morally significant break physiologically indicated by the appearance of the primitive streak: "The primitive streak stage is a vitally important landmark in development because it marks the onset of individuality. [...] Once the primitive streak has formed, we can for the first time recognise and delineate the boundaries of a discrete coherent entity, an individual, that can become transformed through growth and differentiation into an adult human being. If I had to point to a stage and say ‚This was when I began being me', I think it would have to be here" (McLaren 1984). Before the appearance of the primitive streak identical twinning is still possible, chimeara can be created and there is little evidence of an intrinsic unity. For the proponents of this position the early embryo is either more an aggregation than a unity or an individual or it is not the same individual entity as after the point in development when twinning is not any more possible.

Others argued in favour of fertilisation as the onset of a human being. "A change in organism was seen", John Noonan explained in 1970, "to occur at the moment of fertilization which distinguished the resultant from the components. It was easier to mark this new organism off from the living elements which had preceded it than it was to mark it off from some later stage of its organic growth in the uterus. If a moment had to be chosen for ensoulment, no convincing argument now appeared to support Aristotle or to put ensoulment at a later stage of fetal life" (Noonan 1970, 38).

Those who argued in favour of an early onset of human identity or even personhood made use of arguments that the stages of development are linked. "We can say", Norman Ford stated in 1991, "the human person is a living individual with a human nature, i.e. a living ontological individual that has within itself the active capacity to maintain, or at least to begin, the process of the human life-cycle without loss of identity" (Ford 1991, 84–85). In analyzing this sentence and other positions philosophers distinguished a species argument, an identity argument, a continuity argument and a potentiality argument which all were carefully discussed (Damschen and Schönecker 2003; Deutsches Referenzzentrum für Ethik in den Biowissenschaften: In focus: Research with human embryonic stem cells).

In Germany the basic attitudes towards the worthiness of protection to be granted to the human embryo in vitro are similarly heterogeneous like those found in other European countries and in the Anglo-Saxon countries in particular. In the attempts to qualify the necessity to protect life only for some of the first phases of development involved – alongside consistency arguments, such as reference to the legality of nidation inhibitors – the search for caesura in the development of the human organism from fertilisation to birth played an equally important role as in other countries (Rager 2009). Whereas the argument that individuation has not ceased as long as there are residual possibilities of polyembryony played a major role in the Anglo-Saxon discourse of the 80s, the German discussion was more concerned with the criterion of nidation, since – as was said – it is only with nidation that essential nutritional and morphogenetic factors on the mother's side enter into the genetic programme of the fertilised egg cell (Nüsslein-Volhard 2001, see also Heinemann and Honnefelder 2002). Moreover the Constitutional Court made use of the difference between the nasciturus in utero and the embryo in vitro. Since the German Basic Law does not give a definition of a human being and no answer to the question of the status of the early embryo the court declared in two decisions that the right to life extends to the unborn (1975) and affirms that the unborn human life is already entitled of human dignity (1993). Nevertheless in both decisions the court left the question explicitly open if this right to live and to be protected applies already for the embryo before nidation or individuation (1975) although it is argued that insights from medical anthropology might suggest that human life arises prior the pregnancy "with the fusion of egg and sperm cell" (1993).

The individual opinions within the spectrum of positions in Germany break down into two basic patterns of argument, in much the same way as we have seen in other countries (House of Lords 2002; Føllesdal 2006; De Gracia 2012; German Bundestag. Study Commission on Law and Ethics in Modern Medicine, 2001). What both patterns of argument have in common is that they start with a clear assumption that the born human being must be ensured protection, whether on grounds of its status as a moral subject or on the basis of its facility for reason, whether as the holder of preferences of a special kind or as the image of God, or whether simply on grounds of divine command. Both patterns of argument proceed from this fundamental understanding by drawing conclusions as to the status of human beings in the phases prior to birth. The two patterns occur because the development process can be seen, on the one hand, as a process of emerging and, on the other, as a process of growing. One side emphasises that the relevant characteristics and prerequisites of being a person are successively added, while the other side stresses identify and continuity between the embryo and the born human being. This difference comes sharply into focus in the respective understandings of "potentiality". While potentiality is seen on one side as a purely logical or material possibility, the other side regards the entity that has in itself the potential a power of action "leading to the fulfilment of the potential" (Holm 1998, 43). If we would give up the conception of potentiality, the second group argues, we would risk not being able to insist on the protection of persons that are sleeping or in coma (Føllesdal 2006, 70).

Looking at the debate in Germany as a whole, we should note, however, that there have been very few participants in this discourse who favour the option of permitting embryo harvesting for research or therapeutic purposes (Fuchs 2011, 124–129). Additional concerns have been voiced about the possibility that this might occur by means of therapeutic cloning, i.e. via procedures to transplant the nucleus (Fuchs 2003).

This stance reflects a tendency, already visible in the deliberations leading up to the framing of the Embryo Protection Act, that if embryo-consuming research is, even for a therapeutic objective, to be permitted at all, then only when carried out on orphaned supernumerary embryos. Such embryos have become available – although only in small numbers – even under scope of the Embryo Protection Act. This is not due to infringements of the law. For, although the Embryo Protection Act seeks to create a framework for the use of in vitro fertilisation in which such embryos do not occur, it cannot and does not wish to give guarantees that an artificially produced embryo is implanted. Rather, we have to consider the possibility that the mother may fall ill or die, or that the mother – and the Embryo Protection Act also accepts this possibility – may refuse implantation. In general the discussions did make clear that the Embryo Protection Act, although appearing rigid by international comparison, does not only pursue the aim of preventing the unregulated practice of artificial insemination and research on embryos for uses other than the well-being of the embryo, but is also designed to offer protection for, in addition to the embryo, the family with its traditional parents-child structures (Kirchhof 2002, 22–24). Only keeping in mind this dual purpose we can understand why no legal framework was created for embryo adoptions and why, in order to enhance the efficiency of artificial insemination, the production and implantation of more than one embryo is permitted (to a maximum of three). In other words, the debate in the 1980s concerns, as it does today, the question of whether those embryos that, as far as anyone can tell, have no chance of being carried to full term and becoming a child, i.e. can be said to be doomed, could not be put to good use rather than simply allowed to die. Not only utilitarian arguments but also general altruistic intuitions might be used to justify such a use of surplus embryos. The arguments *against* their use are not based on any residual uncertainty about the fate of the embryo. Rather, they seem to be based on a distinction, which can be made from a particular perspective, between a requirement to protect life, which is agreed to be no longer possible, and a requirement to protect dignity, which in a certain sense remains valid in these cases. The idea here is that by allowing the embryo to die we show greater respect than by using it for extraneous purposes, i.e. other than its own well-being.

Those who accept that the production of embryos is a violation of human dignity but have no objection to the use of orphaned embryos argue that the question of instrumentalization, which entails their use, differs between the first case and the second. If one assumes that the early human being has the potential within itself to become a person, i.e. to develop *itself* into a person, then the deliberate production of embryos for extraneous uses will always amount to improper instrumentalization violating human dignity; however, the use of doomed embryos does not automatically have to be seen as such an act of instrumentalization.

And, irrespective of the legal arrangements favoured by the experts in each case, the overwhelming majority in Germany would seem to adopt such an ethical approach that makes a moral distinction between producing embryos for research purposes and using surplus embryos.

14.2 Moral Assessment of Stem Cell Research

For a moral assessment of stem cell research several issues have to be examined, namely questions about the status of the cells and about the status of their source but also questions about available alternatives and the moral evaluation of the goals of research. Producing human embryos in a culture medium for research or therapeutic purposes was an option that – at least through the 1990s – appeared in Germany to be ruled out on ethical grounds. So the debate in Germany essentially revolved around the question of how Germany should respond to developments in other European countries with a more permissive stance, such as in Great Britain or Belgium. Was an erosion of our own ethical standards to be feared if Germany agreed to a middle position under international agreements or sign up to minimal requirements far less restrictive than ones own rules? Such considerations lay in part behind Germany's refusal to sign the Convention on Human Rights and Biomedicine of the Council of Europe, (de Wachter 1997) which Germany has still not yet signed.

14.3 Legal Regulation in Germany: From the Embryo Protection Act to the Amendment of the Stem Cell Act

Indeed, the course chosen in Germany – after an intensive interdisciplinary discussion between legal experts, scientists, medical professionals, philosophers and theologians (Bundesministerium für Forschung und Technologie 1985) and a subsequent parliamentary debate which accentuated their proposals – was to regulate all conceivable options opened up by in-vitro fertilisation under a criminal law with the adoption of the Embryo Protection Act. Passed in 1990, the act only allows the production of human embryos for the purpose of bringing about pregnancy. Other, abusive, applications of reproductive techniques are threatened with serious punishment, as is artificial modification of the human germ-line, reproduction by means of cloning techniques or the creation of chimeras and hybrid beings. Selection according to sex or fertilisation using sperm from someone who has died are also treated as criminal offences. As for research that is not intended to benefit the embryo affected, the law does not provide for any legitimising exemptions.[1]

---

1   Günther et al. (2014). The legal regulation of preimplantation genetic diagnosis in § 3a does not change the prohibition of research with embryos: "With the law of regulation of the preimplantation genetic diagnosis, which got approved by the German parliament on November 21st 2011, and the change of embryo protection law related to it, and despite its fundamental prohibition, the genetic examination of the pluripotent cells of the embryo in vitro, before its intrauterine transfer, within exceptions and tight limits, is declared not illegal. Hence there is an explicitly legal regulation of PGD for the first time. Applying PGD on the basis of the new law is however only permitted once the regulation on the legitimate implementation of preimplantation genetic diagnosis is legally valid." (Deutsches Referenzzentrum für Ethik in den Biowissenschaften 2016, In focus (▶ http://www.drze.de/in-focus/preimplantation-genetic-diagnosis/legal-aspects?set_language=en)

The debate took on a new urgency when, in November 1998, an American-Israeli research group headed by the American embryologist James A. Thomson reported the first ever successful cultivation of human embryonic stem cells (Thomson et al. 1998). It was generally assumed that the possibility of keeping embryonic stem cell lines in a culture medium was a key prerequisite for developing a wider understanding of the differentiation process of human cells. There was also a very widespread view that this would at least open up good prospects for successful transplantation of tissues and perhaps even whole organs. The debate in Germany was driven forward above all by opinions presented by the Deutsche Forschungsgemeinschaft (DFG), which is the self-governing body of the sciences and humanities in Germany, funded by the German federal government and the governments of the 16 Länder. As the conviction increasingly emerged in the scientific community that primordial germ cells (EG cells) do not show the same potential as embryonic stem cells (ES cells), the Deutsche Forschungsgemeinschaft adviced in 2001 (Deutsche Forschungsgemeinschaft 2001), also for ethical reasons, a gradual acceptance of ES cell research with the importing of human embryonic stem cells as a first legally and morally legitimate step to be followed in the medium-term by further ethical clarification and, if necessary, policy changes.

In fact the DFG managed to trigger a national debate along these lines. The issues were considered by two national ethics committees instituted by constitutional bodies (Nationaler Ethikrat 2002; and Enquete Kommission "Recht und Ethik der modernen Medizin" 2002), a debate in parliament (Deutscher Bundestag 2002), a widely-heeded sceptical speech by the federal President, and a very intensive and controversial ethical discussion, especially in the national newspapers. At the end of the debate it was decided not to amend the Embryo Protection Act for the time being, but to pass a law to regulate the importing of human embryonic stem cells. In taking this course, Parliament, as the legislature, was actually following the minority opinion among the experts and parliamentarians sitting on the German Bundestag's Study Commission on Law and Ethics in Modern Medicine (2001).

German law as it stands (Gesetz zur Sicherstellung des Embryonenschutzes im Zusammenhang mit Einfuhr und Verwendung menschlicher embryonaler Stammzellen (Stammzellgesetz – StZG)) does not, however, permit the production of stem cells from supernumerary embryos. The import of embryonic stem cells is only legal under certain conditions. In particular, the cells in question must have been derived before the date set by the act, and the intended research must be without alternative and of high priority. The stem cells to be imported must have been taken from orphaned embryos. Each individual research proposal must be subject to expert appraisal, above all with regard to lack of alternatives and priority status, by a central ethics commission convened for this purpose and a decision must be reached by a special committee at the Robert Koch Institute. None believes that this legal arrangement marks the end of the discussion. Nevertheless, it does represent a compromise by providing a middle way between the opposing positions. It also follows the proposal put forward by a group within the Study Commission of the German Bundestag as its own minority position: "Even for the position that regards the harvesting of stem cells from 'supernumerary' embryos as ethically unjustifiable, some differentiation is necessary between the method of derivation and the act of using the stem cell lines, with regard to the weight of the ethical problem. Also of importance is the question of whether such use relates to existing stem cell lines or whether it gives rise to the derivation of additional stem cell lines and therefore to the destruction of further 'supernumerary' embryos" (German Bundestag. Study Commission on Law and Ethics in Modern Medicine 2001, 6).

Alongside these ethical considerations, we also find that an understanding in the perspective of German constitutional law has played a considerable role in the decision to provide for exemptions from the ban on imports and to prevent a 'slippery slope': "Following the deliberations of the Study Commission it seems doubtful whether a complete ban on the importation of human embryonic stem cells derived from embryos abroad can be established on the basis of constitutional and European law. The importation of human embryonic stem cells is therefore to be tolerated under strict conditions. Adherence to these conditions is to be monitored by a state-authorised control body whose operations are open to scrutiny" (German Bundestag. Study Commission on Law and Ethics in Modern Medicine 2001, 14)

Actually the Stem Cell Act (Stammzellgesetz) was amended in 2008[2] and the cut off date postponed from the first January 2002 to the first May 2007. From the point of view of some leading scientists this amendment seemed to be required, since the quality of the stem cell lines produced after 2002 was significantly higher compared to the older stem cell lines. Nevertheless it was discussed if new key date would become object for further amendments in the future. In the meantime, more than 140 projects applications to the Robert Koch Institute for research involving imported human embryonic stem cells have been approved, having been ethically evaluated by the central Ethics Commission for Stem Cell Research, a body established under the Federal Stem Cell Act.

14.4   IPS-Cells and the Question of Totipotency

This half way position to resolve the question of importing embryonic stem cells (ES cells) links an understanding of the moral status of the human embryo with an evaluation of research purposes. For a moral assessment of stem cell research several issues had been taken into consideration: the status of the cells and the status of their source, the moral evaluation of the goals of research and the question about available alternatives. At the beginning of the millennium it was difficult to predict what direction the discussion of the high-priority and no-alternative criteria in the Stem Cell Act would take and what kind of research practice would ensue. Both the public debate and the ethical discourse changed in 2007, when two groups of researchers published data explaining techniques to reprogram human somatic cells so that they showed characteristics of embryonic stem cells. With these induced pluripotent stem cells (iPS-cells) an alternative seemed to be available both in research as for therapeutic applications. Researchers argued that embryonic stem cells would still be necessary as a gold standard for pluripotency. In the first years there were some doubts that iPS technology could be applicable for therapies in humans. A philosophical and ethical question came up if there could be a guarantee that iPS cells cannot become totipotent.

---

2   The first amendment of the Stem Cell Act in 2008 shows that joint European research programmes have brought up the question of harmonisation of the legal situation across Europe and that opponents of restrictive legislation could succeed to reverse the cut-off date. This is because scientists broadly agree that stem cell lines produced before this cut-off date are unsuitable for such purposes.

Stem cells are generally characterised by their high potential for differentiation. In other words, they are not, or not yet, specialised in the same way as other cells. We know that embryonic cells at the stage of the very first cell divisions have the capacity to develop into a complete organism. On the other hand, the adult stem cells have the ability to continue differentiating within a particular tissue. Whereas cells of the former type are referred to as totipotent, the latter are called multipotent. Research into adult stem cells is aimed at showing what the possibilities are for transdifferentiation and reprogramming. If the cell shows the potential to act as the type of cell associated with other sorts of tissue, they must be called pluripotent. Some cells even have the ability to develop into any type of cell found in the body, and it is proposed that these be designated as omnipotent. But there is some disagreement over definitions and the classifications. In particular, the expression "totipotent" is used by some scientists for those properties designated above as "omnipotent". It is also unclear whether a single totipotent cell is necessarily able to form a whole and whether the said whole will necessarily comprise the embryoblast and the trophoblast or whether the embryoblast should be regarded as a sufficient archetype of the living being.

It is hardly surprising that a clarification of definitions has been demanded from German scientists in particular. The reason, however, does not lie in their penchant for conceptual clarity but in the significance assumed by the term totipotency in the Embryo Protection Act. In its own legal definitions, the act considers not only the embryo from the zygote stage onwards, but also each totipotent cell taken from the embryo to be an embryo, i.e. an early human being. Thus, if an embryonic stem cell which had been extracted from the blastocyst turned out to be totipotent, it would require the protection afforded under the Embryo Protection Act.

Hans-Werner Denker, who drew attention at an early stage in the debate in Germany (Denker 2002) to the uncertainty in the definition of the potential of embryonic stem cells, refers in his contributions to findings by the research team around Thomson (presented before their aforementioned publication on the human embryonic stem cells) from experiments on marmosets (Callithrix jacchus). Denker cites the reports of Thomson et al. (1996) on the astounding differentiation achieved by ES cell lines they had harvested from embryos of this South American monkey species. They found that it was sufficient to let the cultures of these cells grow in very close proximity for the spontaneous formation of "embryoid bodies", which as they reported, were amazingly similar to embryos in postimplantation stages and might even be equated with them. To the extent that it was examined, their structure was found to be virtually indistinguishable from that of normal embryos occurring in vivo and implanted in the uterus at the stage of the blastocyst with the primitive streak. Thomson et al. (1996) and Thomson and Marshall (1998) emphasise that these spontaneous developments are not an isolated phenomenon, but occur regularly. Denker takes the view that such ordered developments cannot be excluded for other primates like humans.

It is indeed surprising that while the ethical significance of the totipotency criterion is firmly asserted, the scientific community and the research institutions make no effort to investigate clarification of the relevant uncertainties in development biology. On the other hand, it must also be conceded that even if the worthiness of protection due to ES cells is clarified, no plausible practical conclusions concerning existing ES cell cultures are available.

What would be the ethical conclusion if researchers would find out that iPS cells are totipotent or could become totipotent under certain conditions? Would we than come up with a distinction between naturally totipotent cells and reprogrammed totipotent cell? Would this distinction be relevant for the ontological status? Would it be relevant for the moral status? On the one hand we are well advised not to give up the concept of totipotency in discussions on the status of organisms and parts of organisms. On the other hand it seems to be absurd to take even a somatic cell taken from an adult as if it would have the same status as an entire organism.

14.5 Translational Stem Cell Research. The Question of Patenting

Since the beginning of the stem cell debate in Germany Patent Courts in Germany are confronted with the request if methods for generating specific cells out of human embryonic cells should be regarded as contrary to public order and should be excluded from patentability. In 2009 the German Federal Court of Justice referred to the European Court of Justice with this question of patentability. It was expected that the European View would be more permissible that the national view. But the European Court came to the conclusion that Article 6 (2) of the European Directive 98/44 would exclude the use of human embryos for industrial or commercial purposes from patentability. It argues that any invention has to be excluded from patentability "where the technical teaching which is the subject-matter of the patent application requires the prior destruction of human embryos".

Although this legal argumentation only concerns patent law and has no direct impact on other parts of the legal system the question of patentability has some consequences for the translation of fundamental research into specific applications. Exclusions from patentability might even be an argument in political and parliamentary discussions on public funding of basic research.

14.6 Outlook

Is there a chance to overcome the dissent on the moral value of the embryo? Most participants in the discussion think that there is no such chance? But why? Does the disagreement show that moral questions are purely subjective? In dead we can learn from the long-lasting discussion that a variety of rational arguments is presented. It is not just a question of feelings or subjective opinions. As far as moral principles are concerned we even have considerable consensus. The disagreement is more about the ontological and anthropological framing of terms like unity, identity or individuality than about moral concepts like respect, dignity, utility and so on. There is no categorical reason why these disagreements could not be overcome. Nevertheless practical reasoning has to do with decisions that have to be taken even when there is no complete consensus what the best option might be.

After 20 years of debate many participants are convinced that we are in a situation of moral uncertainty. Some philosophers call this situation a rational disagreement. Different legal solutions show a way to cope with this situation. They try to find ways to protect the embryo without closing the door for advancing new therapeutic options.

Take Home Message

For a moral assessment of stem cell research several issues have to be examined, namely questions about the status of the cells and about the status of their source but also questions about available alternatives and the moral evaluation of the goals of research.

Since taking ES cells from embryos in the blastocyst stage implies the destruction of the embryo the disagreement about the status of the early embryo is central for the debate. In Germany after an intensive public and interdisciplinary discussion parliament decided to regulate all conceivable options opened up by in-vitro fertilisation under criminal law. Under certain conditions, importing embryonic stem cells is allowed according to the Stem Cell Act.

References

Bundesministerium für Forschung und Technologie. (1985). Gentechnologie. Chancen und Risiken 6: In-vitro-Fertilisation, Genomanalyse und Gentherapie. Bericht der gemeinsamen Arbeitsgruppe des Bundesministers für Forschung und Technologie und des Bundesministers der Justiz, München.

Damschen, G., Schönecker, D., & Dieter (Eds.) (2003). Der moralische Status menschlicher Embryonen: pro und contra Spezies-, Kontinuums-, Identitäts- und Potentialitätsargument, Berlin u.a.

De Gracia, D. (2012). Creation ethics. In *Reproduction, genetics, and quality of life.* New York: Oxford University Press.

de Wachter, M. A. M. (1997). The European convention on bioethics. In *Hastings Center Report* (pp. 13–23).

Denker, H-W. (2002). Forschung an embryonalen Stammzellen: eine Diskussion der Begriffe Totipotenz und Pluripotenz. In F. S. Oduncu, U. Schroth, & W. Vossenkuhl (Eds.) *Stammzellenforschung und therapeutisches Klonen* (pp. 19–35). Göttingen.

Deutsche Forschungsgemeinschaft. (2001). Empfehlungen der Deutschen Forschungsgemeinschaft zur Forschung mit menschlichen Stammzellen. *Jahrbuch für Wissenschaft und Ethik, 6,* 349–385.

Deutscher Bundestag. (2002). Bericht der 214. Sitzung des Deutschen Bundestages (30.01.2002), Plenarprotokoll 14/214, in: Enquete Kommission "Recht und Ethik der modernen Medizin" (Ed.: Deutscher Bundestag. Referat Öffentlichkeitsarbeit), Stammzellforschung und die Debatte des Deutschen Bundestages zum Import von menschlichen embryonalen Stammzellen, Berlin.

Deutsches Referenzzentrum für Ethik in den Biowissenschaften. (2016). In focus http://www.drze.de/in-focus/preimplantation-genetic-diagnosis/legal-aspects?set_language=en.

Deutsches Referenzzentrum für Ethik in den Biowissenschaften. In focus: Research with human embryonic stem cells. Accessed 23 Nov 2018. http://www.drze.de/in-focus/stem-cell-research/ethical-discussion?set_language=en.

Faden, R. R., & Beauchamp, T. L. (1986). *A history and theory of informed consent.* New York: Oxford University Press.

Føllesdal, D. (2006). The ethics of stem cell research. *Jahrbuch für Wissenschaft und Ethik, 11,* 67–77.

Ford, N. (1991). *When did I begin? Concept of the human individual in history, philosophy and science.* Cambridge et al.

Fuchs, M. (2003). Herstellung menschlichen Lebens zu Heilungszwecken. In: *GAIA. Ecological perspectives in science, humanities and economics 12, 3* (pp. 224–225).

Fuchs, M. (2011). Bioethics in Germany. Historical background, mayor debates, philosophical features. In *The Turkish Annual of the Studies on Medical Ethics and Law, 2009–2010,* (Vol. 2–3, Istanbul 2011, pp. 117–139).

German Bundestag. Study Commission on Law and Ethics in Modern Medicine. (2001). Summary report. Supplement to the interim report on stem cell research focusing on importation problems. Research in imported human embryonic stem cells.

Günther, H-L., Taupitz, J., & Kaiser, P. (2014). Embryonenschutzgesetz: juristischer Kommentar mit medizinisch-naturwissenschaftlichen Grundlagen, 2. neu bearb. Aufl., Stuttgart.

Heinemann, T., & Honnefelder, L. (2002). Principles of ethical decision making regarding embryonic stem cell research in Germany. *Bioethics, 16*(6), 532–543.

Holm, S. (1998). Art. 'Embryology, ethics of'. In R. Chadwick (Ed.) *Encyclopedia of applied ethics* (Vol. 2, pp. 39–45).

House of Lords. (2002). *Stem cell research – Report.* http://www.parliament.the-stationery-office.co.uk/pa/ld200102/ldselect/ldstem/83/8301.htm (*Summary of Conclusions and Recommendations* reprinted in: Jahrbuch für Wissenschaft und Ethik, Humana Press. vol. 7, pp. 483–488).

Hug, K., & Hermerén, G (Eds.) (2011). Translational stem cell research. Issues beyond the debate on the moral status of the human embryo. Humana Press (Springer).

Kirchhof, P. (2002). Genforschung und die Freiheit der Wissenschaft, in: Höffe, Otfried, Honnefelder, Ludger et al., Gentechnik und Menschenwürde. An den Grenzen von Ethik und Recht, Köln, 9–35.

Lockwood, M. (1985). When does life begin? In M. Lockwood (Ed.) *Moral dilemmas in modern medicine* (pp. 9–31). New York: Oxford University Press.

McLaren, A. (1984). Where to draw the line? In *Proceedings of the Royal Institution*, G.B. 56, pp. 101–121.

Meissner, A., & Jaenisch, R. (2005). Generation of nuclear transfer-derived pluripotent ES cells from cloned Cdx2-deficient blastocysts. *Nature, 439*, 212–221.

Merkel, R. (2001). Rechte für Embryonen?: die Menschenwürde lässt sich nicht allein auf die biologische Zugehörigkeit zur Menschheit gründen; eine Antwort auf Robert Spaemann und ein Vorschlag wider das Geläufige, in: Die Zeit 56, 5.

Nationaler Ethikrat. (2002). Stellungnahme zum Import menschlicher embryonaler Stammzellen, and Enquete Kommission "Recht und Ethik der modernen Medizin" (2002): Schlussbericht der Enquete-Kommission "Recht und Ethik der modernen Medizin".

Noonan, J. (1970). An almost absolute value in history. In J. Noonan (Ed.) *The morality of abortion. Legal and historical perspectives* (1–59). New York: Oxford University Press.

Nüsslein-Volhard, C. (2001). Wann ist ein Tier ein Tier, ein Mensch kein Mensch?: eine wunderbare Symbiose; die Befruchtung ist nur der halbe Weg zur Entwicklung des Individuums, in: Frankfurter Allgemeine Zeitung 53.

Rager, G. (2009). *Beginn, Personalität und Würde des Menschen*. München: Freiburg.

Rau, J. (2001). "Wird alles gut?; für einen Fortschritt nach menschlichem Maß": Berliner Rede 2001 in der Staatsbibliothek zu Berlin.

Singer, P., & Kuhse, H. (1985). Should the baby live? The problem of handicapped infants. New York: Oxford University Press.

Spaemann, R. (2001). Gezeugt nicht gemacht: wann ist der Mensch ein Mensch?; das britische Parlament hat den Verbrauch von Embryonen erlaubt, und Kulturminister Julian Nida-Rümelin verteidigt diese Genehmigung; sie ist aber ein Anschlag auf die Menschenwürde, in: Die Zeit, 56, 4.

Gesetz zur Sicherstellung des Embryonenschutzes im Zusammenhang mit Einfuhr und Verwendung menschlicher embryonaler Stammzellen (Stammzellgesetz – StZG).

Tachibana, M., Amato, P., Sparman, M., Gutierrez Nuria, M., Tippner-Hedges, R., Ma, H., Kang, E., Fulati, A., Lee, H.-S., Sritanaudomchai, H., Masterson, K., Larson, J., Eaton, D., Sadler-Fredd, K., Battaglia, D., Wu, D., Jensen, J., Patton, P., Gokhale, S., Stouffer, R. L., Wolf, D., & Mitalipov, S. (2013). Human embryonic stem cells derived by somatic cell nuclear transfer. *Cell, 135*, 1228–1238.

Thomson, J. A., & Marshall, V. S. (1998). Primate embryonic stem cells. *Current Topics in Developmental Biology, 38*, 133–165.

Thomson, J. A., Kalishman, J., Golos, T. G., et al. (1996). Pluripotent cell lines derived from common marmoset (Callitus jacchus) blastocysts. *Biology of Reproduction, 55*, 254–259.

Thomson, J. A., Itskovitz-Fldor, J., et al. (1998). Embryonic stem cell lines derived from human blastocysts. *Science, 282*(6 Nov), 1145–1147.

Further Reading

Heinemann, T. (2005). Klonieren beim Menschen. Analyse des Methodenspektrums und internationaler Vergleich der ethischen Bewertungskriterien, Berlin, New York (Studien zu Wissenschaft und Ethik, Bd. 1).

Höffe, O., Honnefelder, L. et al. (2002). Gentechnik und Menschenwürde. An den Grenzen von Ethik und Recht, Köln.

Honnefelder, L. (2008). Embryonic stem cell research – Arguments of the ethical debate in Germany. In L. Østnor (Ed.), *Stem cells, human embryos and ethics* (pp. 177–185). Berlin, Heidelberg, New York: Interdisciplinary Perspectives.

Steinbock, B. (2007). Moral status, moral value, and human embryos. Implications for stem cell research. In B. Steinbock (Ed.), *The Oxford handbook of bioethics* (pp. 416–440). Oxford: Oxford University Press.

Printed in the United States
By Bookmasters